RURAL BUILDINGS

VOLUME THREE

Construction

INTERMEDIATE TECHNOLOGY PUBLICATIONS 1995

Published by Intermediate Technology Publications Ltd,
103–105 Southampton Row, London WC1B 4HH, UK
Under licence from TOOL, Sarphatistraat 650, 1018 AV, Amsterdam, Holland

© TOOL

A CIP catalogue record for this book
is available from the British Library

ISBN 1 85339 320 7

Printed in Great Britain by SRP, Exeter, UK

PREFACE

This official text book is designed purposely to meet the needs of trainees who are pursuing rural building courses in various training centres administered by the National Vocational Training Institute.

The main aim of this book is to provide much needed trade information in simple language and with illustrations suited to the understanding of the average trainee.

It is the outcome of many years of experiment conducted by the Catholic F.I.C. brothers of the Netherlands, and the German Volunteer Service instructors, in simple building techniques required for a rural community.

The National Vocational Training Centre is very grateful to Brothers John v. Winden and Marcel de Keijzer of F.I.C. and Messrs. Fritz Hohnerlein and Wolfram Pforte for their devoted service in preparing the necessary materials for the book; we are also grateful to the German Volunteer Service and the German Foundation For International Development (DSE) - AUT, who sponsored the publication of this book.

We are confident that the book will be of immense value to the instructors and trainees in our training centres.

DIRECTOR: National Vocational Training Institute, Accra

© Copyright
by Stichting Kongregatie F.I.C.
Brusselsestraat 38
6211 PG Maastricht
Nederland

Alle Rechte vorbehalten
All rights reserved.

INTRODUCTION TO A RURAL BUILDING COURSE

Vocational training in Rural Building started in the Nandom Practical Vocational Centre in 1970. Since then this training has developed into an official four year course with a programme emphasis on realistic vocational training.

At the end of 1972 the Rural Building Course was officially recognised by the National Vocational Training Institute. This institute guides and controls all the vocational training in Ghana, supervises the development of crafts, and sets the examinations that are taken at the end of the training periods.

The Rural Building programme combines carpentry and masonry, especially the techniques required for constructing housing and building sanitary and washing facilities, and storage facilities. The course is adapted to suit conditions in the rural areas and will be useful to those interested in rural development, and to farmers and agricultural workers.

While following this course, the instructor should try to foster in the trainee a sense of pride in his traditional way of building and design which is influenced by customs, climate and belief. The trainee should also be aware of the requirements of modern society, the links between the old and new techniques, between traditional and modern designs -- and how best to strike a happy medium between the two with regard to considerations like health protection, storage space, sewage and the water supply. The trainee should be encouraged to judge situations in the light of his own knowledge gained from the course, and to find his own solutions to problems; that is why this course does not provide fixed solutions but rather gives basic technical information. The instructor can adapt the course to the particular situation with which he and the trainee are faced.

This course is the result of many years of work and experimentation with different techniques. The text has been frequently revised to serve all those interested in Rural Development, and it is hoped that this course will be used in many vocational centres and communities. It is also the sincere wish of the founders of this course that the trainees should feel at the completion of their training that they are able to contribute personally to the development of the rural areas, which is of such vital importance to any other general development.

We are grateful to the Brothers F.I.C., the National Vocational Training Institute and the German Volunteer Service for their assistance and support during the preparation of this course.

> Bro. John v. Winden (F.I.C.)
> Wolfram Pforte (G.V.S.)
> Fritz Hohnerlein (G.V.S.)

LAY-OUT OF THE RURAL BUILDING COURSE

The Rural Building Course is a block-release-system course, which means that the trainee will be trained in turn at the vocational centre and at the building site. The period of training at the centre is called "off-the-job" training, and the period on the building site is called "on-the-job" training. Each will last for two years, so that the whole course will take four years and will end with the final test for the National Craftsmanship Certificate.

BLOCK RELEASE SYSTEM

YEAR	TERM 1	TERM 2	TERM 3
1	X	X	X
2	O	O	O
3	O	X	O
4	X	O	X

X = OFF-THE-JOB TRAINING
O = ON-THE-JOB TRAINING

The total "off-the-job" training period is approximately 76 weeks, each week 35 hours. During this training about 80% of the time is spent on practical training in the workshop. The remaining 20% of the time is devoted to theoretical instruction.

The total "on-the-job" training period is approximately 95 weeks, each week 40 hours. During this period the trainee does full-time practical work related to his course work. In addition some "homework" is assigned by the centre and checked by the instructors.

A set of books has been prepared as an aid to the theoretical training:
- A - Rural Building, Basic Knowledge (Form 1)
- B - Rural Building, Construction (Forms 2, 3, 4)
- C - Rural Building, Drawing Book (Forms 1, 2, 3, 4)
- D - Rural Building, Reference Book

All these books are related to each other and should be used together. The whole set covers the syllabus for Rural Building and will be used in the preparation for the Grade II, Grade I, and the National Craftsmanship Certificate in Rural Building.

CONTENTS

BUILDING PRELIMINARIES — 1
- Site selection / Location plan / Working drawings / Plot and site clearing / Site organization / Water supply

SETTING OUT — 7
- Interfering objects / Uneven ground / Sloping sites / Profiles

FOUNDATIONS — 15
- Types of foundations / Functions of the foundation / Dimensions for strip foundations

NON-CONCRETE FOUNDATIONS — 17
- Rubble and boulder foundations / Ashlar masonry foundations / Bonding principles / Laterite rock foundations / Artificial stone foundations

CONCRETE FOUNDATIONS — 29
- Excavating the foundation trenches / Measurements of stepped foundations / Levelling the trenches and marking the foundation depth / Mix proportions / Preparations before casting / Casting / Levelling the foundations / Curing

FOOTINGS — 43
- Functions of the footings / Materials and measurements

SPECIAL BONDS — 45
- Different wall thicknesses at a quoin / Different wall thicknesses at a T-junction / Different wall thicknesses at a cross junction

HARDCORE FILLING — 59
- Functions of the hardcore filling / Materials and arrangement / Compaction of the hardcore

PLINTH COURSE — 60

POSITIONS OF DOOR AND WINDOW FRAMES — 61

DOOR FRAMES — 63
- Types of door frames / Joints for door frames / Heads of door frames / Measurements for door frames / Concrete shoe / Installing door frames

WINDOW FRAMES — 77
- Types of window frames / Measurements of window frames / Steel cramps / Joints for window frames / Cills of window frames / Heads of window frames / Installing window frames

WALLING UP BETWEEN FRAMES	87
MOSQUITO PROOFING	87
BURGLAR PROOFING	89
– Types of burglar of burglar proofing for windows	
DOORS	91
– Types of doors / Sizes of doors / Position of the door / Finishing treatment of the door / Ledged and battened doors / Ledged, braced and battened doors / Panelled doors / Flush doors / Large doors	
WINDOWS	109
– Types of casements / Battened casements / Panelled casements / Glazed casements / Flush casements / Joints / Louvre windows	
WORKING SCAFFOLDS	117
– Jack scaffold / Blocklayers scaffold / Ladder scaffold / Independent scaffold	
SUPPORTING SCAFFOLD	125
PROTECTIVE SCAFFOLDING	127
ARCHES	128
– Technical terms / Stability of arches / The segmental arch / Types of centring / Setting out the turning piece / Positions of the archstones / Setting up the centring / Skewback template / Laying the archblocks	
REINFORCED CONCRETE LINTELS	145
– Formwork / Reinforcement / Reinforcement systems / Corrosion in reinforced concrete	
WALLING ABOVE FRAMES	155
FORMWORK SYSTEM FOR RING BEAM	155
ROOFS IN GENERAL	157
– Roof types / Size of the roof / Roof pitch / Effective length of the sheets / Outline design / Loads	
LIGHT-WEIGHT STRUCTURES	165
– Basic shape of the roof construction / Technical terms	
LENGTHENING JOINTS FOR LIGHT-WEIGHT STRUCTURES	169
– Where to place a lengthening joint / Fish joints / Other lengthening joints	

CONSTRUCTION DETAILS	175

- Lean-to roof / Pent roofs / Gable roof / Designing built-up trusses / Joints for the truss / Marking out built-up trusses / Assembly of built-up trusses / Erecting a roof structure for a gable roof / Purlins / Fascia boards / Verandahs / Anchorage

ROOF COVERING	201

- Alignment of the sheets / Sidelaps / Ridge caps / Helpful hints for roof covering

HEAT PENETRATION - INSULATION	207
LIGHTENING	207
HIP ROOFS	209

- Setting out and constructing the half truss / Position and measurements of the hip truss / Constructing the hip truss / Erecting the hip of the roof / Covering a hip roof

HIP AND VALLEY ROOF	219

- Position and measurements of the hip and valley roof / Construction of a hip and valley truss / Covering a hip and valley roof

PLASTER AND RENDER	225

- Composition of plaster or render / Preparing the wall for plastering / Application techniques / Sequence of operations for plastering a straight wall / Positions of edge boards for plastering corners / Positions of edge boards for plastering around frames / Positions of edge boards for plastering the tops of gables / Positions of edge boards for plastering the top of a parapetted pent roof / Positions of the edge boards for plastering openings

FLOOR CONSTRUCTION	241

- Sequence of operations for one-course work / Verandah floors

STAIRS	249

- Requirements

FORMWORK	251

- Formwork for coping / Formwork for pillars / Formwork for slabs / General hints for formwork

REINFORCED CONCRETE SLABS	255

- Construction

HANGING DOORS AND WINDOWS   257
- Hanging a door with butt hinges

LOCKS AND FITTINGS   261
- Mortice lock / Rim lock

CEILINGS   267
- Construction, parts and sizes of plywood ceilings

PAINTING   268
- Preparation of surfaces for painting / How to paint / How to
  apply timber preservatives

APPENDIX

WATER PURIFICATION   273
- Small water filter / Large water filter

HAND-DUG WELLS   277
- Lining wells / Tools and materials needed / Construction

GRAIN STORAGE SILO   283
- Construction of a silo / Using the silo / Circular masonry
  work / Sequence of operations

PIT LATRINE   289
- Capacity / Location / Design and construction of the squatting
  slab / Construction / Toilet room / Toilet partitions /
  Permanent latrine

BUCKET LATRINE   299

AQUA PRIVY SYSTEMS   299

SOAKAWAYS   301

MANHOLES   301

SEPTIC TANKS   301

BOOK INTRODUCTION

Rural Building Construction is your main construction book. It is built up in the same logical way that a house is built up: starting with the preliminaries of setting out and ending with the roof construction, hanging the doors, and the finishing.

Because of the structure of the training course, it might sometimes be necessary for the instructors to rearrange this sequence and treat certain working methods either earlier or later in the course (for example, plastering is done late in the construction of a building, but it might have to be treated quite early in the actual Centre training). In any case the instruction should follow the current Rural Building syllabus, and the instructor should pick out the chapters which need to be covered according to the syllabus.

There is room for the instructor to add his own ideas, knowledge and experiences concerning the local ways of building, and to adapt the lessons to the local circumstances. The trainee should be able to adapt the knowledge he has gained to the requirements of the rural areas.

In order to work with this text book, you should be familiar with drawing techniques so that you can understand the sketches and drawings. Don't be afraid to make notes and sketches in the book: the notes made here can be helpful to you again and again in your future building career.

- The tools and materials mentioned in this book are described and explained in the Rural Building Reference Book. Some figures essential for the Rural Builder are given in the Tables of Figures at the end of the Drawing Book.

- All measurements, figures and constructions given and explained in this book are made in accordance with the standard sizes of timber, steel, etc. which are commonly used in Ghana.

The appendix gives some basic designs and construction information for water filters, wells, sanitation systems, and a silo for improved grain storage.

BUILDING PRELIMINARIES

SITE SELECTION

When choosing the location of the planned building, the responsible builder will strongly advise his client to avoid building on valuable farm land, if possible.

Most Rural Builders come from farming communities; they are building for farmers, and they also do farming themselves. Thus they are very conscious of the land and know that good land should never be wasted by building on it.

Moreover, the Rural Builder is aware that there are a number of advantages in building on higher, stony areas not suitable for farming.

Some of the advantages are:

- The higher the building is situated, the healthier and more comfortable it will be to live in, because of the better ventilation and the reduced danger from dampness rising from the ground.

- Building materials such as stones and laterite are often available on the spot, and do not need to be brought from far away.

- The soil in higher areas is more likely to be the desired type to form a firm base for the house. This makes the construction of the foundations easier, and reduces the amount of building materials required, keeping the costs lower.

NOTES:

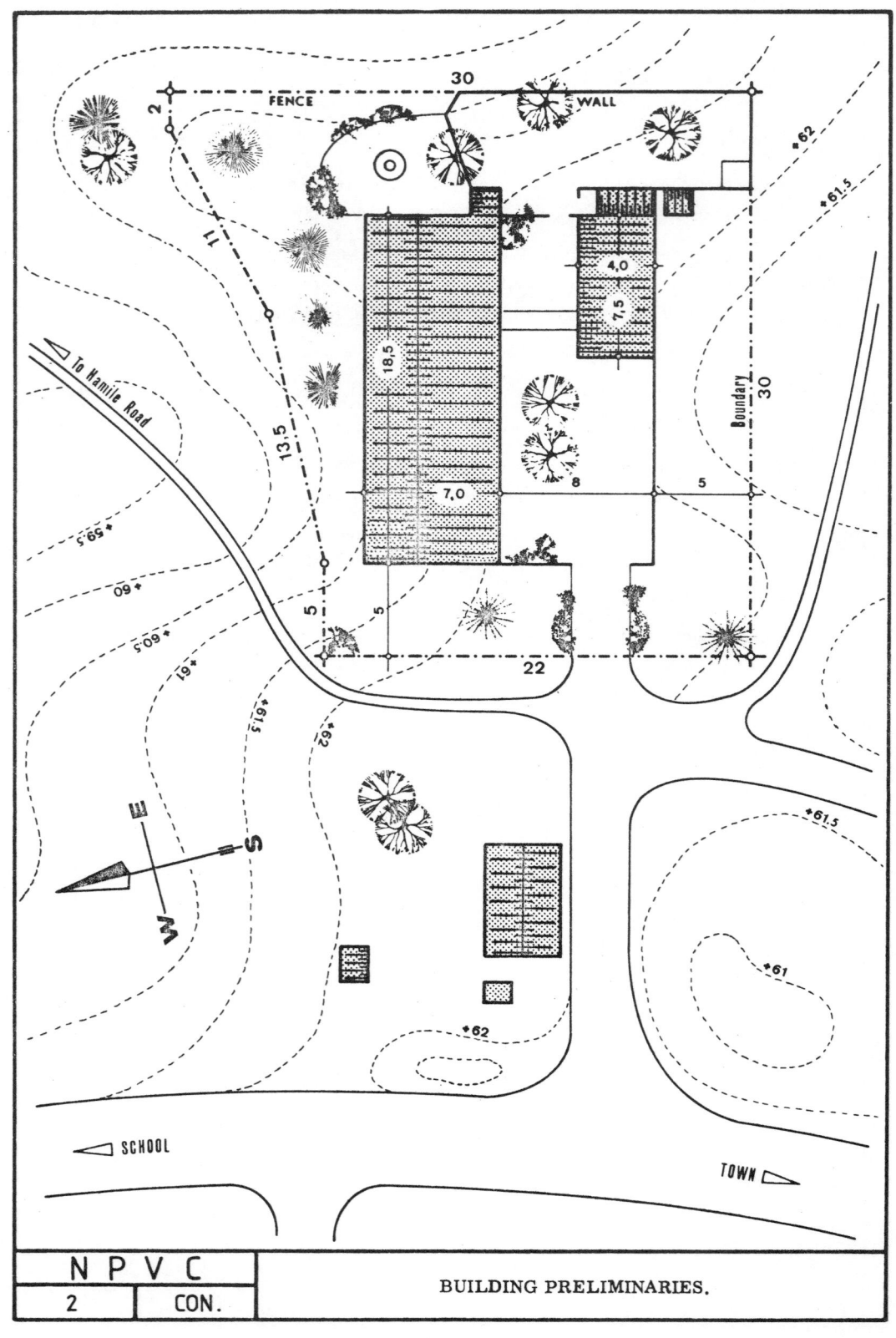

BUILDING PRELIMINARIES.

LOCATION PLAN

When building a house or any other structure one must have certain information available, in order to arrive at the best and most economical result.

The most basic information is the location, size and nature of the plot. This is contained in the location plan, which shows the plot and the immediate surroundings in scale. The scale can be from 1:200 up to 1:1000, depending on the size of the project.

The plan outlines the shape of the plot and the dimensions of its boundaries, as well as the location of the future building. It should also show the nature of the area, because it is very important to know whether the site is sloping or if the ground is uneven.

Roads, drive ways and the positions of the bigger trees are also marked on the location plan (Fig. 1).

In Rural Building, location plans are often not available, so in order to get the necessary information, both the client and the builder have to examine the plot thoroughly. Important matters such as the best position for the future building are discussed and decided on the spot.

When the site has been examined thoroughly and all the measurements and particulars have been obtained, the drawings for the house can be prepared.

- NOTE: Before the building can be started, all the required drawings have to be ready, and at the disposal of the foreman. Before you can prepare any drawings, you need to have all the particulars of the site.

NOTES:

BUILDING PRELIMINARIES.

N	P	V	C
CON.			3

WORKING DRAWINGS

The working drawings are the drawings which the builder uses before the construction starts and during the construction; to plan for materials requirements, to plan the work; and finally to carry out the construction according to the directions contained in the drawings.

The drawings include "plans", "cross sections", "elevations", and "detail drawings". They are all prepared in scales which are suitable to the particular drawing. The first three types of drawing have a scale of between 1:50 and 1:20.

- PLANS: A plan is a view that shows a certain layer or horizontal section of a building as if it were looked at from above (see Drawing Book, page 55). The plans usually include the "floor plan", which shows the walls, positions of the doors and windows, etc. (Drawing Book, page 54); the "foundation plan" gives the dimensions of the foundations and footings; and the "roof plan" which shows the shape of the roof and its dimensions. The floor plan and the foundation plan are often combined into one drawing.

- CROSS SECTIONS: A cross section is a drawing which shows the inside of the structure, as if the building were cut into two (see the Drawing Book, pages 61 and 62). The exact position where the cross section is taken has to be shown on the plan.

- ELEVATIONS: These are the views of the building as it would look from the outside (see pages 59 and 60 in the Drawing Book).

- DETAIL DRAWINGS: These show members or portions of the structure in a larger scale than the other drawings; such as 1:10, 1:5, 1:2,5, etc. Detail drawings are made when:

  - There is not enough space in the other drawings to clearly indicate all the required measurements.

  - The member of the structure is too small to be properly shown in the other drawings.

  - The member has a complicated shape and more views or cross sections are needed to explain it (see Drawing Book, page 84).

  - Important construction hints have to be pointed out.

  - The member is built up from several different materials.

The working drawings have to show the various materials used in the structure (see Drawing Book, pages 49 to 50). This enables the builder to make a list of

N P V C	BUILDING PRELIMINARIES.
4 CON.	

the materials that will be required and the amount of each material (see the Reference Book, Tables of Figures, pages 234 to 240.

When this is done the builder can estimate the cost of the building and order the materials he will need. Ordering materials has to be done in advance so that the materials are there when they are needed.

PLOT AND SITE CLEARING

Once the planning work has been completed, the plot and site both have to be prepared for the setting out. The location plan shows exactly from which areas the trees, bushes, grass and stones must be removed. The ground is levelled. The part of the plot which is cleared will be the actual site that the future building will occupy, including a space of about 5 m all around the building.

One very important measure is to remove all the tree roots from the site area. If the roots remain, they will sometimes grow again and might damage the structure. This is particularly true with the roots of the neem tree.

Clearing does not mean that all the trees on the whole plot are removed. Beyond the 5 m clear space, as many trees as possible should be allowed to remain, because they will provide shade for the people using the building or living there.

SITE ORGANIZATION

The first step in organizing the site is deciding where to make the driveway.

Next, choose the area where the building materials will be stored and arrange the storage facilities. Certain materials like sand and stones may have to be located, excavated, and transported to the site. This can be done as soon as the storage areas are located.

The building materials should not be stored too far away from the working place, nor should they be too close.

Space has to be provided for making blocks, and for mixing concrete and mortar. This also should not be too close to the future building.

The same applies to any temporary work sheds or storage sheds that are erected.

NOTES:

BUILDING PRELIMINARIES.

NPVC
CON. 5

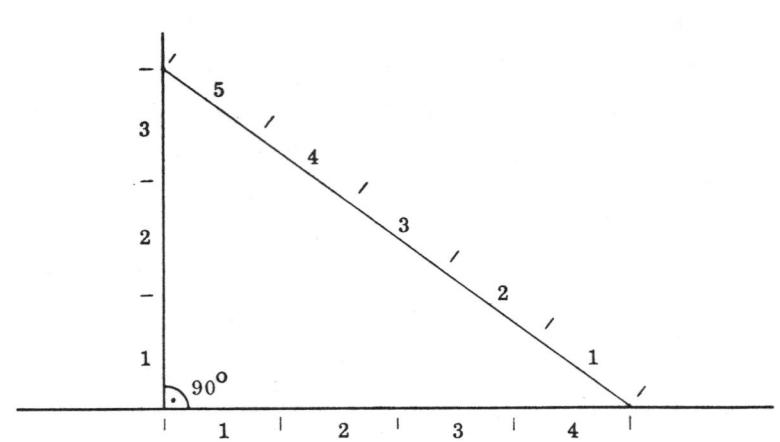

Fig. 1   3 - 4 - 5 METHOD

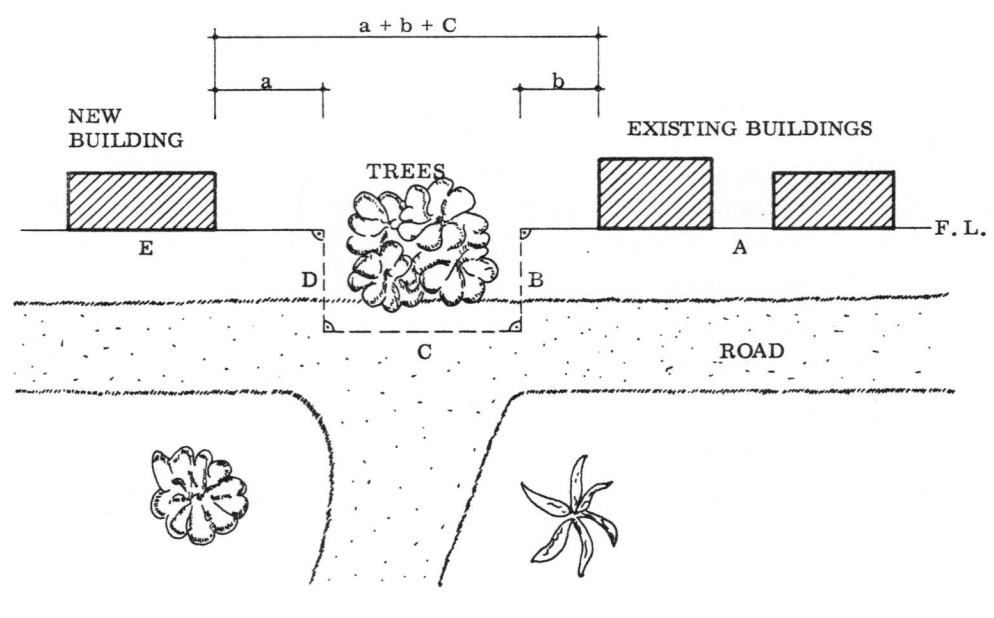

Fig. 2

BUILDING PRELIMINARIES & SETTING OUT.

WATER SUPPLY

Any kind of building would be impossible without water. Therefore one of the main tasks of the builder is to provide a guaranteed supply of clean water.

Usually piped water is not available, so it is advisable to build a simple water tank, or else dig a well if this is possible (see the Appendix, pages 277 to 281).

The water tank should be located so that after construction is completed it can be used to store rain water collected from the roof. The best site will be close to either gable end of the house so that the rainwater from the gutters can be fed into the tank directly.

Investigate the possibilities for digging a well, as this would eliminate the need to transport water and thus cut down on unnecessary effort and cost.

SETTING OUT

Once the plot and site clearing is completed, the setting out can be done: first mark the frontline of the building and from there mark all the other lines using the 3-4-5 method described in the Basic Knowledge book, page 142.

However, not all sites are conveniently flat and level, and the Rural Builder will frequently face more difficult situations. While the construction of a right angle remains the same, the measurement of distances and the determination of directions might sometimes be difficult.

There is also the case where there are trees or other buildings which should remain, but which are in the way when we are making a particular measurement.

INTERFERING OBJECTS

Provided that you have mastered the 3-4-5 method, it is relatively easy to by-pass interfering trees or buildings in setting out: it only requires a bit more time to construct the additional angles (Fig. 1).

Fig. 2 illustrates an example: the frontline of the existing buildings must be maintained in the new building, but it is difficult to set out directly because there are trees which are in the way. Rather than cut the trees down, we can by-pass them.

From the frontline of the existing buildings (line A), set out line B, using the 3-4-5 method, perpendicular to line A. Then construct lines C and D using the same method, followed by line E which is the front line of the new building. Check that line C plus the distances "a" and "b" all add up to the total planned distance between the new building and the existing buildings (Fig. 2).

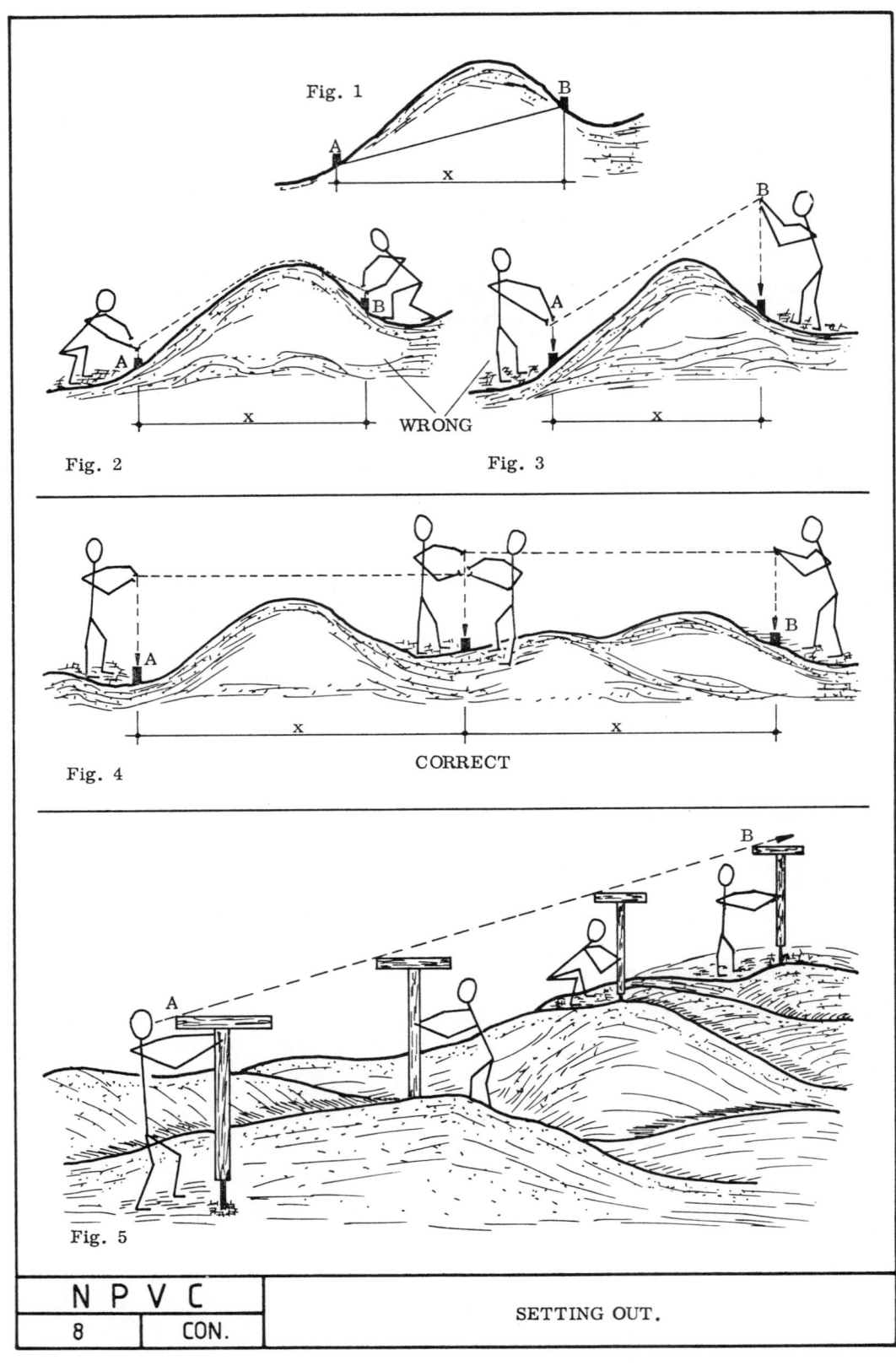

UNEVEN GROUND

Setting out on uneven ground, and particularly measuring distances, requires you to apply some simple geometry.

When we measure distances in setting out, we are actually looking for the horizontal distance between two points (Fig. 1, x). We don't measure the distance along a slope, because the house we want to build will not slope; it will have level floors and walls.

Since the ground is not flat, and the points are at different heights (point A is lower than point B), the horizontal distance between them has to be measured indirectly.

Fig. 2 shows two men trying to measure the distance between pegs A and B along uneven ground. Their result cannot be correct because the line they are holding is neither straight nor is it horizontal (measure "x" and compare it to the length of their line).

The men in Fig. 3 also fail to get the correct measurement. Their line is stretched taut and is therefore straight, but it is still not horizontal (measure "x" and compare it to the length of their line).

In order to find the true horizontal length "x", the line or tape measure has to be held horizontally and stretched taut so it is straight. Both ends of the line are kept vertically above the pegs A and B by means of plumb bobs (Fig. 4). This method as shown is a good rough method for short distances.

If a larger distance has to be measured, the work is carried out in intermediate steps of suitable lengths (Fig. 4).

For very large distances, the use of boning rods (Reference Book, page 25) can help to make sure that the different steps are in line and the total length measured is straight (Fig. 5). With the boning rods and a water level (Reference Book, page 25) you can also make sure that the whole distance is horizontal.

Use the water level to level between two boning rods; then any points in between can also be levelled by simply sighting along the boning rods (Fig. 5).

NOTES:

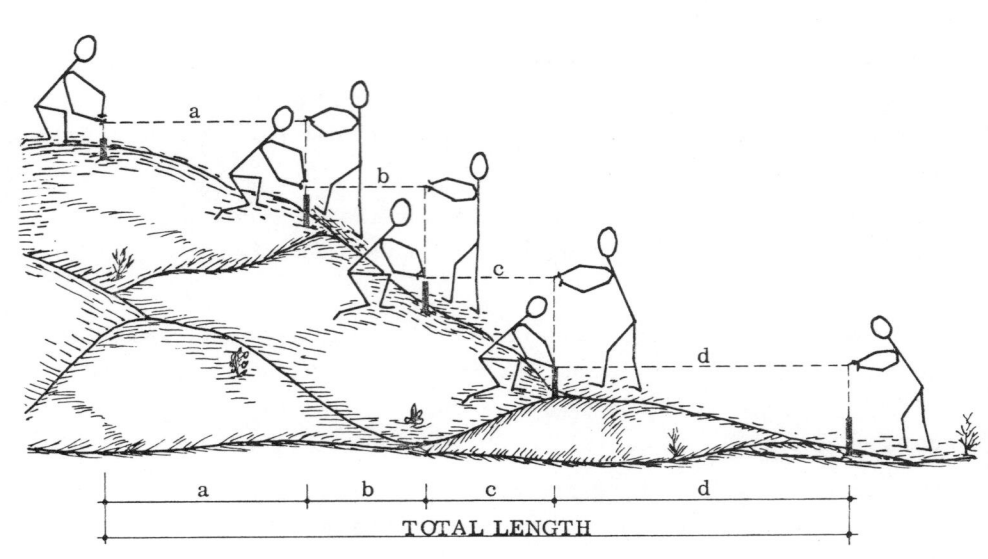

Fig. 1   MEASURING A DISTANCE ON A SLOPING SITE

Fig. 2   3 - 4 - 5 METHOD ON A SLOPING SITE

N P V C		SETTING OUT.
10	CON.	

SLOPING SITES

Setting out and measuring distances on sloping sites is simply the same procedure as explained in the section on uneven ground.

The work is carried out in steps (Fig. 1), with the line always held horizontally when measurements are taken.

The length of each section depends on how steep the slope is. The steeper the slope, the shorter the section, while slightly sloping areas allow longer sections (Fig. 1).

The construction of a right angle according to the 3-4-5 method is the same on a slope, as long as all the lines involved are straight. To see that this is true, take your try square and turn it around in your hands. You will observe that whatever the position of the try square, the angle is always the same, because the sides of the try square remain straight.

Even on a site which slopes in two or more directions (Fig. 2) it is possible to construct a right angle as long as the lines are straight and do not touch the ground. The pegs may have to be different lengths, in order to keep the lines off the ground.

Keep in mind that on sloping sites all lines stretched between the pins do not represent the horizontal length (unless they are also levelled) but only the future positions and directions of the walls, etc. (Fig. 2, see also Figs. 1 to 4 on the previous page).

The Rural Builder knows that the decision to build on a sloping site can have expensive consequences. There are some advantages such as easier sewage disposal, but the disadvantages are usually greater (Drawing Book, page 103).

NOTES:

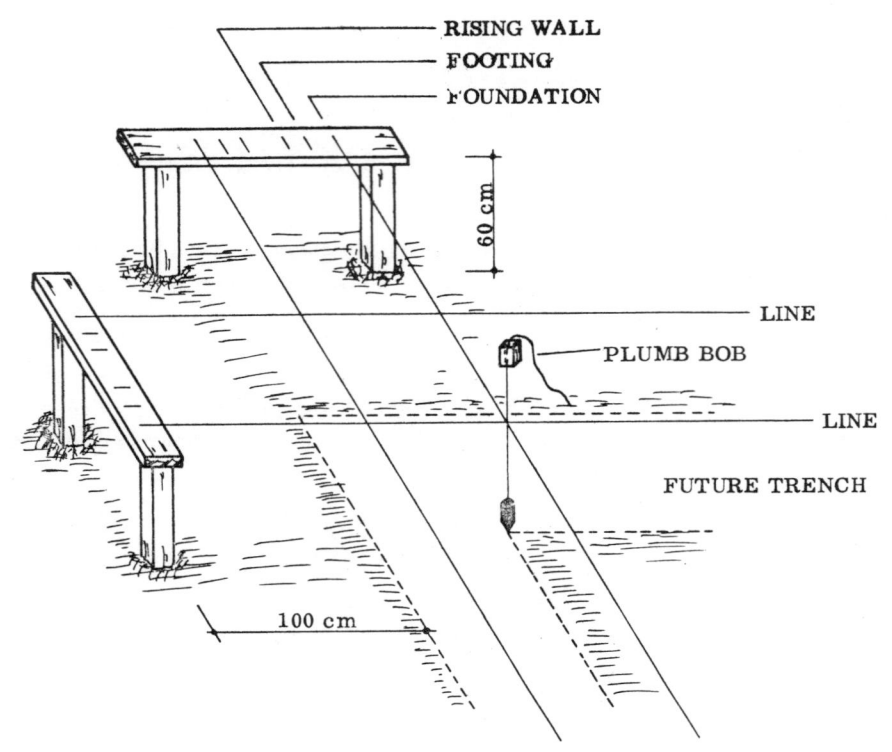

Fig. 1　　MARKING THE POSITION OF THE FOUNDATION ON THE GROUND

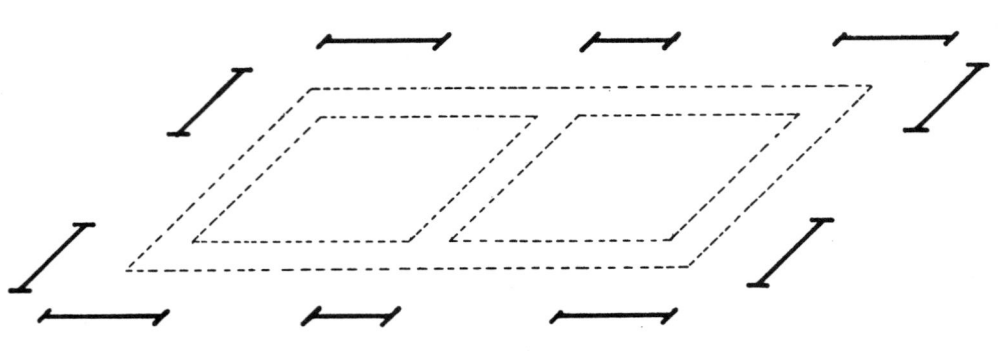

Fig. 3　　POSITIONS OF THE PROFILE BOARDS

NPVC		SETTING OUT.
12	CON.	

PROFILES

When the positions of the corners of the building are known, and the distances between them, then we can mark the positions and widths of the foundations as well as of the footings and plinth course.

This marking should be carried out in a relatively permanent way, so that it is accurate for a longer period. We do this by using profiles. A profile is a simple, temporary structure which maintains the correct locations of the various marks.

The profile consists of a board nailed flatwise on top of two pegs which are set in the ground, at a height of about 60 cm (Fig. 1). This height is necessary to lift the line well above the footings, so that later the plinth course can be marked from the profile.

If the soil is too hard to drive the wooden pegs into, then specially made iron pegs designed to receive a profile board can be used (Fig. 2). If these are not available, plain iron rods can be used, although they are less accurate since the lines are fixed directly onto the rods.

At the corners of the building, two boards are used, to mark in two directions (Fig. 3). To mark off the dividing walls, one board is used at each end of the future wall (Fig. 3).

The profiles are erected at a distance of about 1 m from the outside edge of the foundation (Fig. 1). Permanent divisions are marked on the boards to indicate the width of the foundations and the thicknesses of the footings and rising walls. The marks may be either saw-cuts or short nails, so that lines can easily be fixed to them as they are needed.

The positions of the foundations are then marked on the ground by plumbing down from the lines stretched between the profiles (Fig. 1).

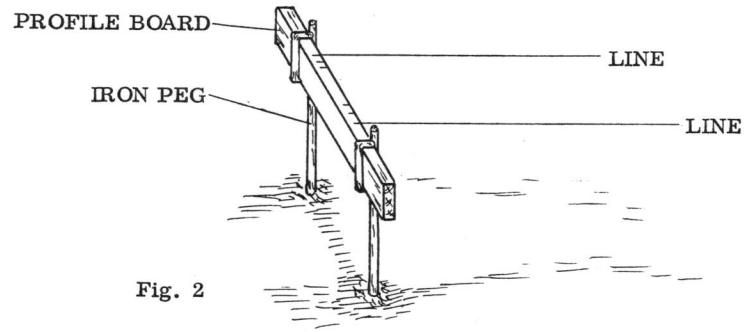

Fig. 2

SETTING OUT.

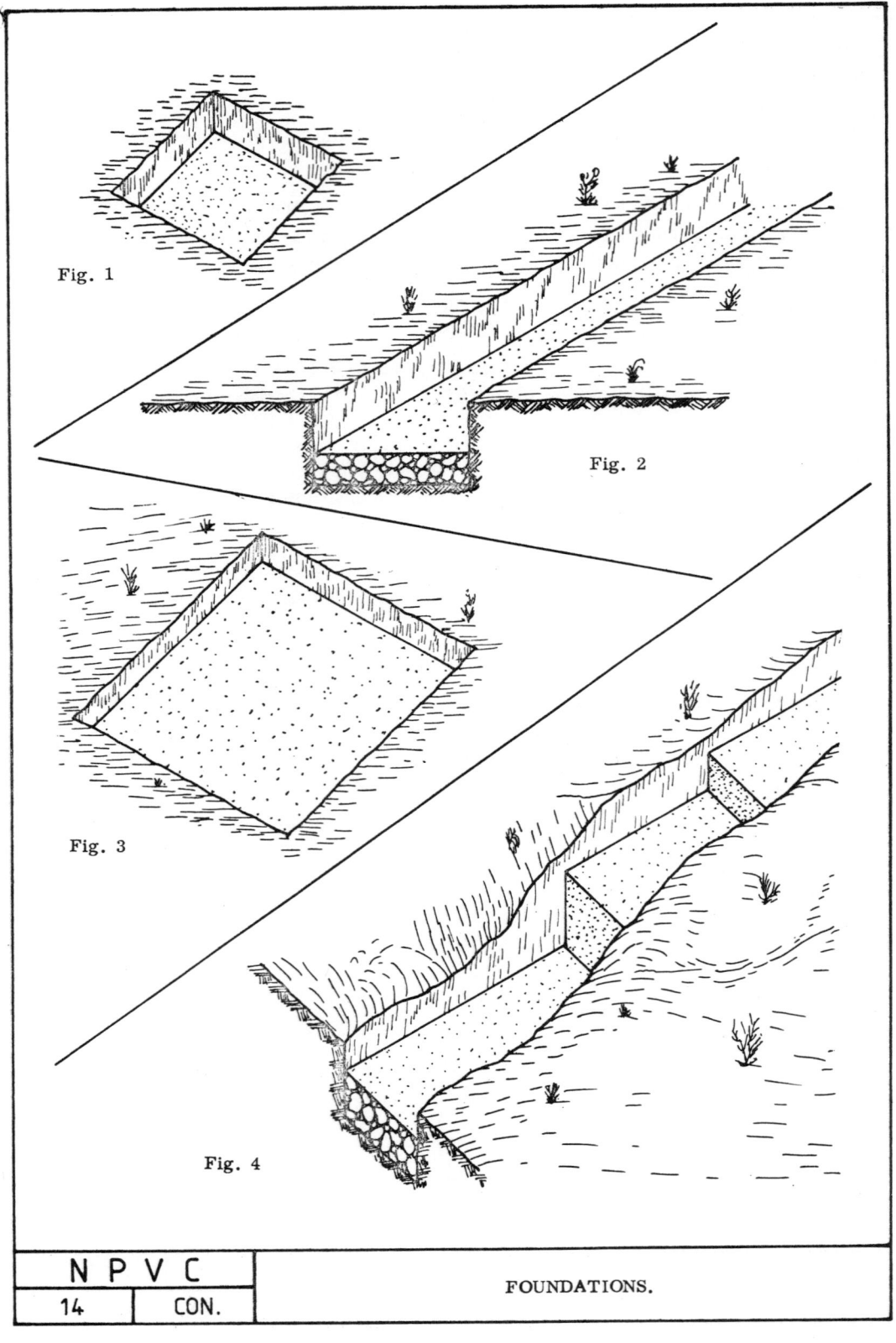

Fig. 1

Fig. 2

Fig. 3

Fig. 4

N P V C
14 CON.

FOUNDATIONS.

FOUNDATIONS

The stability of all buildings depends largely on the load-bearing capacity of the ground under them. As far as Rural Building is concerned, all hard soils are suitable for building. Soft soils, and those soils which become very soft and turn into mud when wet, are not suitable.

If a building is erected on the wrong kind of soil or if the foundations were constructed incorrectly, the building might settle unevenly, tilt or slide, or even collapse.

This can be avoided by selecting the right site for the building and by adapting the foundation construction to the soil conditions and the nature of the building.

TYPES OF FOUNDATIONS

The most common forms for foundations in Rural Building are:

- SINGLE FOUNDATION: This is for columns, pillars and poles, if they are detached from the building (Fig. 1).

- STRIP FOUNDATION: This is the most widely used foundation for walls (Fig. 2).

- SLAB FOUNDATION: This is used for water tanks and septic tanks (Fig. 3).

- STEPPED FOUNDATION: On sloping sites the so-called stepped foundation must be used, which is in fact just a special form of the strip foundation (Fig. 4).

NOTES:

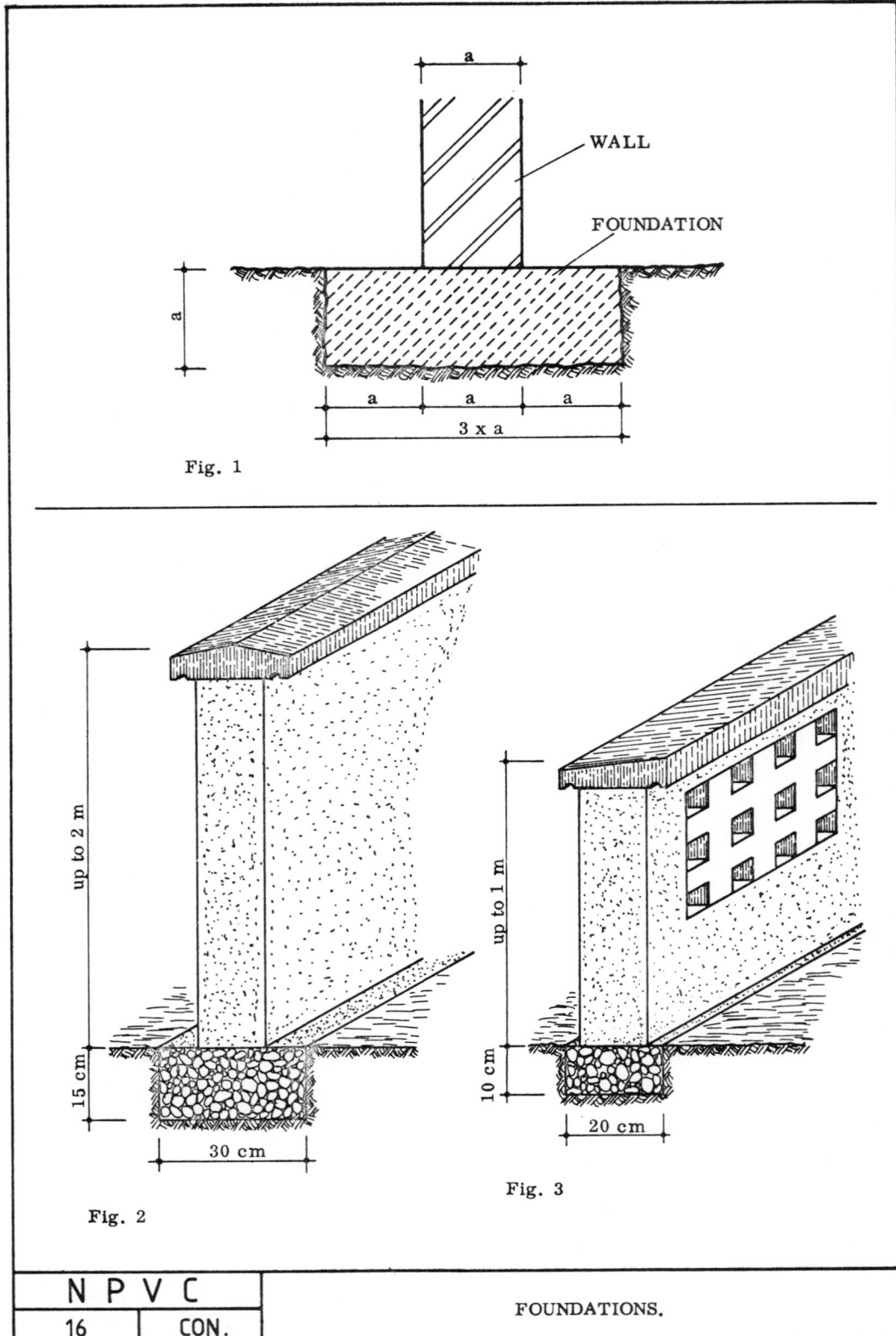

Fig. 1

Fig. 2

Fig. 3

N P V C		FOUNDATIONS.
16	CON.	

FUNCTIONS OF THE FOUNDATION

A foundation has to be constructed in a way that it can fulfill the demands that the building structure places on it. These are:

- To provide a solid, level base for the building

- To receive loads from the structure above

- To distribute the loads onto the ground over a larger area

- Thus, to prevent uneven settling of the building.

The above demands are met through the correct choice of dimensions, materials, and constructions for the foundation.

DIMENSIONS FOR STRIP FOUNDATIONS

Many years of building experience in northern Ghana have proven that the dimensions for concrete foundations shown in Fig. 1 are sufficient, provided that the structure consists of only a ground floor and it is built upon firm soil.

A simple rule is: the depth of the foundation should be no less than the thickness of the rising wall; while its width should be no less than 3 times the thickness of the rising wall (Fig. 1).

Lower walls (not exceeding 2 m in height) which do not carry loads can be built on smaller foundations; these should be at least 30 cm wide and 15 cm deep (Fig. 2).

Foundations which carry very low walls (less than 1 m high) such as decorative openwork screen walls enclosing verandahs, may be reduced to no less than 20 cm wide and 10 cm deep (Fig. 3).

NON-CONCRETE FOUNDATIONS

Although most foundations in Rural Building are made of concrete, there are situations where the choice of another material would be practical and economical. Some of the common alternatives to concrete foundations are:

- Rubble and boulder foundations

- Ashlar masonry foundations

- Laterite rock foundations

- Artificial stone (brick and sandcrete block) foundations.

In certain areas of the world, the first two types of masonry were formerly widely

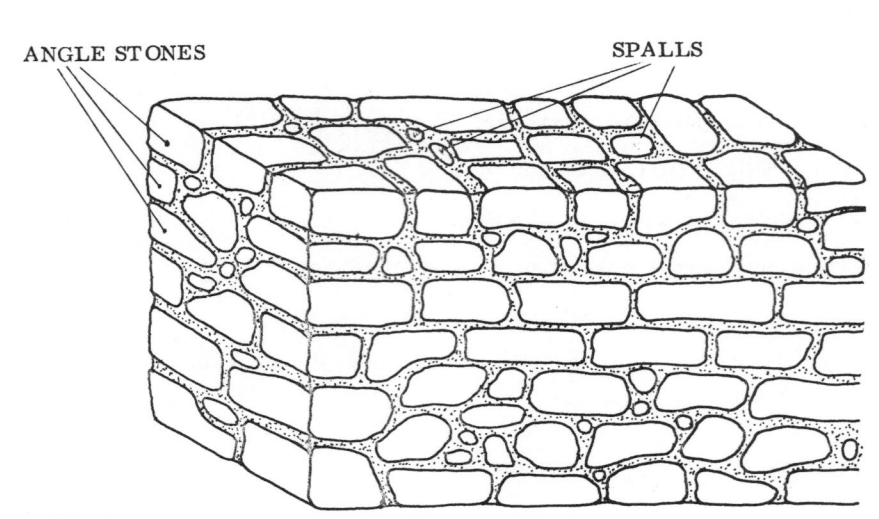

Fig. 1

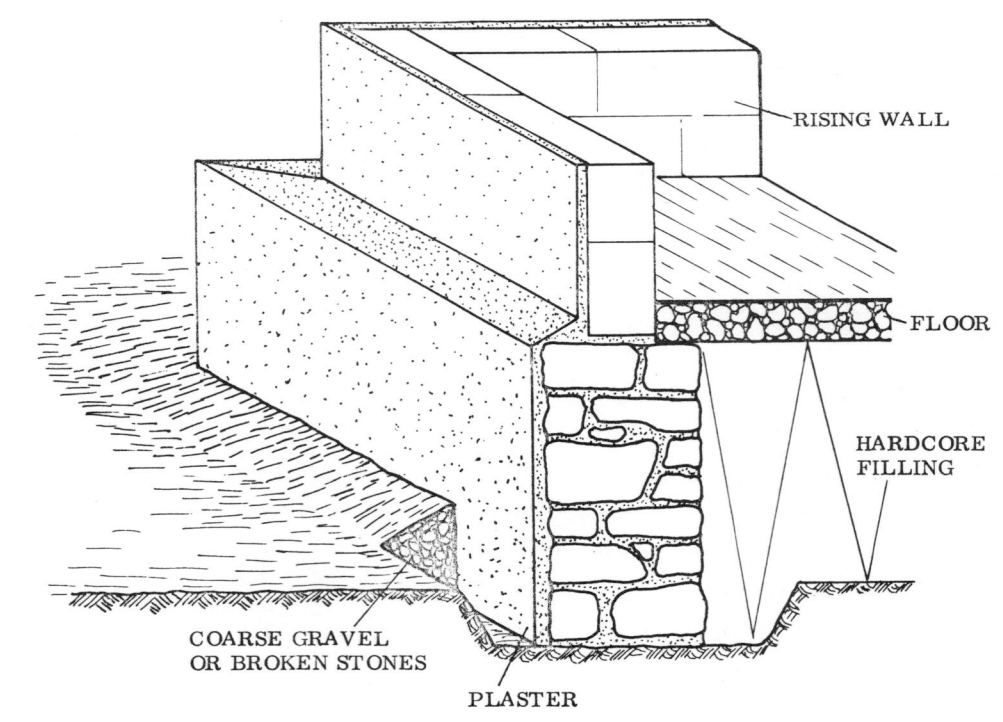

Fig. 2

NON-CONCRETE FOUNDATIONS.

used: not only foundations but also footings, plinths and even rising walls were made out of stones. Due to the increased use of cement worldwide, and the improved concrete technology, stone masonry became uneconomical to build with because it takes more time and labour. Eventually the use of some of these types of masonry disappeared.

However, cement can be very expensive and it makes sense to return to the use of local materials when possible.

RUBBLE AND BOULDER FOUNDATIONS

This kind of foundation is preferred on stony sites where rubble, rocks and boulders are found in large quantities, and where the soil is firm enough to permit its use.

This foundation saves building materials such as cement and timber for formwork, but it takes more time to construct and requires some skill.

As shown in Fig. 1, as far as possible very large stones are used, and the spaces that remain between them must be filled up with spalls. Spalls are smaller stones used to fill up the voids left in boulder masonry, in order to save cement mortar.

The stones are used in their raw state, but the boulders forming the quoin should be shaped a bit more regularly to form a secure corner bond. The first course is laid in mortar, not directly onto the bare ground. It is usually impossible to construct regular courses because of the irregular shapes of the stones.

If the foundation is made of a porous rock like laterite rock, it is advisable to plaster both the inside and outside faces to prevent dampness from penetrating. Otherwise only the outside face is plastered.

The excavation for the trench is made a little wider than the foundation (Fig. 2) so that the plaster can be applied right down to the bed of the first course. Coarse gravel or broken stones can be used later to refill the space when the plastering is finished. This prevents the erosion of soil along the foundations, especially under the eaves of an overhanging roof.

NOTES:

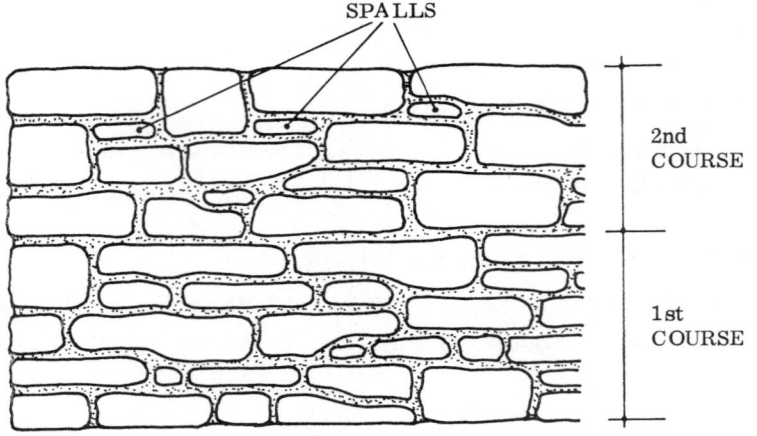

Fig. 1   ROUGH STONE MASONRY

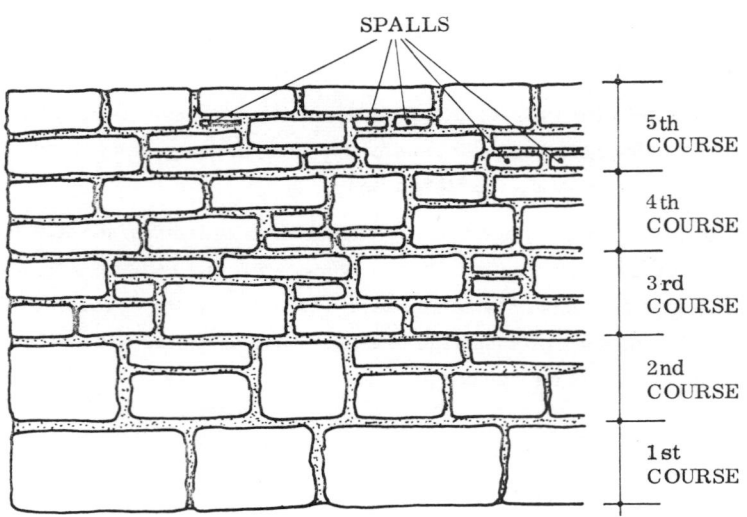

Fig. 2   HAMMER-DRESSED ASHLAR MASONRY

N P V C		NON-CONCRETE FOUNDATIONS.
20	CON.	

ASHLAR MASONRY FOUNDATIONS

In this type of masonry, the stones are "dressed" before they are used in the structure. To dress stone means to cut and shape it, and it is done to make the stones fit together better. There are four different types of ashlar masonry, depending on how much dressing is done and how the stones are put together. These are:

- Rough stone masonry

- Hammer-dressed ashlar masonry

- Broken range masonry

- Range masonry.

These are listed according to the increasing amounts of stone dressing and stone arrangement required for each method.

- ROUGH STONE MASONRY: This sort of masonry consists of natural stones which are shaped only slightly along their bed faces, or not shaped at all. As in boulder masonry, regular courses are not seen because of the irregularly shaped stones (Fig. 1).

- HAMMER-DRESSED ASHLAR MASONRY: As the name implies, the stones used for this type of masonry are roughly shaped with a hammer, so that the stretcher and header faces are approximately square to each other.

  The stones are laid in regular courses but the thickness of the stones may vary within one course (Fig. 2).

NOTES:

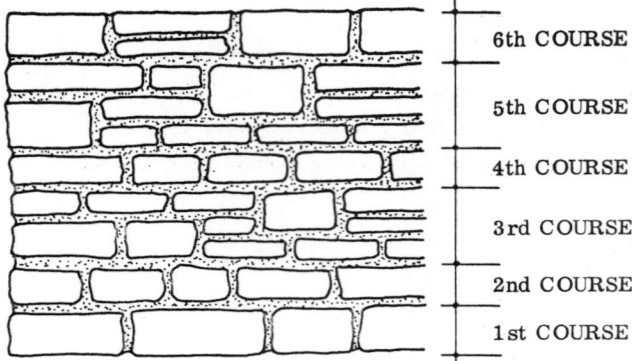

Fig. 1   BROKEN RANGE MASONRY

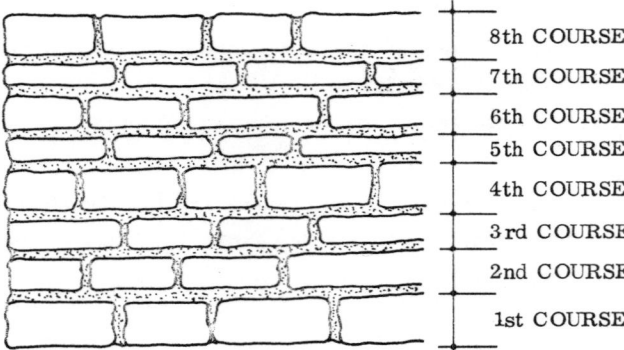

Fig. 2   RANGE MASONRY

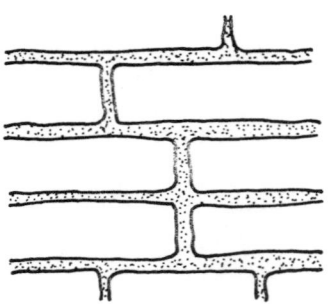

Fig. 3   WRONG!

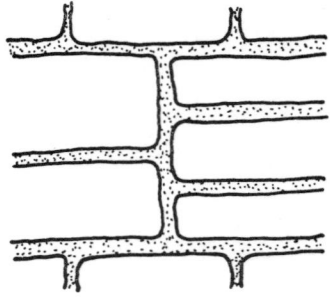

Fig. 4   WRONG!

NON-CONCRETE FOUNDATIONS.

- BROKEN RANGE MASONRY: The stones of this masonry are accurately shaped with the club hammer and cold chisel (Reference Book, page 15) so that all the faces are square to each other. The bond should not contain joints more than 3 cm thick. The height of the stones may vary within a course, and the height of the courses may also vary, with the result that the courses are continuous for only short distances (Fig. 1).
- RANGE MASONRY: The accurately squared stones are laid in courses, and each course is uniformly thick throughout its length. However, the courses are not all necessarily all the same thickness (Fig. 2).

BONDING PRINCIPLES

The following rules are observed for all four ashlar masonry techniques, regardless of how much dressing is done on the stones.

- Observe the structure of the rock, and if possible lay the stone in the way it has "grown". For example, if the stone appears to have horizontal layers, it should be laid so that the layers are flat.
- Never lay stones edgewise.
- The stones should overlap each other as far as possible; avoid making any continuous cross joints between two courses (Fig. 3).
- Also avoid making a group of continuous cross joints in one course (Fig. 4), as this creates an unpleasant appearance and an impression of a separation. A better arrangement is shown in Fig. 1.
- Fill up the bigger voids between stones with spalls, to save mortar.
- If ashlar masonry is used as a foundation, the first course is always laid in mortar, not directly on the ground.
- In case the foundation is combined with the footings and the plinth course, all ashlar masonry must be coursed and levelled at a height not exceeding 1,5 m, and preferably every 50 cm.

NOTES:

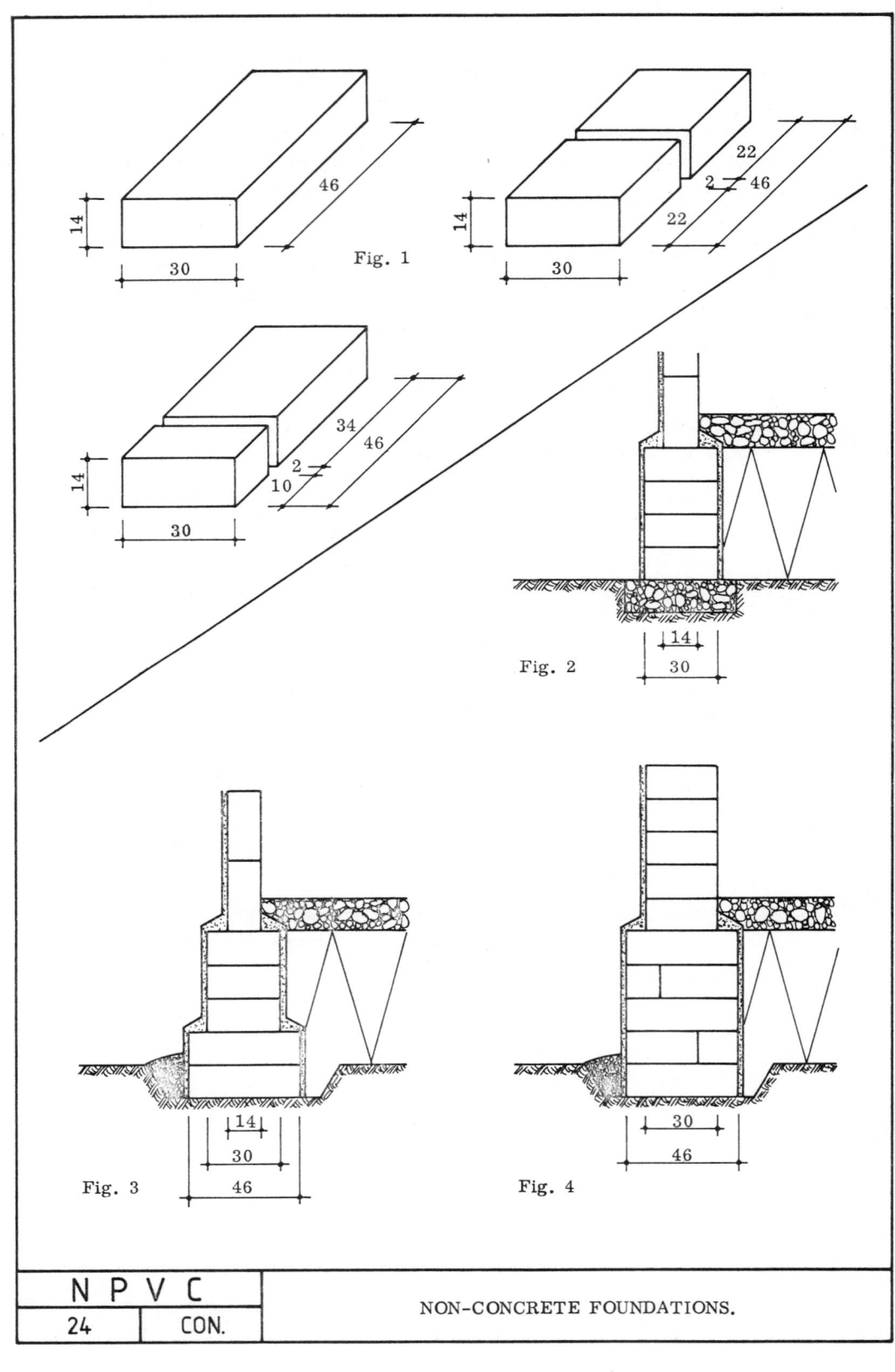

LATERITE ROCK FOUNDATIONS

Many areas of northern Ghana offer the opportunity to work with a highly suitable building material; namely laterite rock.

In structure and colour it appears similar to laterite soil, and although it is rather porous and comparatively soft when freshly dug from the earth, it gradually becomes hard and rock-like when it is exposed to the air.

Laterite stone has been used for many sorts of masonry because it is readily available, easily shaped, and strong.

Its only disadvantage compared with concrete is that it takes time to excavate the stone and shape it into blocks.

However, taking all the factors into consideration, there are some cases where laterite rock represents the most economical choice for building material.

Although the stone can be given any convenient shape, the dimensions of blocks made from laterite stone should preferably be 14 x 30 x 46 cm, which allows proper bonding with all sorts of masonry. Accordingly, the dimensions of half-blocks are 14 x 30 x 22 cm; for 3/4 blocks, 14 x 30 x 34 cm; and for 1/4 blocks, 14 x 30 x 10 cm (Fig. 1).

When blocks of laterite stone are used for footings, they are laid in stretcher bond, flatwise, which results in a footing which is 30 cm thick instead of 23 cm when sandcrete blocks are used (Fig. 2).

Foundations can be built out of laterite stone blocks; the blocks are laid in header bond, with a minimum of two flatwise courses (Fig. 3). Foundation and footings can be combined as shown in Fig. 4, which allows the erection of stronger outside walls.

Since the laterite stone is porous, the foundations should be plastered on both the inside and outside faces.

NOTES:

NON-CONCRETE FOUNDATIONS.

NPVC CON. 25

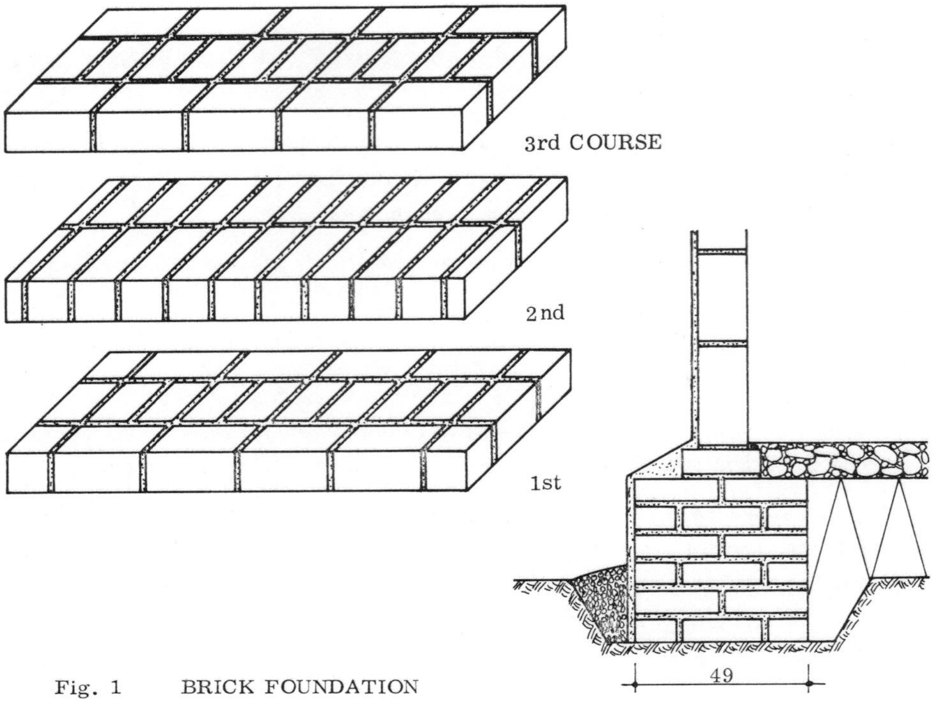

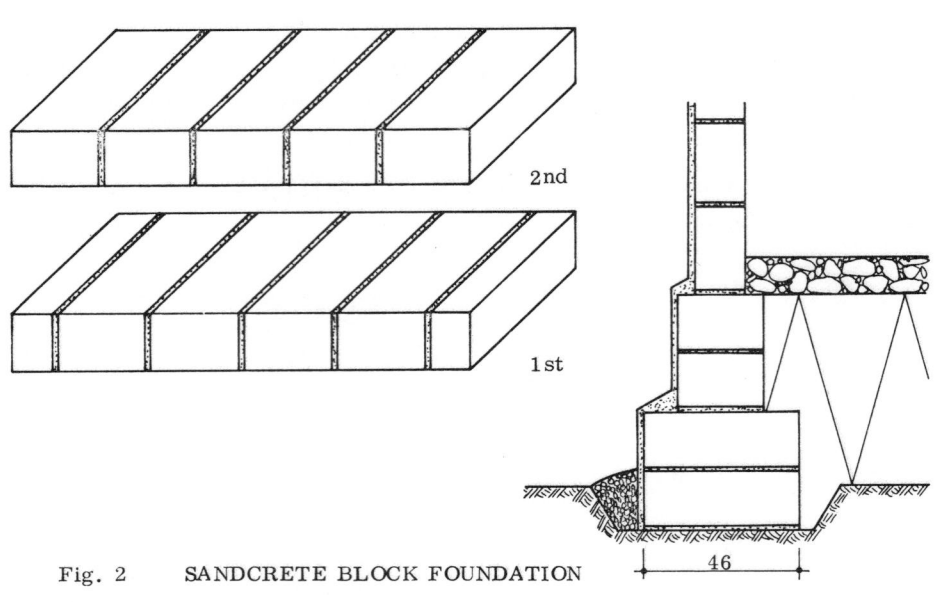

Fig. 1 BRICK FOUNDATION

Fig. 2 SANDCRETE BLOCK FOUNDATION

N P V C		NON-CONCRETE FOUNDATIONS.
26	CON.	

ARTIFICIAL STONE FOUNDATIONS.

As far as Rural Building is concerned, there are only two artificial stone products of importance for foundations: bricks and sandcrete blocks.

Both products are seldom used for foundations in northern Ghana. We don't use bricks because, while the clay is available to make them, it would take too much firewood or other fuel to fire the bricks. Concrete or laterite stone blocks are usually more economical to use than sandcrete blocks.

- BRICK FOUNDATIONS: Brick is one of the oldest known artificial stones; it is made from clay which is shaped, dried, then fired in a kiln. The usual dimensions of a brick are 7,1 x 11,5 x 24 cm, a size which results in a masonry containing many joints, but which allows the construction of most complicated bonds.

  Foundations made out of bricks must be at least 5 courses high and carefully laid in cross bond or English bond (Fig. 1). Note the different arrangements of the alternating courses.

- SANDCRETE BLOCK FOUNDATIONS: In an area where there is a lot of sand available but no broken stones, gravel or laterite rock, etc., one may be forced to use sandcrete blocks to build the foundations. The mix proportion for these blocks should be no less than 1:8, and the blocks themselves should be laid in header bond, as seen in Fig. 2.

NOTES:

Fig. 1

Fig. 2

Fig. 3

Fig. 4

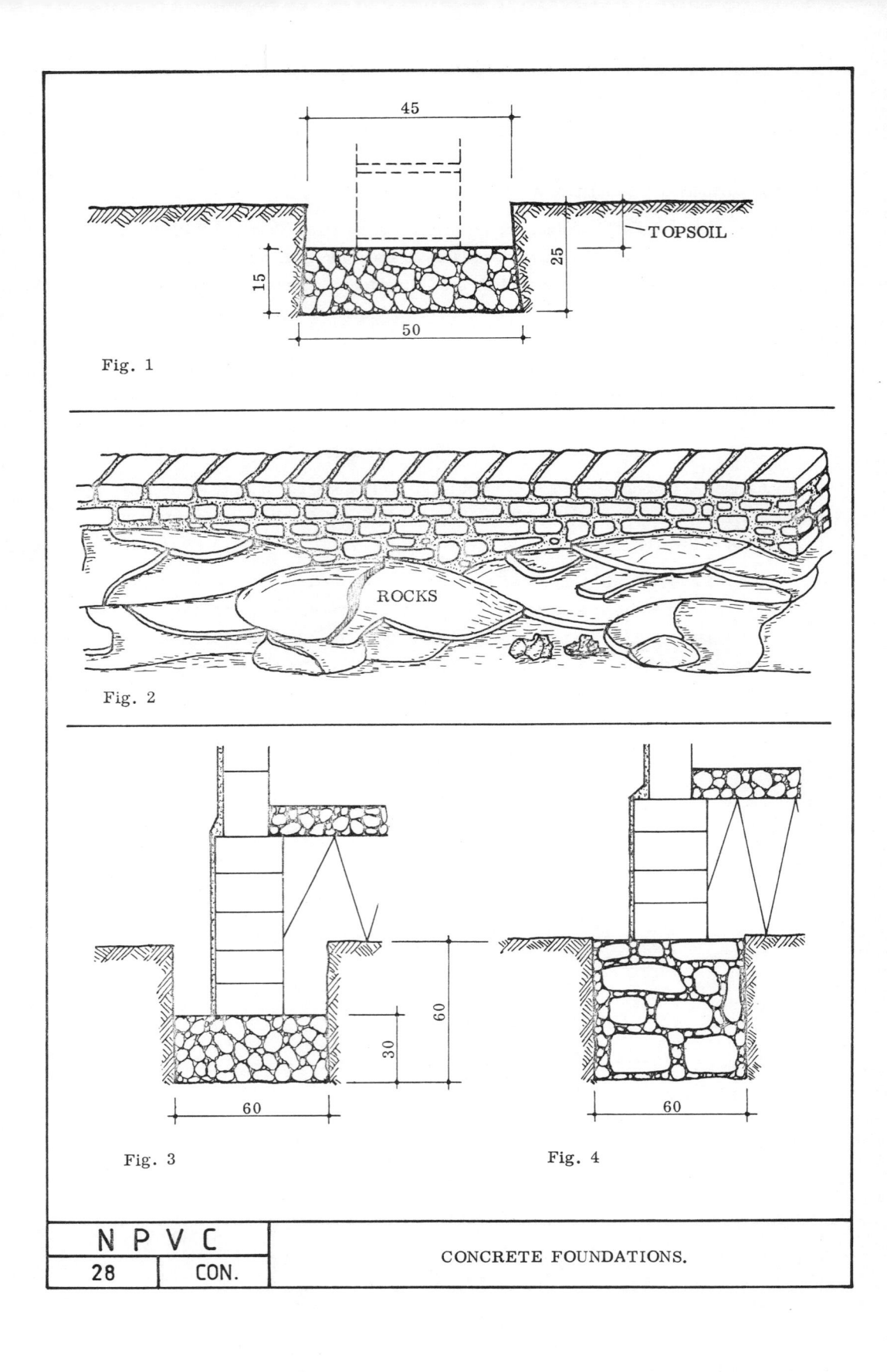

N P V C		CONCRETE FOUNDATIONS.
28	CON.	

CONCRETE FOUNDATIONS

EXCAVATING THE FOUNDATION TRENCHES

After the setting out is complete, including marking the positions and widths of the foundations on the ground, the lines are removed from the profiles or pins and the excavation work is started.

All the topsoil and as much as possible of the soft soil and light soil is removed, and brought to a place where it can either be used immediately or kept for a future dry season garden.

In most cases soils of a firm consistency and good load bearing capacity are found at depths ranging from 15 to 25 cm, which is deep enough for the construction of foundations for a rural dwelling place (Fig. 1).

There are of course exceptional cases where the rock is close to the surface, or even at the surface, which makes the work easier; or the opposite situation of very deep soft soil, which is a problem.

In areas where there is light or heavy rock close to the surface, the thickness of the foundation can be reduced, but to no less than 5 cm thick. The surface of the rocks should be roughened if necessary to provide a good grip, and cleaned before the concrete is cast. These areas are usually rich in stones, so one should check out the possibility of making boulder foundations. The boulder foundations can also be carried out as a stepped foundation if necessary, following the contours of the rocky surface (Fig. 2).

Very deep soft soil is a problem, and it requires special measures. In general, all excavations should be continued until a layer of good firm soil is found. When a depth of 60 cm is reached and the soil is still too soft, the excavation work is stopped.

Instead of digging deeper, widen the trench to at least 60 cm. This increases the total area on which the structure rests, with the result that the pressure on the soil underneath is distributed over a wider area.

Fig. 3 shows that the thickness of the widened foundations must be increased to 30 cm, because of how the foundation material distributes the pressure of the wall above.

In order to save concrete, put bigger stones, rubble or boulders into the foundation; it should preferably be flush to the ground surface (Fig. 4).

CONCRETE FOUNDATIONS. NPVC CON. 29

SECTION THROUGH THE LENGTH

CROSS SECTION

IRON ROD ⌀ 6 mm

IRON RODS ⌀ 12 mm

Fig. 1

50 cm

50 cm

SOFT SOIL

Fig. 2

Fig. 3 WRONG

Fig. 4 CORRECT

Fig. 5 CORRECT

N P V C	
30	CON.

CONCRETE FOUNDATIONS.

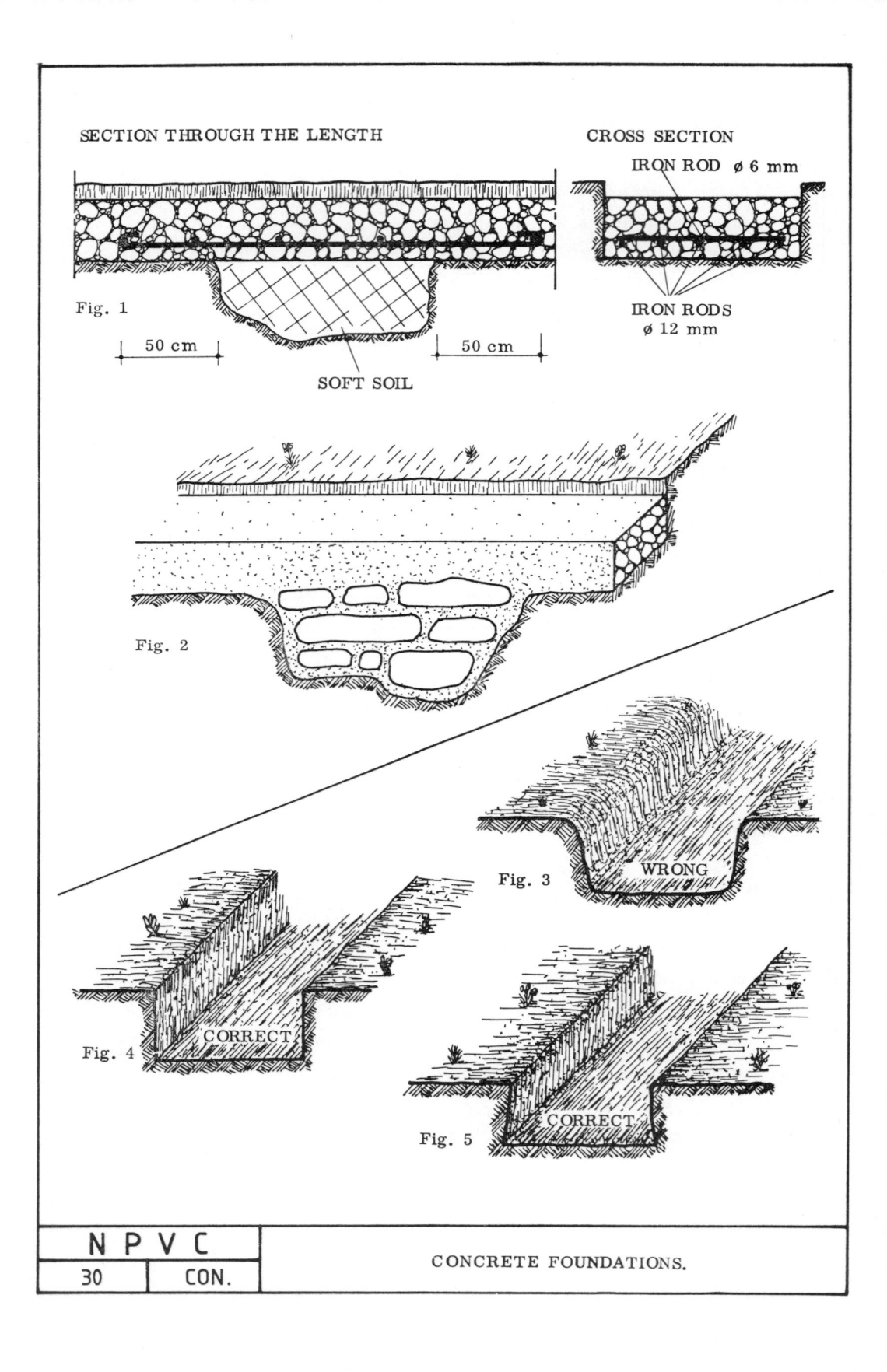

- SOFT POCKETS: During the excavation work it sometimes happens that the character of the soil changes unexpectedly: spots or pockets of soft soil are found in the firm soil. These may be natural deposits or man-made holes that have gradually filled in. The load bearing capacity of the soil is reduced in these places.

If the area is relatively small and runs across the trench, it is advisable to reinforce the concrete foundation covering the pocket with a network of crossed iron rods (Fig. 1). The rods should be long enough to bridge the soft part and project past it on both ends to anchor securely in the foundation which is supported by firm soil.

In case the area is too wide to be reinforced in this way, the soft soil is removed and the hole is refilled with compacted layers of boulders, stones, and sharp sand; or with concrete, thus improving the load bearing capacity (Fig. 2).

When the foundation has to cross a very wide section of soft soil, that section of the foundation has to be made wider and deeper than the rest (see the previous page).

- SHAPE OF THE TRENCHES: Although it is generally accepted that the ideally shaped foundation trench has walls that are at right angles to the levelled bottom of the trench (Fig. 4), it is often observed that unskilled or even skilled workers tend to dig trenches which are incorrectly shaped. Special attention should be given to this problem during excavations in harder soils, because the more difficult the digging, the more likely it is that incorrectly made trenches will result (Fig. 3).

Experience has shown that it is best to ask the workers to make the trenches a bit wider at the bottom, to ensure that the minimal width of the foundation is maintained everywhere (Fig. 5).

NOTES:

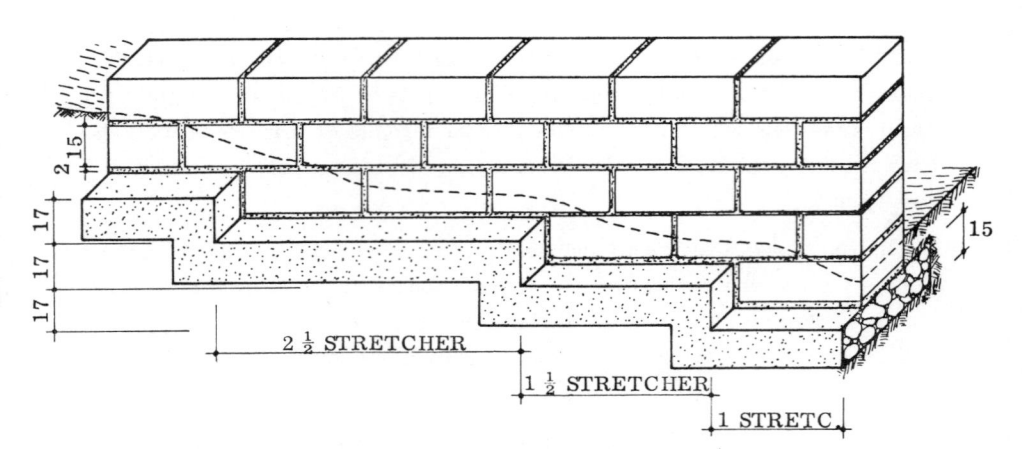

Fig. 1   STEPPED FOUNDATION WITH SANDCRETE FOOTING

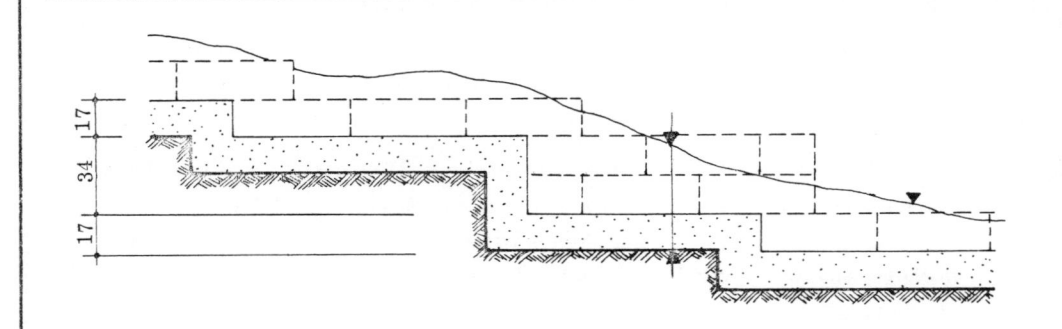

2½ STRETCHER
1½ STRETCHER
1 STRETC.

Fig. 2   STEPPED FOUNDATION WITH 2 COURSES ON ONE STEP

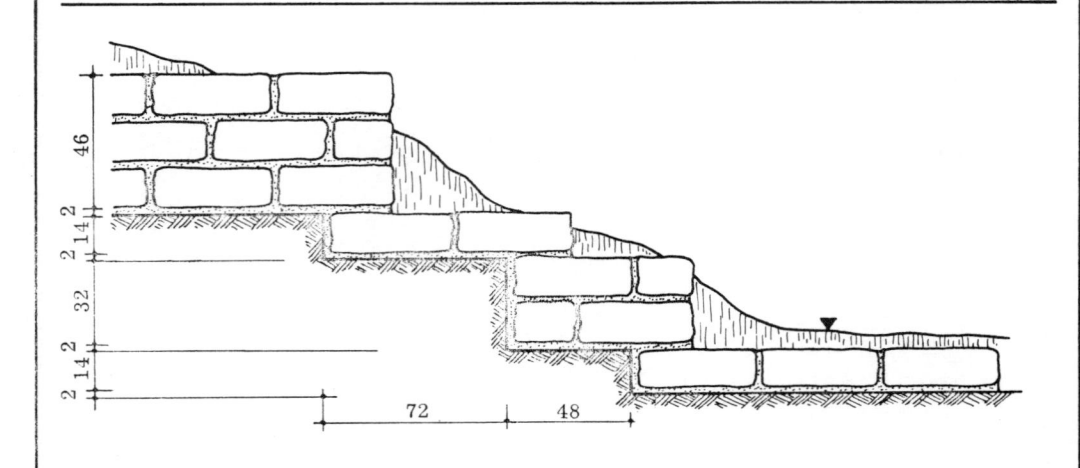

Fig. 3   STEPPED FOUNDATION WITH LATERITE STONE

N P V C		
32	CON.	CONCRETE FOUNDATIONS.

MEASUREMENTS OF STEPPED FOUNDATIONS

The number, length and height of the steps in the foundation depend on the shape and steepness of the ground contours. To make it easier to construct the footings later and to save materials, the skilled builder will adapt the length and height of the steps to fit with the type of blocks that will be used for the footings.

Fig. 1 shows a common stepped foundation made of concrete, and footings built of sandcrete blocks. It can be seen that the concrete thickness is 15 cm throughout, while the steps are different lengths.

- HEIGHTS OF THE STEPS: The height of each concrete step is always the thickness of a sandcrete block, plus the thickness of the mortar bed; making 15 cm plus 2 cm equals 17 cm.

    In case the steps have to be steeper, the height of the step is always a full multiple of 17 cm: 34, 51 cm, etc. so that 2 or 3 courses of sandcrete blocks can fit on the step (Fig. 2).

- LENGTHS OF THE STEPS: All masonry is supposed to be built in half-block bond. Accordingly, the minimum length of a step is half a block plus the joint: 22 cm plus 2 cm equals 24 cm; or for longer steps, a full multiple of this: 48 cm, 72 cm, 96 cm and so on.

If the foundation is to be constructed with laterite stone blocks (Fig. 3), the height of one step is reduced to 16 cm, that is 14 cm for the block plus 2 cm for the joint. The lengths of the steps are the same as for sandcrete blocks.

The actual construction of a stepped foundation and its footings must always be started from the lowest step, to prevent bonding mistakes.

NOTES:

CONCRETE FOUNDATIONS.

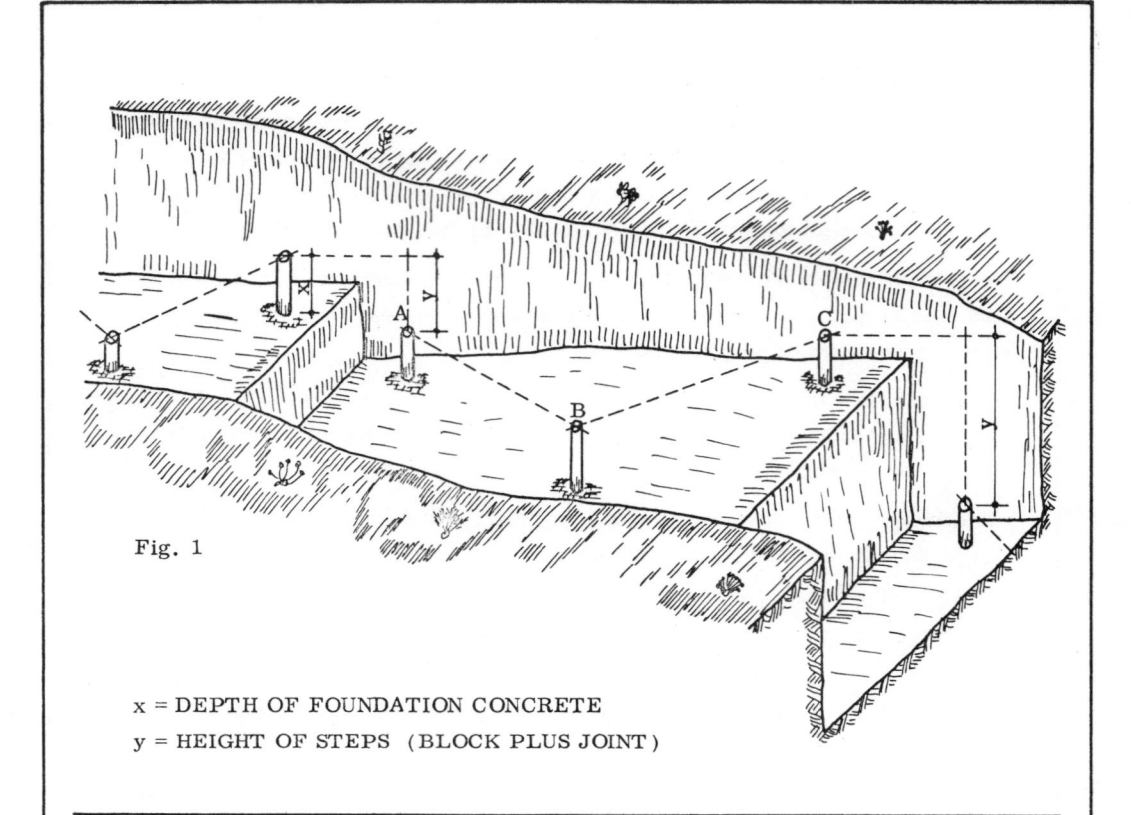

Fig. 1

x = DEPTH OF FOUNDATION CONCRETE
y = HEIGHT OF STEPS (BLOCK PLUS JOINT)

NOTES:

N P V C		
34	CON.	CONCRETE FOUNDATIONS.

LEVELLING THE TRENCHES AND MARKING THE FOUNDATION DEPTH

The bottoms of all foundation trenches must be perfectly level, not only in the length but also across the width. Otherwise the building may settle unevenly, which can cause cracks in the structure or even a complete collapse.

This is why a foundation on a sloping site must be constructed in steps (Fig. 1) instead of taking a course parallel to the slope of the ground.

An improperly constructed foundation can result in a building that is a danger to the people living in it, and also a possible loss of valuable property.

When the trench has been excavated and roughly levelled, the next step is to determine the depth of the foundation concrete and at the same time, to level the bottom of the trench more exactly.

This is done by inserting iron or wooden pegs in the trench so that they project from the bottom by a distance equal to the depth of the foundation (Fig. 1). The distance between the pegs should be no further than the straight edge can bridge, so that it is possible to level between them using the straight edge and spirit level. Trenches can be levelled more accurately over longer distances by using the water level.

The triangular pattern of the pegs is necessary to obtain a level surface: if the distances A-B and B-C are level, then A-C must also be level (Fig. 1).

When the tops of the pegs are level, we can measure how far each peg sticks out from the bottom of the trench, and thus check whether the trench bottom is also level. High areas are skimmed off with a shovel, while hollows are filled with concrete. Never refill hollow areas with excavated soil, as this might lead to uneven settling of the foundation.

NOTES:

CONCRETE FOUNDATIONS. N P V C CON. 35

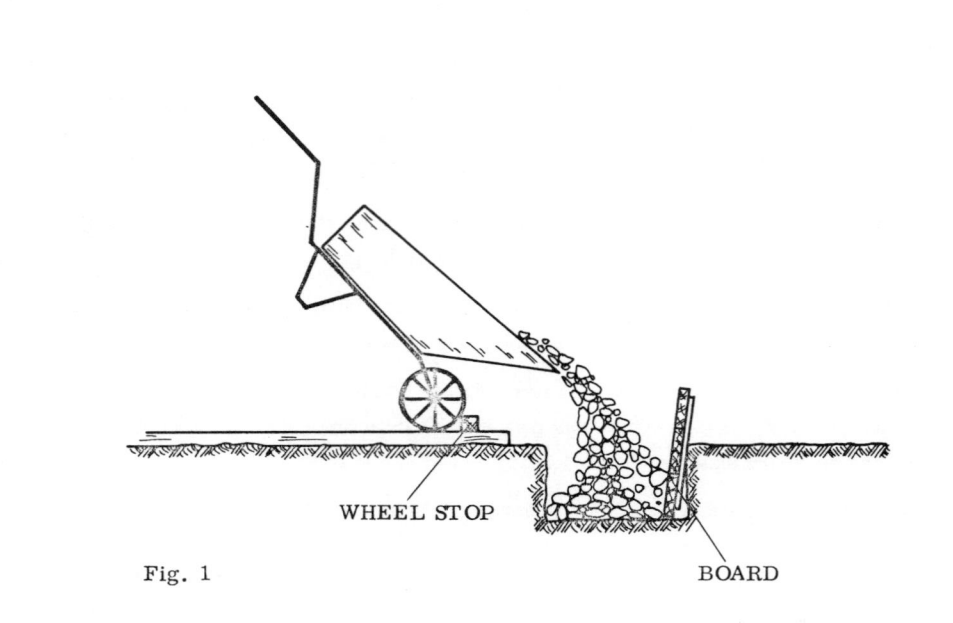

Fig. 1

WHEEL STOP

BOARD

NOTES:

CONCRETE FOUNDATIONS.

MIX PROPORTIONS

The correct mix proportions for foundation concrete should generally be determined through the calculations of an engineer. Since this is usually not possible in Rural Building, the Rural Builder can decide the mix proportion with the help of the Tables of Figures, Reference Book, page 234. The data given there and the following hints are approximate values and are intended as a guide. The general range for the mix is from 1:10 to 1:15.

- For small projects like single dwellings, stores, etc., a mix proportion of 1:15 is sufficient because the total weight on the foundation is not so great.

- Bigger buildings like two-storeyed houses, halls or churches require a better mix proportion of 1:12.

- Elevated water storage tanks, bell-towers and other large heavy structures should be based on foundations mixed in a proportion of 1:10.

PREPARATIONS BEFORE CASTING

If the ground is dry, the foundation trenches have to be wetted down before the concrete is cast. This helps to reduce the absorption of moisture from the concrete by the soil. Take care that the sides of the trenches are also thoroughly wetted.

After the foundation concrete is mixed (Reference Book, pages 161 to 163), it has to be transported to the trenches, either in buckets, headpans or wheelbarrows. All transportation of the concrete has to be done without too much delay or vibration, as this could lead to the aggregates becoming separated. If wheelbarrows are used, the paths should be covered with boards to reduce vibration and to make it easier to push the wheelbarrows.

It is a good idea to provide a wheel stop (Fig. 1) beside the trench. This helps to ensure that the concrete is emptied right in the middle of the trench and does not take along dust and dirt from the sides of the trench. A board can be used to protect the opposite side of the trench (Fig. 1). This is not necessary when the concrete is picked up with a shovel from the wheelbarrow.

The containers in which concrete is transported should be kept wet, so that no concrete sticks to the container when it is poured.

NOTES:

CONCRETE FOUNDATIONS.

N P V C
CON. 37

Fig. 1 — LEVELLING, COMPACTING, POURING, RAMMER

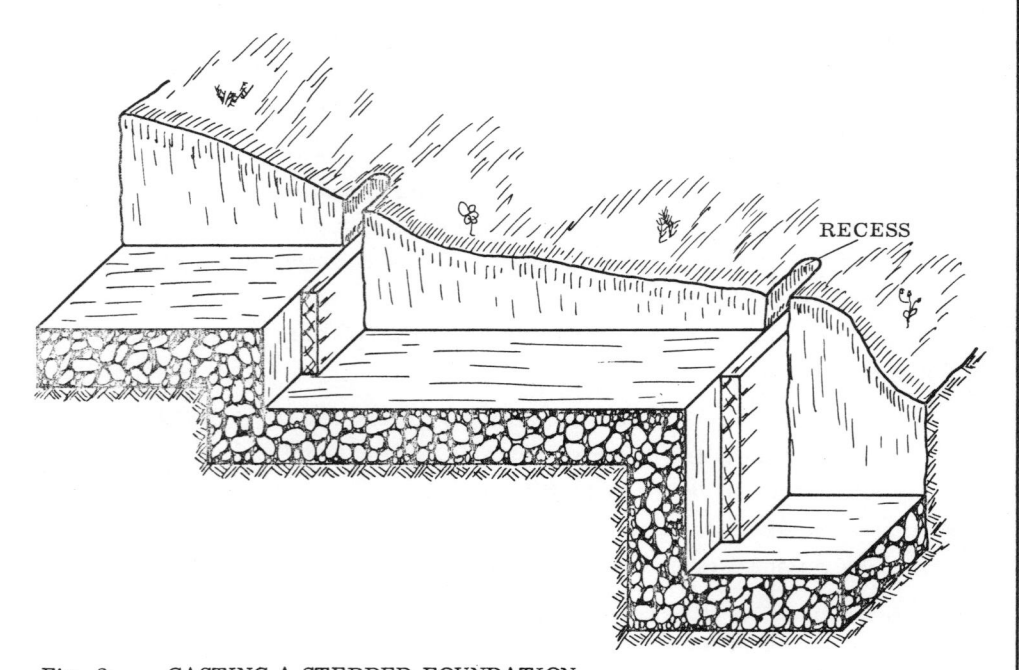

Fig. 2   CASTING A STEPPED FOUNDATION — RECESS

NPVC 38 CON.   CONCRETE FOUNDATIONS.

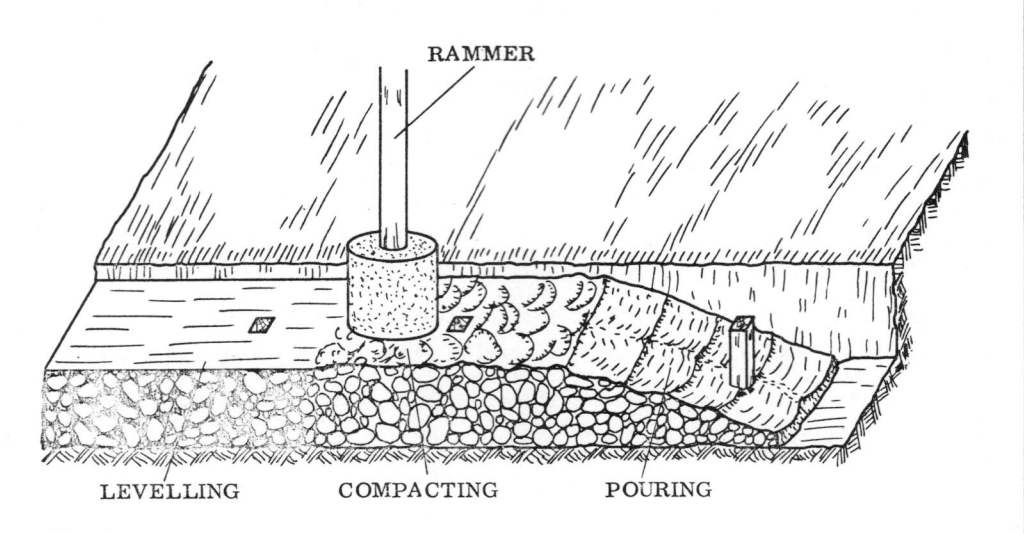

CASTING

The concrete must be cast systematically so that the compacting and levelling follow immediately after it is poured in the trench. The headpan loads or wheelbarrow loads are deposited in an orderly way, not just dumped anywhere in the trench (Fig. 1).

Do not cast foundation concrete in sections with spaces in between, as this can lead to disturbances of the hardening process when the gap is filled up later. Always start from the far end of the trench and work your way closer to the mixing area. Two groups of workers can cast foundations on different sides of the building at the same time.

- COMPACTION: The concrete must be well compacted so that no air voids remain. Since foundation concrete has a rather stiff consistency, to compact it requires the use of heavy rammers (Reference Book, page 19).

  The concrete is applied in layers no more than 15 cm deep, and each layer is compacted with the rammers. Pay special attention to compacting the corners and the outside edges. Stiff concrete should be compacted until its surface becomes wet.

  Do not be tempted to compact the concrete just roughly, or to add water to ease the work. The production of a good foundation is a hard job, but only hard work will result in a good quality foundation.

- STEPPED FOUNDATION: In casting the vertical parts of a stepped foundation, pieces of board are used as shown in Fig. 2. These are fitted into recesses cut in the side of the trench. The width of the board is the same as the height of the step (Fig. 2).

NOTES:

Fig. 1

Fig. 2

Fig. 3

Fig. 4

N P V C		
40	CON.	CONCRETE FOUNDATIONS.

LEVELLING THE FOUNDATIONS

As soon as the foundation is compacted, the top has to be levelled flush with the top of the pegs. For this purpose a strike board is used.

If the compaction has been carried out correctly, the levelling should be fairly easy as there is not much concrete left to strike off.

All wooden pegs have to be removed and the holes filled with concrete, while the iron pins may remain if these were used.

As soon as the hardening process starts, as can be seen by the dull and dry looking surface, the area where the footing course will be set must be slightly roughened with the blade of a trowel, to provide a good grip for the mortar (Fig. 1).

CURING

If the concrete is disturbed during the hardening process, serious defects may be produced.

Freshly cast concrete must therefore be covered with empty cement bags, straw, mats, boards or moist sand to protect it against:

- The rain, which can wash out the cement paste, leaving the non-bound aggregate behind (Fig. 2).

- The sun, which can "burn" the surface of the concrete, so that although the concrete looks cured, under the surface it has not set hard enough (Fig. 3).

- The wind, which can dry up the surface, resulting in cracks due to excessive dryness and shrinkage: especially during the harmattan season (Fig. 4).

- The dirt, which may get on the surface and interfere with the grip of the footings.

Any vibration near the hardening concrete could cause cracks and destroy the internal structure of the concrete. Don't continue with any excavation work, etc. near the just-completed foundation.

Once the surface has set sufficiently hard, the concrete must be wet down and kept wet for three days.

NOTES:

BUILDING WITHOUT FOOTING

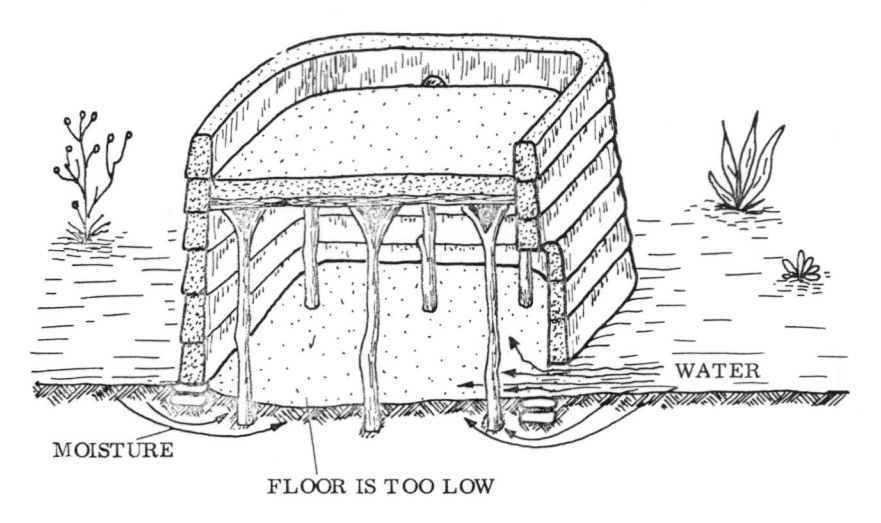

MOISTURE

FLOOR IS TOO LOW

WATER

Fig. 1

BUILDING WITH FOOTING

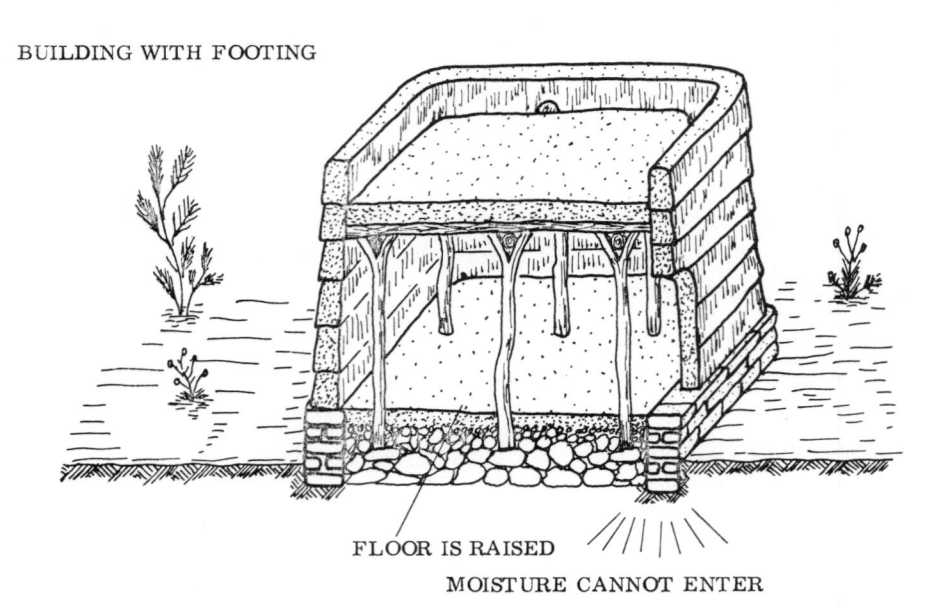

FLOOR IS RAISED
MOISTURE CANNOT ENTER

Fig. 2

N P V C		FOOTINGS.
42	CON.	

FOOTINGS

The projecting courses at the foot of a wall, between the rising wall and the foundation, are called the footings. If there is no wider concrete foundation, the combination foundation/footing is called a footing (see Foundations, page 25).

FUNCTIONS OF THE FOOTINGS

Since the footings form the link between the rising wall and the foundation, the demands on them are similar to those set on the foundations. These are:

- To provide a solid, level base for the walls
- To receive loads from the structure above
- To distribute the loads onto the foundation (or the ground).

In addition, the footings have to:

- Raise the floor high enough above ground level to prevent moisture from rising through to the floor, and to keep the landcrete blocks of the rising wall dry.

As far as Rural Building is concerned the last function is the most important one.

MATERIALS AND MEASUREMENTS

The most common material used for footings is sandcrete blocks, laid flatwise in a half-block bond. The courses are 23 cm thick and 17 cm in height, and they are laid in the centre of the foundation strip.

In order to effectively prevent moisture penetration, three courses should be laid above ground level, thus raising the soffit of the floor to a height of 51 cm above ground level. This is the minimum for all bedrooms and living areas.

Only in cases where the building is not to be used as a dwelling place for people, for example a store or a shelter for animals, may the footings be built to a lower level.

The use of boulder or laterite rock masonry is preferred as this can save considerable amounts of cement. The dimensions of such footings will be the same as those of sandcrete footings, when the laterite rock blocks are cut to the same dimensions as sandcrete blocks.

- COMPARE: Fig. 1 shows a building with no footing. The floor is at ground level and water can easily enter. The building in Fig. 2 rests on a footing and the floor is raised above ground level, so it will be drier and thus healthier to live in, and also less likely to collapse since the mud walls are kept drier.

Fig. 1

Fig. 2

RISING WALL

FOOTING

VERANDAH

Fig. 3

RISING WALL

FOOTING

VERANDAH

Fig. 4

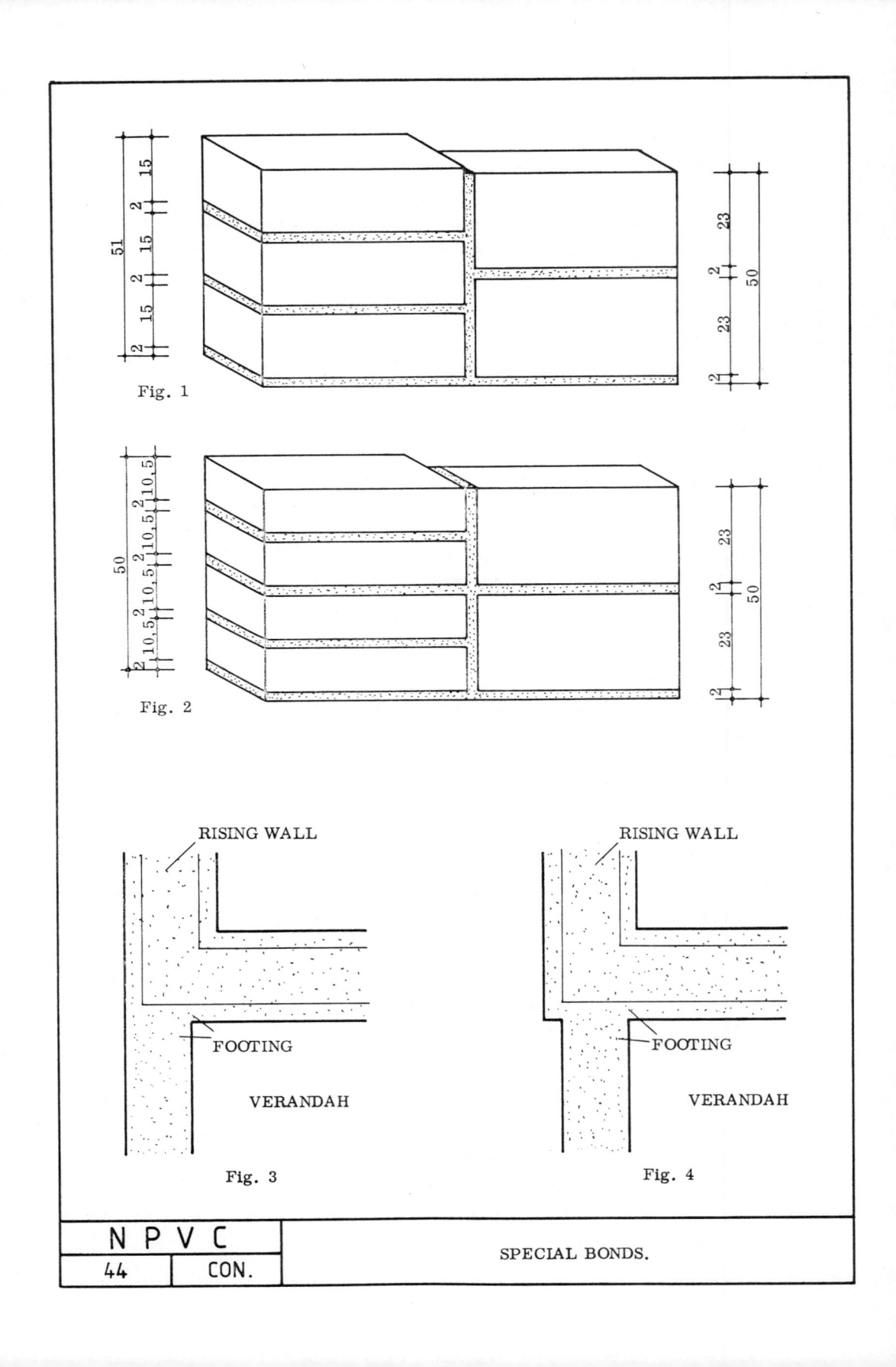

N P V C		SPECIAL BONDS.
44	CON.	

SPECIAL BONDS

Special bonding problems occur at quoins, T-junctions and cross junctions where walls of different thicknesses meet. These situations are frequently found in connection with the footings. They may also occur in other masonry.

An example where footings with different thicknesses meet is in the construction of a verandah. This is because the footing courses under verandahs are laid edgewise instead of flatwise, in order to save material. This is possible because there are no great loads on the verandah footings.

The bonding problem arises from the fact that the dimensions of common sandcrete blocks make it difficult to form a bond between the edgewise courses and those laid flatwise. Fig. 1 shows that two courses laid edgewise do not exactly correspond with three courses laid flatwise: there is a 1 cm difference in height.

Footing blocks can be specially made to dimensions that will show a better correspondence, so that bonding is easier and less block-cutting is needed to make the bonds. Fig. 2 shows four flatwise courses of specially made blocks, 10,5 cm thick, which correspond perfectly with two courses laid edgewise. Thus bonding is made easier.

The verandah footing may be kept in line with the outside face of the wall footing (Fig. 3), or in line with the rising wall (Fig. 4) which results in a recessed footing (see Drawing Book, pages 65, 66, and 90).

Each type of block can be used with either a recessed footing or a flush footing. Therefore there are four possible situations:

- Common sandcrete blocks in a flush footing

- Common sandcrete blocks in a recessed footing

- Specially made sandcrete blocks (10,5 cm thick) in a flush footing

- Specially made sandcrete blocks in a recessed footing.

NOTES:

HOUSE
LOGGIA

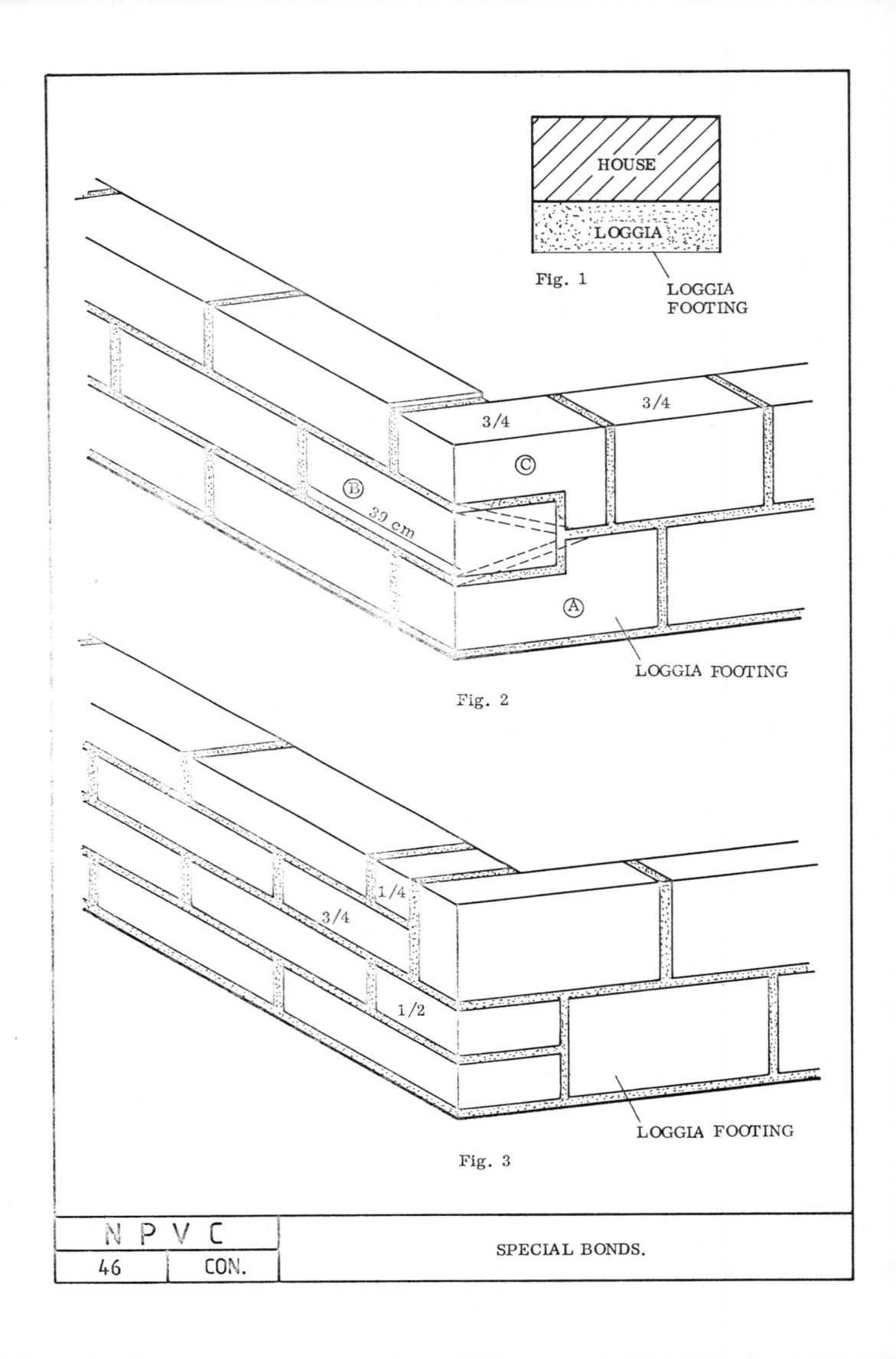

Fig. 1

LOGGIA FOOTING

3/4 3/4
Ⓒ
Ⓑ
39 cm
Ⓐ
LOGGIA FOOTING

Fig. 2

1/4
3/4
1/2
LOGGIA FOOTING

Fig. 3

N P V C	
46	CON.

SPECIAL BONDS.

DIFFERENT WALL THICKNESSES AT A QUOIN

This situation occurs when a verandah is closed at one end with a wall, or where a "loggia" is made. A loggia is a room which has one side open to the courtyard, garden, etc. (Fig. 1). The footings which carry walls have to be regular flatwise footings, while the footing on the open side can be constructed with the blocks edgewise, since it does not carry any wall.

- COMMON BLOCKS, FLUSH TO OUTSIDE: Fig. 2 shows that both quoin stretchers of the edgewise courses (blocks A and C) have to be cut to an odd shape to fit together with the flatwise quoin header of the second course (block B). In order to maintain the half-block bond, block B has to be cut to a length of 39 cm.

- SPECIALLY MADE BLOCKS, FLUSH TO OUTSIDE: If specially made blocks are used, it is not necessary to cut them to odd shapes, and the bond is easy to make as well as more economical (Fig. 3).

A corner bond in which the edgewise courses are set back (recessed) in order to be in line with the outside of the rising wall does not make sense and it is therefore not shown here. It is best to apply one of the arrangements above, or to make the whole structure stronger by laying all the corner blocks flatwise (Fig. 4, below).

- NOTE: It is best to avoid making these bonds with common blocks: use specially made blocks instead if possible, because it is extremely difficult to cut the odd shaped blocks (Fig. 2, blocks A and C) to the ideal shape without breaking them. If specially made blocks are not available, the common blocks can be given a wedge shape where they intersect, as indicated in Fig. 2 by the broken lines. This shape is easier to cut without breaking the blocks.

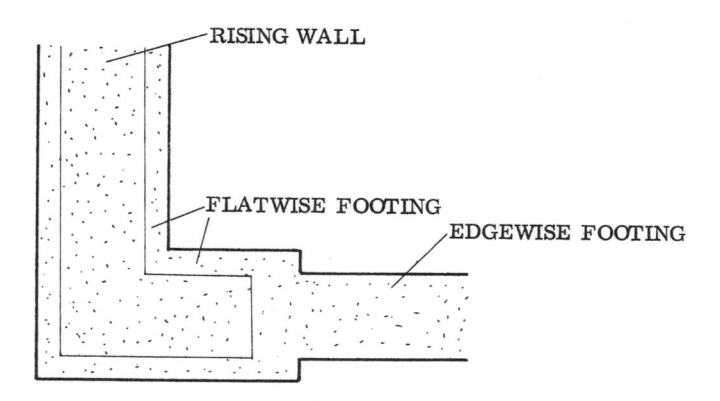

Fig. 4

SPECIAL BONDS.

Fig. 1   SITUATION "A"
COMMON BLOCKS, OUTSIDE FLUSH

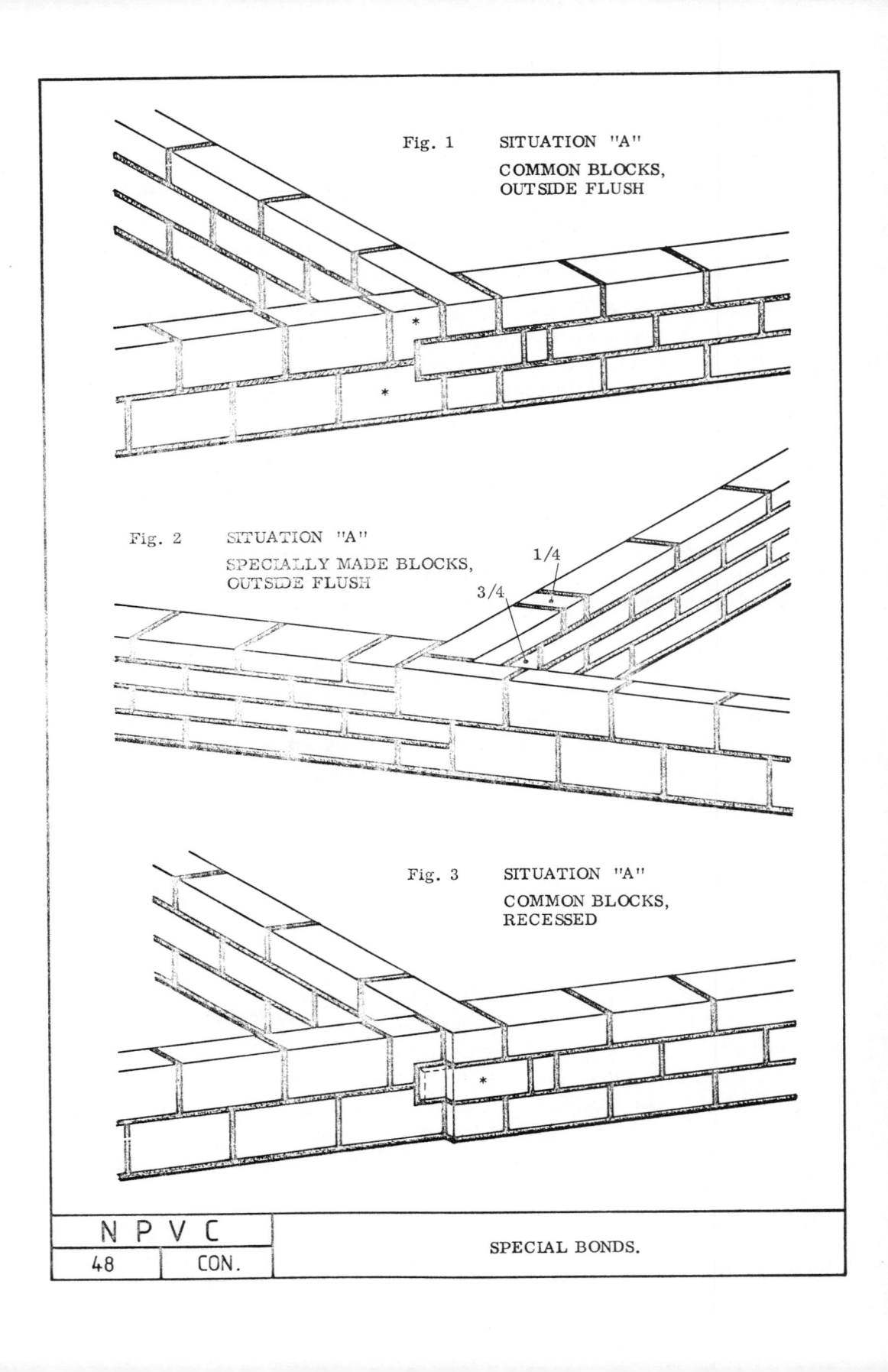

Fig. 2   SITUATION "A"
SPECIALLY MADE BLOCKS, OUTSIDE FLUSH
1/4
3/4

Fig. 3   SITUATION "A"
COMMON BLOCKS, RECESSED

N P V C	
48	CON.

SPECIAL BONDS.

DIFFERENT WALL THICKNESSES AT A T-JUNCTION

In Rural Building we are mostly concerned with those T-junctions in which a course of edgewise blocks meets either a corner ("A" below) or a wall ("B" below) of flatwise blocks.

Situation "A" occurs for example when a verandah is built flush with the corner of the house; while situation "B" will occur if the verandah meets the wall of the house away from the corner.

There are also possible situations where a flatwise course meets a corner or wall of edgewise blocks, but these situations are not common in Rural Building.

Here we discuss four methods which can be used in situation "A", depending on the type of blocks available and on whether the edgewise course will be recessed or flush to the outside of the flatwise courses. Two methods are given to deal with situation "B".

- SITUATION "A", COMMON BLOCKS, OUTSIDE FLUSH: As with the corner bond using common blocks (previous page), this bonding problem can only be solved by shaping the first blocks of both courses of the thin wall so that the second course of the thick wall can be bonded (Fig. 1). This requires precise and careful cutting.

- SITUATION "A", SPECIALLY MADE BLOCKS, OUTSIDE FLUSH: Bonding becomes simpler and more economical when these blocks are used. Fig. 2 shows that there is no material wasted since the 1/4 block which is left from cutting the 3/4 block will be used in the fourth flatwise course.

- SITUATION "A", COMMON BLOCKS, RECESSED: This bond is almost the same as the first method above, except that the stretcher of the second flatwise course projects on both sides in the direction of the thin wall (Fig. 3, broken lines). After the mortar has set hard, the projecting corners are chiselled off.

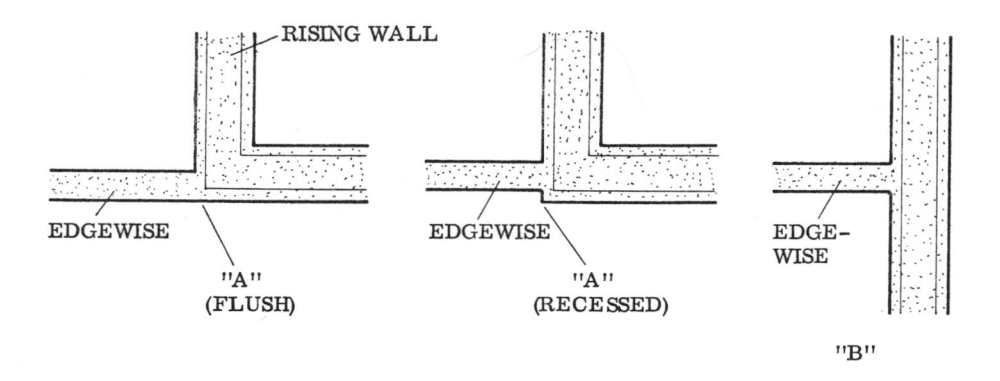

SPECIAL BONDS.

N P V C
CON. 49

Fig. 1   SITUATION "B"
         COMMON BLOCKS

DIFFERENT IN HEIGHT

Fig. 2   SITUATION "B"
         SPECIALLY MADE BLOCKS

N P V C		SPECIAL BONDS.
50	CON.	

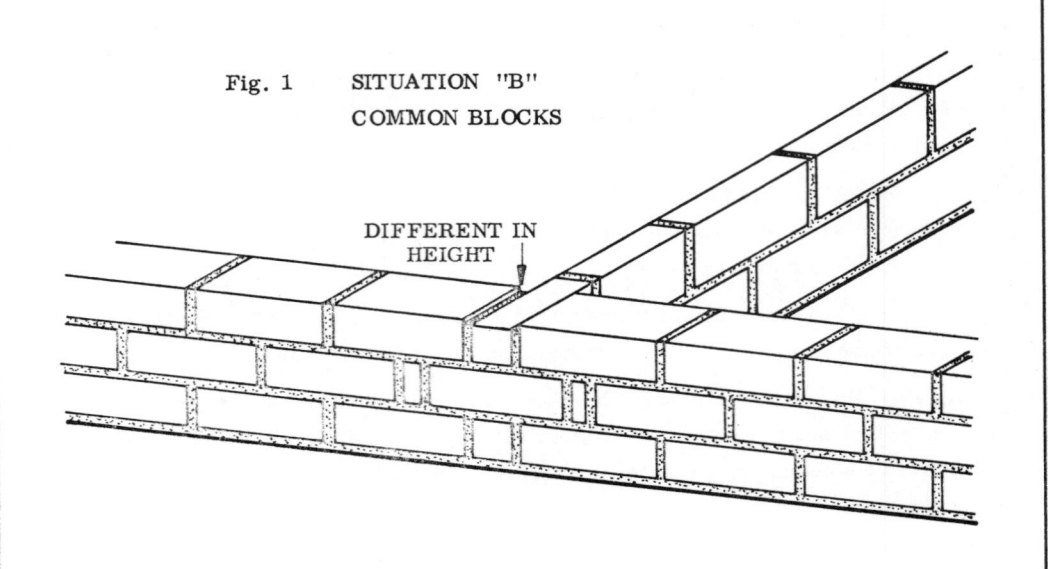

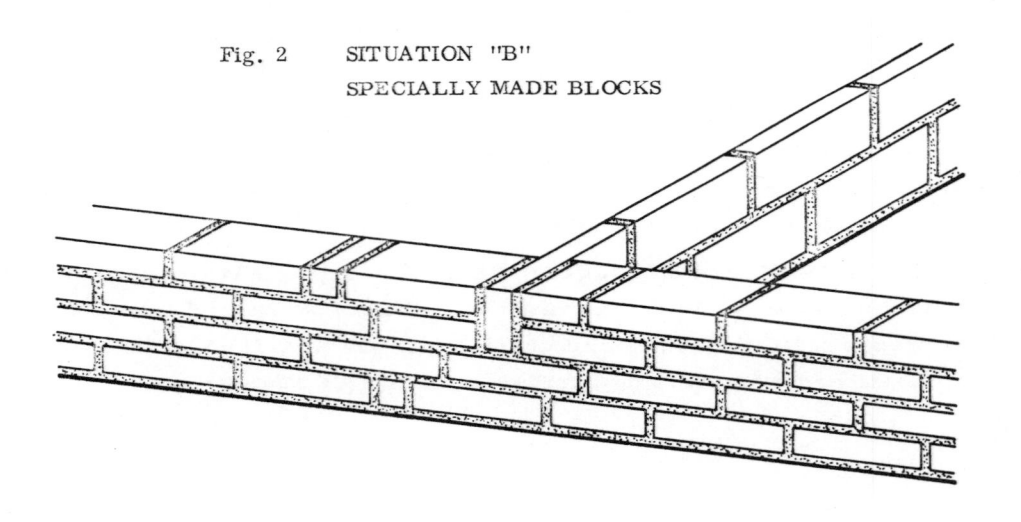

- SITUATION "A", SPECIALLY MADE BLOCKS, RECESSED: In this case we can use the same bond as in the second method (Fig. 2, last page). The 3/4 block and the full block above it have to be shortened by 4 cm. This will reduce the inside overlap of the junction, but that cannot be helped.

- SITUATION "B", COMMON BLOCKS: Again, here it does not make sense to talk about making a recessed thin wall: as at a quoin, the edgewise course will always be flush with the outside face of the flatwise footing course.

    Fig. 1 shows that here again we find the same problem which always arises when common blocks are used in a bond between different wall thicknesses. The first blocks of the two edgewise courses have to be cut to odd shapes.

- SITUATION "B", SPECIALLY MADE BLOCKS: When this bonding method (Fig. 2) is compared with the last method, it is clear that once again the specially made blocks are the best choice.

NOTES:

Fig. 1c
3rd COURSE

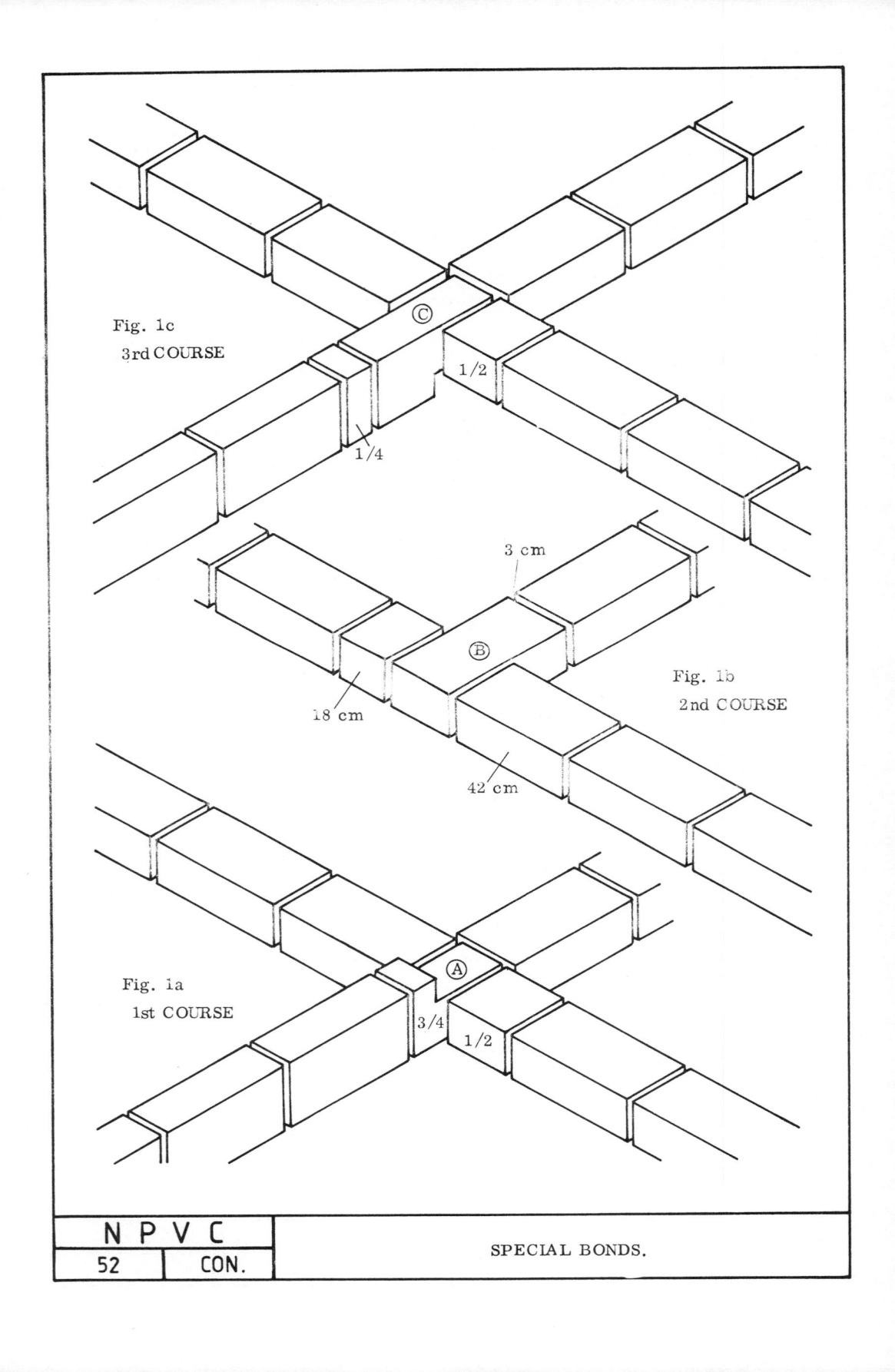

Ⓒ

1/2

1/4

3 cm

Ⓑ

18 cm

Fig. 1b
2nd COURSE

42 cm

Fig. 1a
1st COURSE

Ⓐ

3/4

1/2

N P V C	
52	CON.

SPECIAL BONDS.

DIFFERENT WALL THICKNESSES AT A CROSS JUNCTION

There are four possible situations for a cross junction with different wall thicknesses. The one which is described here is the most common in Rural Building. This situation occurs when the end of a verandah joins the wall of the house and is in line with an inside wall (Fig. 2, below). This means that the footing blocks are laid flatwise in three directions, and edgewise in one direction.

- COMMON BLOCKS: The first course of the thin wall is bonded into the thick one by a 3/4 block (block A) which must be shaped in such a way that the second course of the thick wall can overlap it (block B). Block A is followed on one of its stretcher sides by a 1/2 block, while all other directions are continued with full blocks (Fig. 1a shows the first courses).

In order to maintain the half-block bond, the second course of the thick wall must contain a block which is 18 cm long, and another 42 cm long; the latter overlaps the 1/2 block below. Don't forget that the first cross joint following the stretcher overlapping the thin wall is 3 cm wide, or else the first two cross joints are made 2,5 cm wide (Fig. 1b).

The third course of the thick walls is a repetition of the first one. The second course of the thin wall starts with a full block (block C) which is the same shape as the 3/4 block below it, so that it can be bonded into the thick wall. It is followed by the 1/4 block which was left over from the first course. Note that the thin wall remains 1 cm lower than the other walls, unless the mortar beds in the thin wall are both made 2,5 cm thick (Fig. 1c).

A disadvantage of this bond is that the overlap across the thin wall is rather short. This problem is not fully solved by using the specially made blocks.

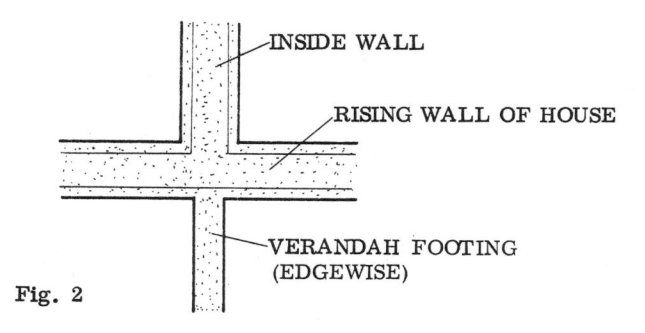

Fig. 2

SPECIAL BONDS.

Fig. 1b
2nd COURSE

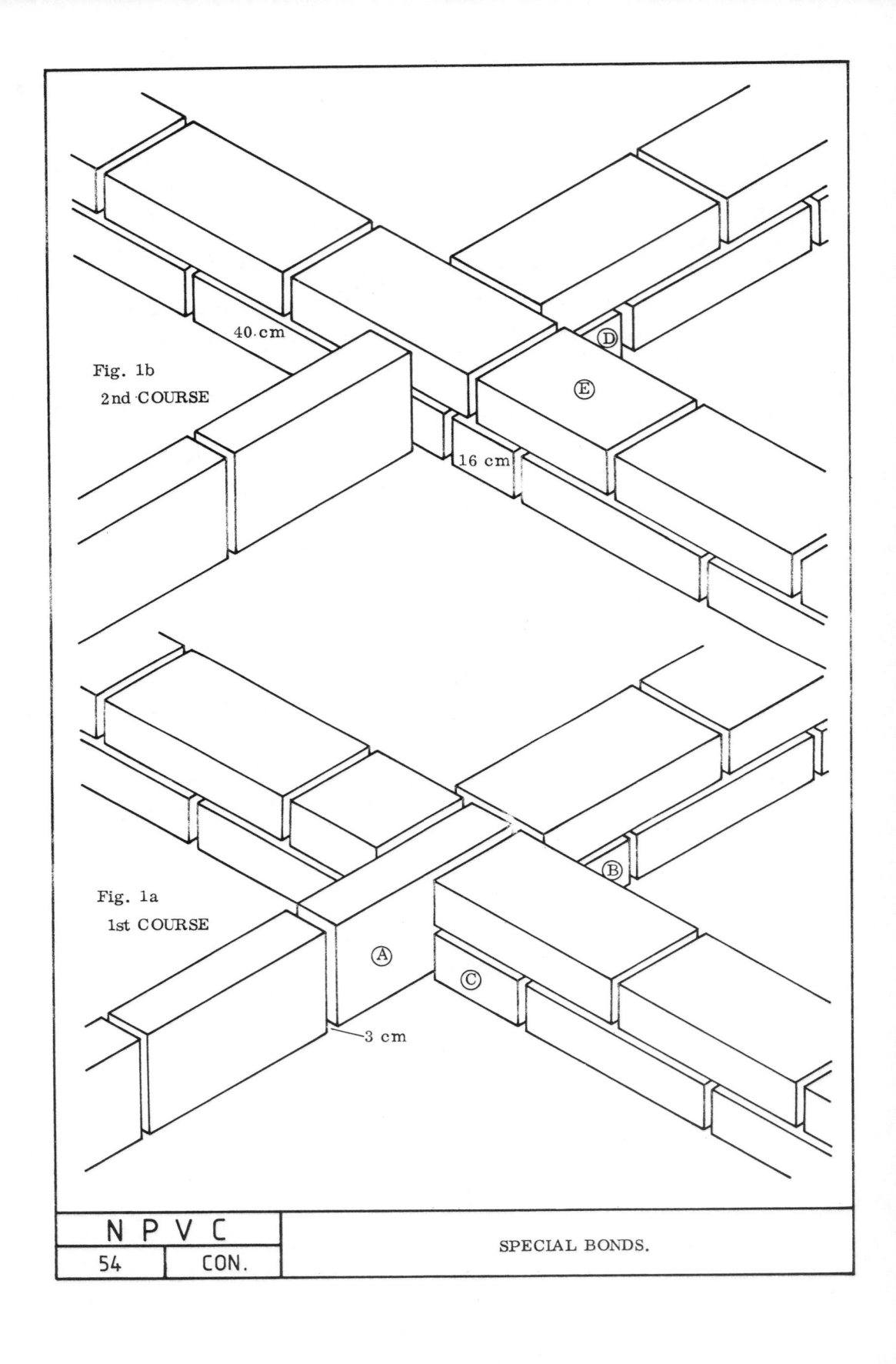

40 cm

Ⓓ

Ⓔ

16 cm

Fig. 1a
1st COURSE

Ⓐ

Ⓑ

Ⓒ

3 cm

N P V C		SPECIAL BONDS.
54	CON.	

- SPECIALLY MADE BLOCKS: The first course of the thin wall is bonded into the first two courses of the thick one by using a full block (block A). The first cross joint following this block is made 3 cm wide, or else the first two cross joints are made 2,5 cm wide. The first course of the thick walls continues with a 1/2 block (block B) in the direction of the thin wall, and with a 1/2 block (block C) either to the right or left side of the cross. The second course of the thick walls starts with full blocks where there are half-blocks below, and with a half-block which is above the full block of the first course (Fig. 1a).

The third course of the thick walls starts with a full block (block D) covering three cross joints. A block of 16 cm in length is placed beside this, above the full block; and a block of 40 cm length is placed on the other side, above the 1/2 block. The second edgewise course and the fourth flatwise course consist of full blocks, with the exception of a 3/4 block (block E) overlapping the 16 cm block below by 10 cm (Fig. 1b).

This cross junction bond, even with the use of the specially made blocks, still does not provide a very good overlap.

NOTES:

SPECIAL BONDS.

NPVC
CON. 55

Fig. 1b
2nd COURSE

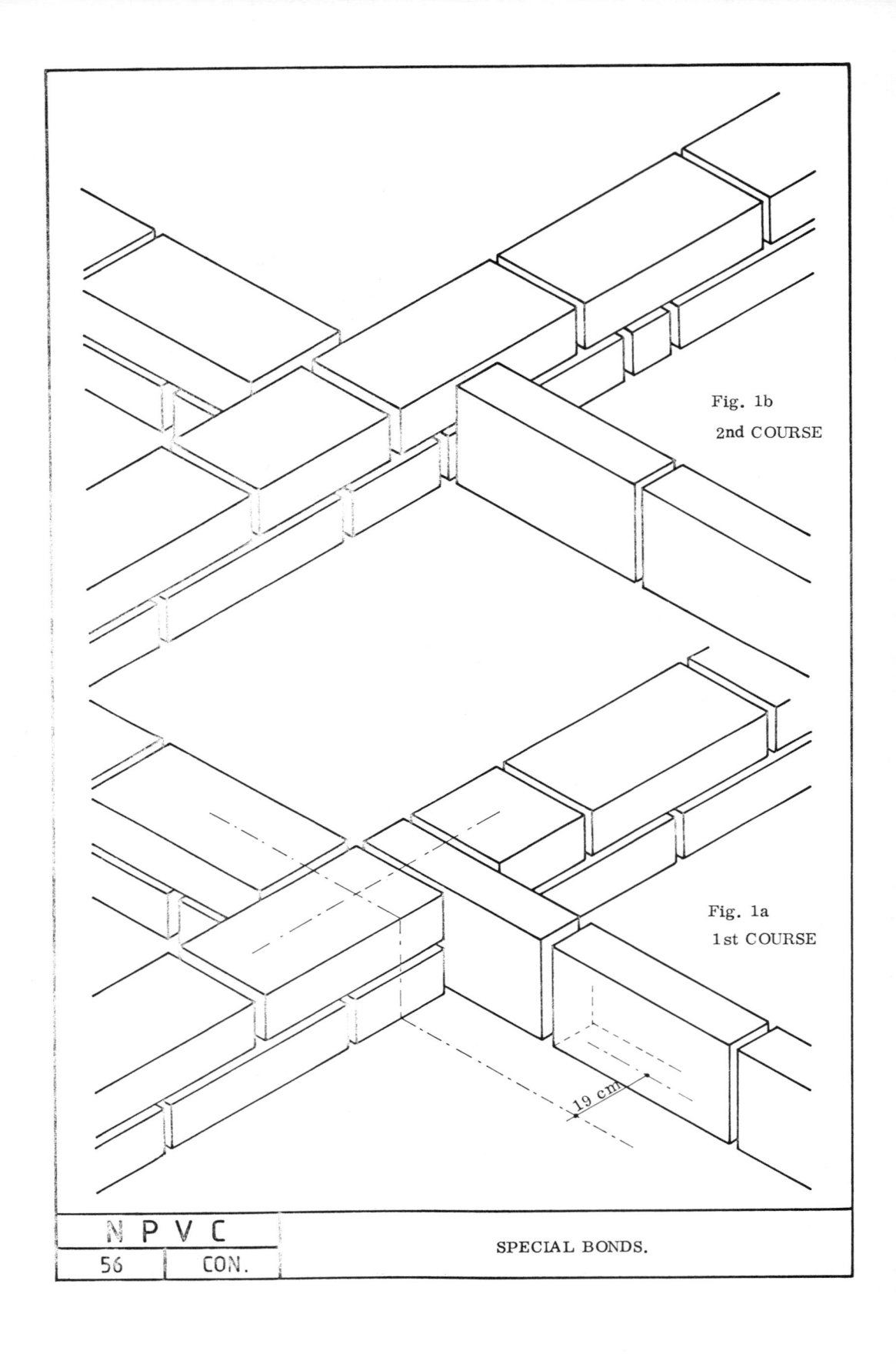

Fig. 1a
1st COURSE

19 cm

N	P	V	C
56		CON.	

SPECIAL BONDS.

- STAGGERED BOND WITH SPECIALLY MADE BLOCKS: A simple but good masonry bond can be achieved by **staggering the bond**. This means that the thin wall is shifted to either side of the crossing thick wall. The distance between the two centre lines of the walls will be about 19 cm (Fig. 1a).

The first course of the thin wall is bonded into the thick one by using a full block. The two courses on the left side of this almost form a quoin, while the courses on the other side are completed in the normal way. Apart from the 1/2 block in the second course, only full blocks are needed (Fig. 1a).

The third and fourth courses of the thick wall are bonded as if in a T-junction, except that the 1/4 block of the third course is placed further away from the junction, to avoid having too many cross joints in the junction area (Fig. 1b).

- NOTE: By planning ahead and doing a little extra work in setting out, you can avoid complicated and costly constructions.

For footings where there will be junctions between flatwise and edgewise blocks, it is always better to use specially made blocks which are 10,5 cm thick. These allow proper bonding, are more economical, and avoid the necessity for constructing junctions between two edgewise courses and three flatwise courses.

Cross junctions between different wall thicknesses should be avoided because the bonding is difficult and the overlap is poor. A staggered bond is preferred in those cases where a simple T-junction cannot be made instead of the cross junction.

NOTES:

RAIN

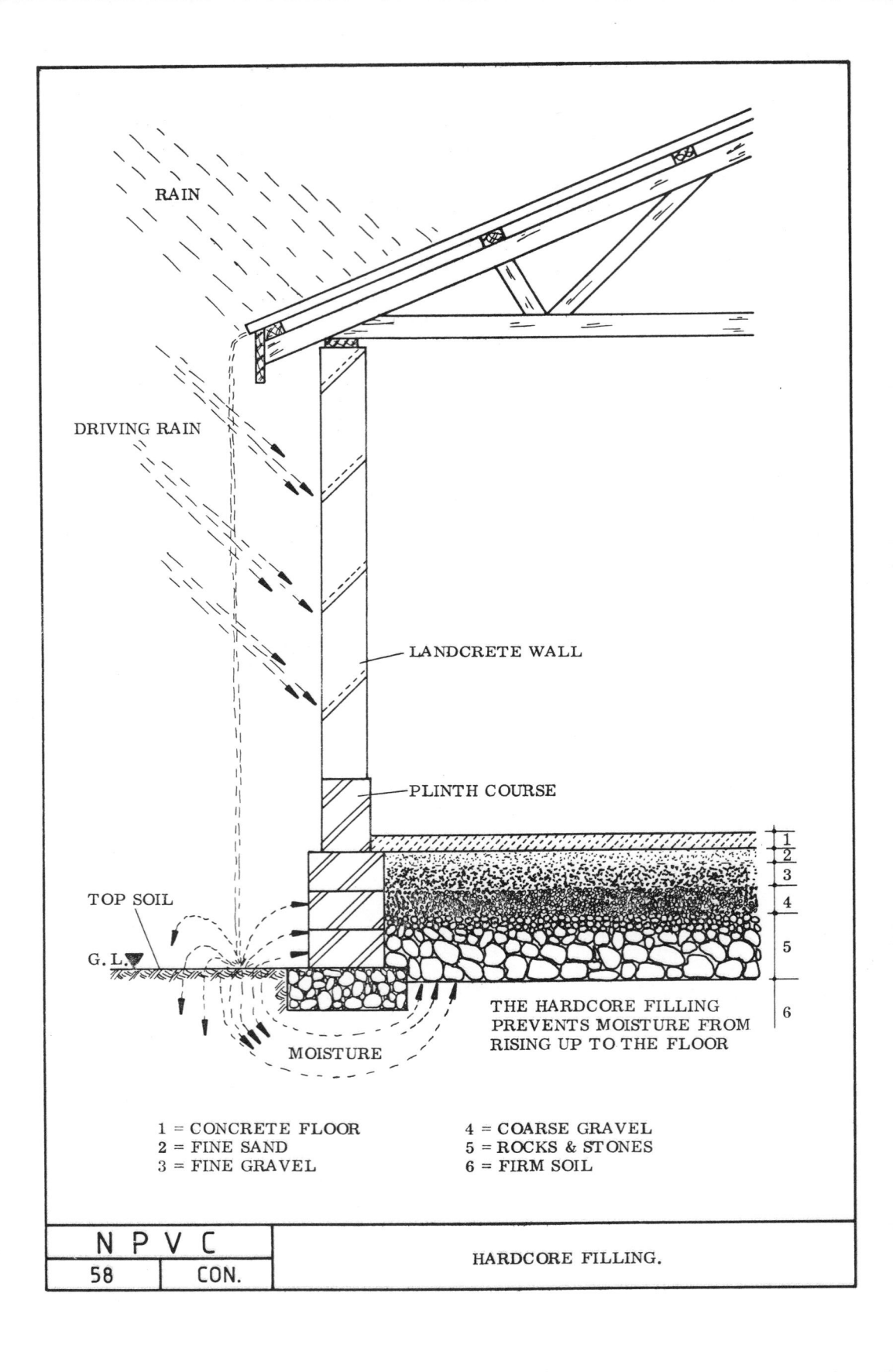

DRIVING RAIN

LANDCRETE WALL

PLINTH COURSE

TOP SOIL

G.L.

MOISTURE

THE HARDCORE FILLING PREVENTS MOISTURE FROM RISING UP TO THE FLOOR

1 = CONCRETE FLOOR
2 = FINE SAND
3 = FINE GRAVEL
4 = COARSE GRAVEL
5 = ROCKS & STONES
6 = FIRM SOIL

N P V C	
58	CON.

HARDCORE FILLING.

HARDCORE FILLING

FUNCTIONS OF THE HARDCORE FILLING

The two main functions of the hardcore filling are:

- To support the concrete floors, including all loads imposed on the floors

- To raise the floors high enough above ground level to prevent dampness from penetrating.

The hardcore has to be made from hard, solid materials which are arranged in such a way that moisture is not able to rise through them to the floor.

MATERIALS AND ARRANGEMENT

All topsoil is removed before the hardcore is applied. Then the filling is applied in layers no more than 15 cm thick. Each layer is compacted thoroughly before the next one is added. The hardcore filling occupies the space between the subsoil and the tops of the footings (Fig. 1).

The first layer consists of rocks, stones and broken sandcrete blocks. These are no smaller than fist-size and no bigger than 15 cm in diameter. The stones for this layer are preferably set into place by hand, according to their shape, in order to fit them together.

For the second layer we use broken stones which are about half the size of a fist. The following layers are formed of smaller stones, coarse gravel, or gravelly laterite soil; up to about 6 cm below the tops of the footings.

The last 6 cm to the top of the footings should be filled with sand or laterite soil to seal the surface of the hardcore, in order to keep the cement paste from the floor from leaking downwards.

- BACK-FILLING: If the foundation consists of stone masonry, the extra space left inside the foundation trenches has to be filled in before the hardcore is done. If the space is more than about 30 cm wide, it is best to use hardcore to fill it. If less, rammed soil will be sufficient.

  The outside of the trench excavation has to be refilled with gravel, to prevent erosion around the foundations (Foundations, page 19).

Never be tempted to fill up the site with soil underneath the hardcore, even on a sloping site: even if the soil is compacted it will still settle and cause the floor above to settle and crack. The hardcore should rest directly upon the subsoil.

COMPACTION OF THE HARDCORE

All the layers are normally compacted by hand with heavy rammers. This exhausting work can be made easier by using heavy iron or concrete rollers, or by driving a tractor several times over the hardcore to compact it.

Wet down the gravel and sand layers as they are filled in. This not only reduces the dust in the air but also makes compaction easier and ensures a dense layer.

Sometimes the building schedule allows construction to halt during the rainy season. If so, then it is a good idea to plan ahead in order to complete the hardcore filling before the rains come. This is done so the hardcore can be left open during the rainy weather to become very well compacted by the rain. However, it is still necessary to compact the layers of hardcore as they are added. The hardcore can be inspected from time to time during the rainy season, and any sunken areas can be filled up as they develop.

- NOTE: It is sometimes observed that a builder uses soil excavated from the foundation trenches to make the hardcore. This is convenient, but it is a bad procedure. The laterite soil acts as a sponge to draw moisture up from the subsoil into the building. The hardcore is made with a rock base because the rocks do not draw moisture upwards, so the building stays drier.

Gravelly laterite soils can be used for the layers just below the final layer, and fine laterite soil may be used for the final layer which seals the hardcore filling.

PLINTH COURSE

The plinth course forms the first course of the rising wall, immediately above the footings. Its sole function is to raise the landcrete blocks or sun-dried laterite blocks of the following courses high enough to protect them from moisture penetration.

The materials used for the plinth course should therefore be moisture resistant to some extent, such as natural stone or sandcrete blocks (Fig. 1, last page). It is known from experience that one plinth course is sufficient in northern Ghana, but there are situations where it is better to use a more durable material up to window cill level, especially in the rain-forest areas of Ghana.

In northern Ghana the rising landcrete walls are erected flush with the outside of the plinth course. In the south, however, it is advisable to lay the landcrete blocks flush to the inside of the plinth, so that thicker render can be applied on the outside to protect against moisture penetration.

POSITIONS OF DOOR AND WINDOW FRAMES

Before the plinth course is laid, the positions of the doors are marked on the footings.

If the door frames have been already made and painted they may now be set into place according to the marks. If the frames are not yet available, the plinth course and the rising walls can be erected anyway. In this case one must remember that the measurements given in the drawing are the outside measurements of the frame. Therefore the opening left in the wall must be about 1 cm wider on all sides, so that later on the frame can fit in (see Fig. 1).

The way the door and window frames are positioned in the surrounding masonry depends on the type of door or window, which way it opens, and whether mosquito-proofing and/or burglar-proofing will be installed.

Fig. 2 shows the most common position in which the frame is set flush to the outside of the render. The frame may be flush to both the inside and outside surfaces (Fig. 3), although the cill may project past the render on the outside face. Keep in mind the thickness of the plaster or render when you set the frame in the unplastered wall. The frame edge should be in line with the future plaster surface.

Vibrations from opening and closing the door or window, as well as the shrinking and swelling of the frame, will cause cracks to appear in the plaster and even cause pieces of plaster to come off. To prevent this and to cover the unavoidable cracks, a V-joint is made wherever plaster or render is flush to the frame. This may be covered with a wooden strip (Fig. 4). It is incorrect to apply plaster or render directly against the frame (Fig. 5, arrow).

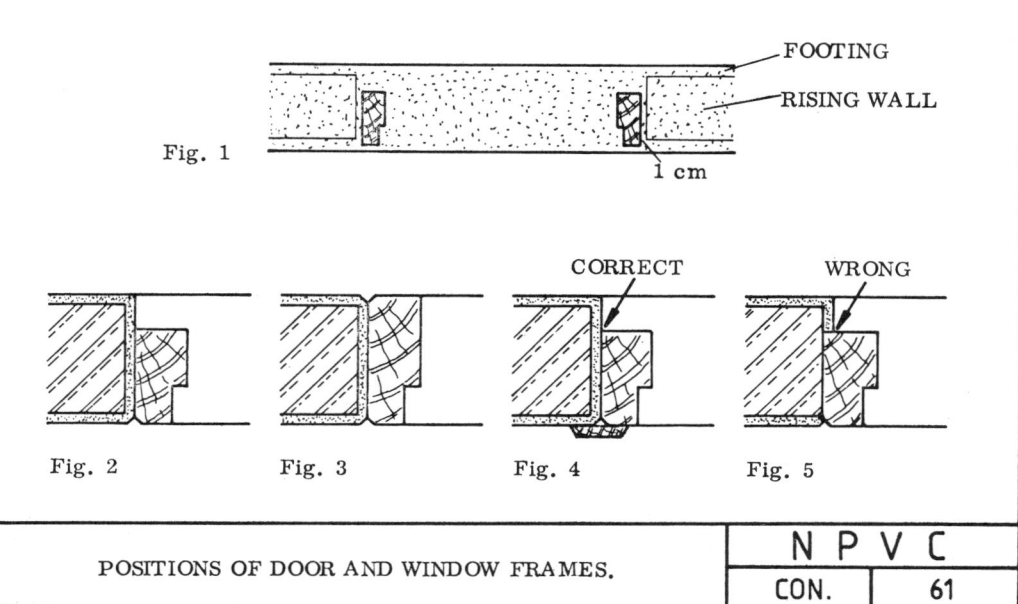

Fig. 1 — FOOTING, RISING WALL, 1 cm

Fig. 2 Fig. 3 Fig. 4 CORRECT Fig. 5 WRONG

POSITIONS OF DOOR AND WINDOW FRAMES. N P V C CON. 61

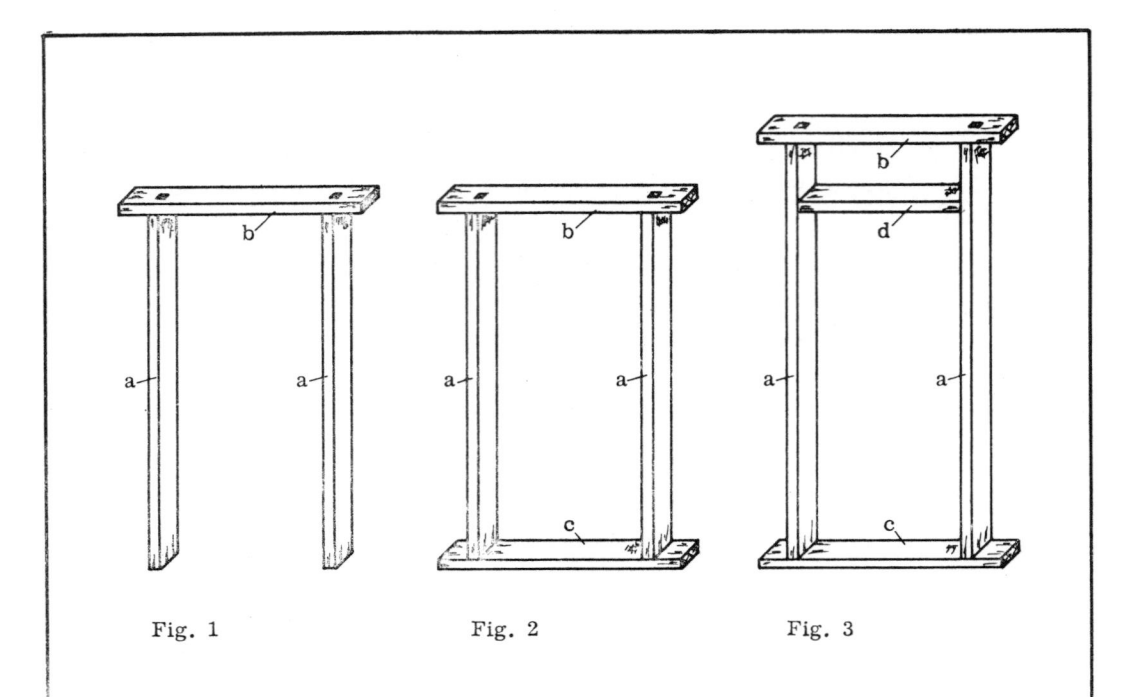

Fig. 1 Fig. 2 Fig. 3

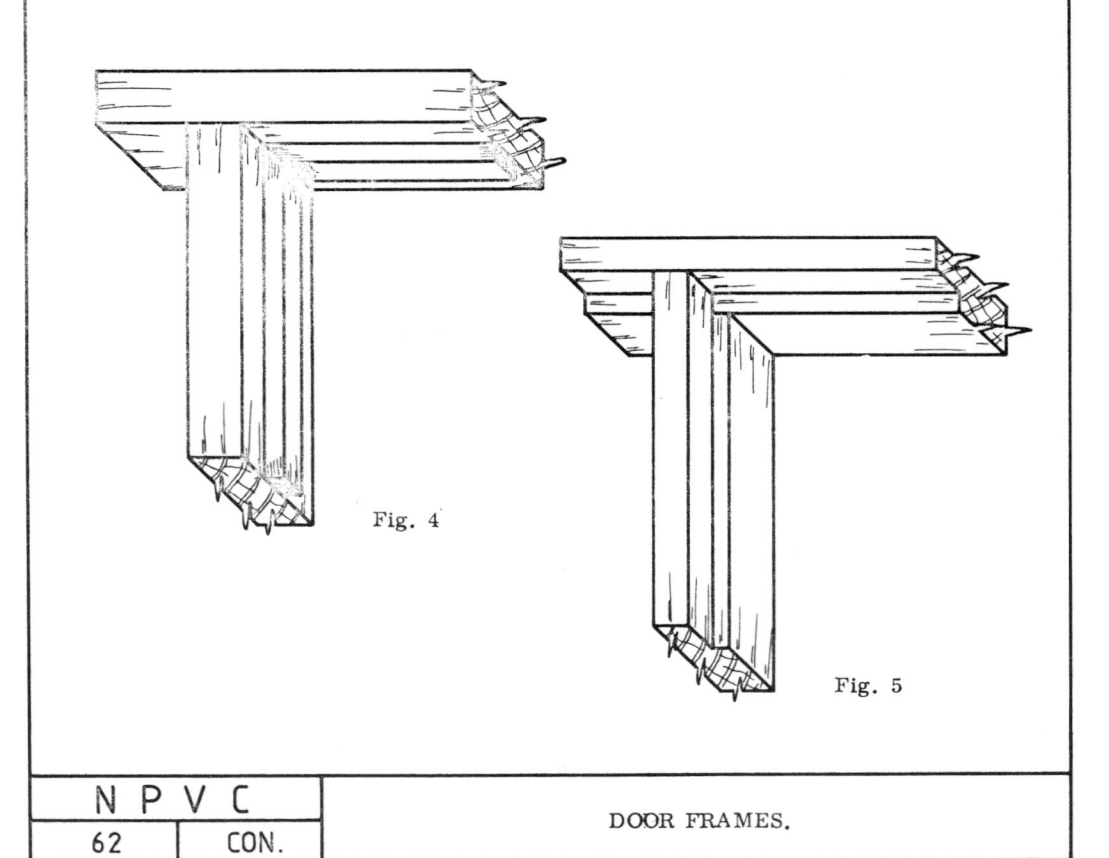

Fig. 4

Fig. 5

NPVC 62 CON. DOOR FRAMES.

DOOR FRAMES

TYPES OF DOOR FRAMES

The door frame provides a secure attachment for the door. Depending on the type of door, its function, and its location, different types of door frames can be constructed. The most common door frame consists of three members (Fig. 1):

- Two vertical members called posts (a)

- One horizontal member called a head (b).

When it is necessary for the door to fit tightly in the frame, it consists of four members (Fig. 2):

- Two posts (a)

- One head (b)

- One member on the floor, called the threshold (c).

When additional light or ventilation is needed for a room, a fanlight is added to the frame structure by adding one more member between the posts, called a transom (Fig. 3, d).

Note that since the threshold is in direct contact with the floor, or even partly embedded in the floor, it is in special danger from attack by termites or fungus (Reference Book, pages 141 to 144).

- BEAD OR REBATE TYPE DOOR FRAMES: Each of the above door frames can be made either with beads (Fig. 4) or a rebated construction (Fig. 5).

    Beads are preferred over a rebate for two reasons: the construction is easier, and it is possible to adjust the beads after the door is hung. Beads are always fixed into place after the frame is installed and the door is hung.

    Rebates, however, do have a better appearance than beads, and no extra wood is required to make a rebated frame.

NOTES:

Fig. 1
WIDTH OF REBATE (x)
1/3 OR LESS

HEAD

POST

Fig. 2
WIDTH OF REBATE (x)
MORE THAN 1/3

HEAD

POST

N P V C		DOOR FRAMES.
64	CON.	

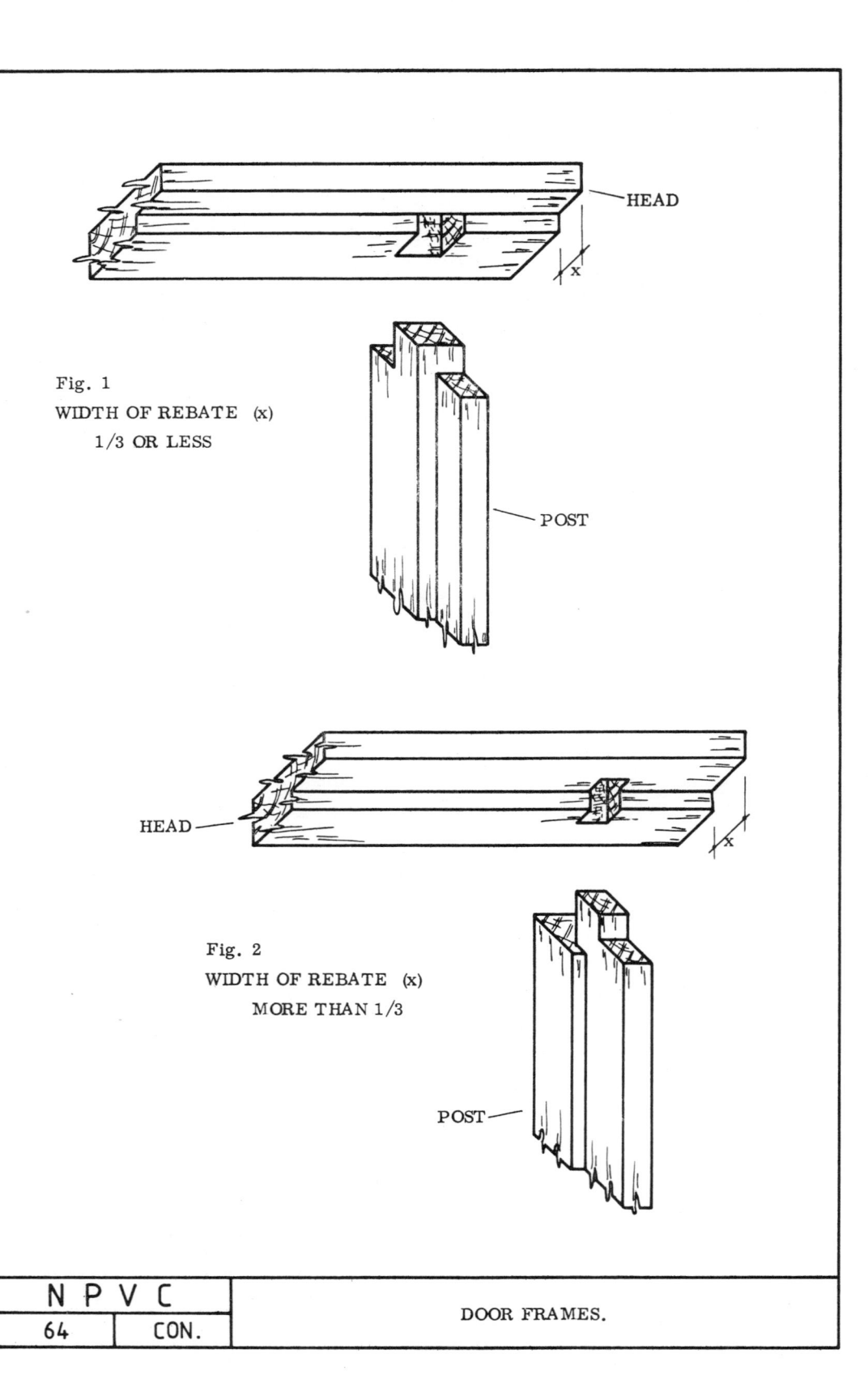

JOINTS FOR DOOR FRAMES

- **JOINTS FOR DOOR FRAMES WITH BEADS:** The joints for a door frame made with beads can be common mortice and tenon joints for box-like constructions. They may be either pegged, wedged or nailed (Basic Knowledge, pages 104 to 106, and 124; also Drawing Book page 43 and pages 67 to 71).

  If a transom is required, it should be installed using a stub tenon joint to prevent water from penetrating the mortice (Basic Knowledge, page 122).

  If a threshold is installed, a common mortice and tenon joint is used for it, but wedges should not be used since they might fall out. Pegs cannot be used either, because the threshold is too thin. This joint should be nailed.

- **JOINTS FOR DOOR FRAMES WITH REBATES:** If the door frame is made with a rebate, take care that the shoulders of the tenons are set out and cut correctly. One shoulder of the tenon is cut longer to fit the rebate (Drawing Book, pages 67 & 68).

  The thickness of the tenon is affected by the size of the rebate. If the rebate is 1/3rd or less of the width of the member, the tenon will not be reduced in size (Fig. 1). If the rebate is wider than 1/3rd of the member, the thickness of the tenon (and accordingly the mortice width) is reduced by the depth of the rebate (Fig. 2).

  Besides these considerations, the construction of these joints will be the same as for any common mortice and tenon joints for box-like constructions.

NOTES:

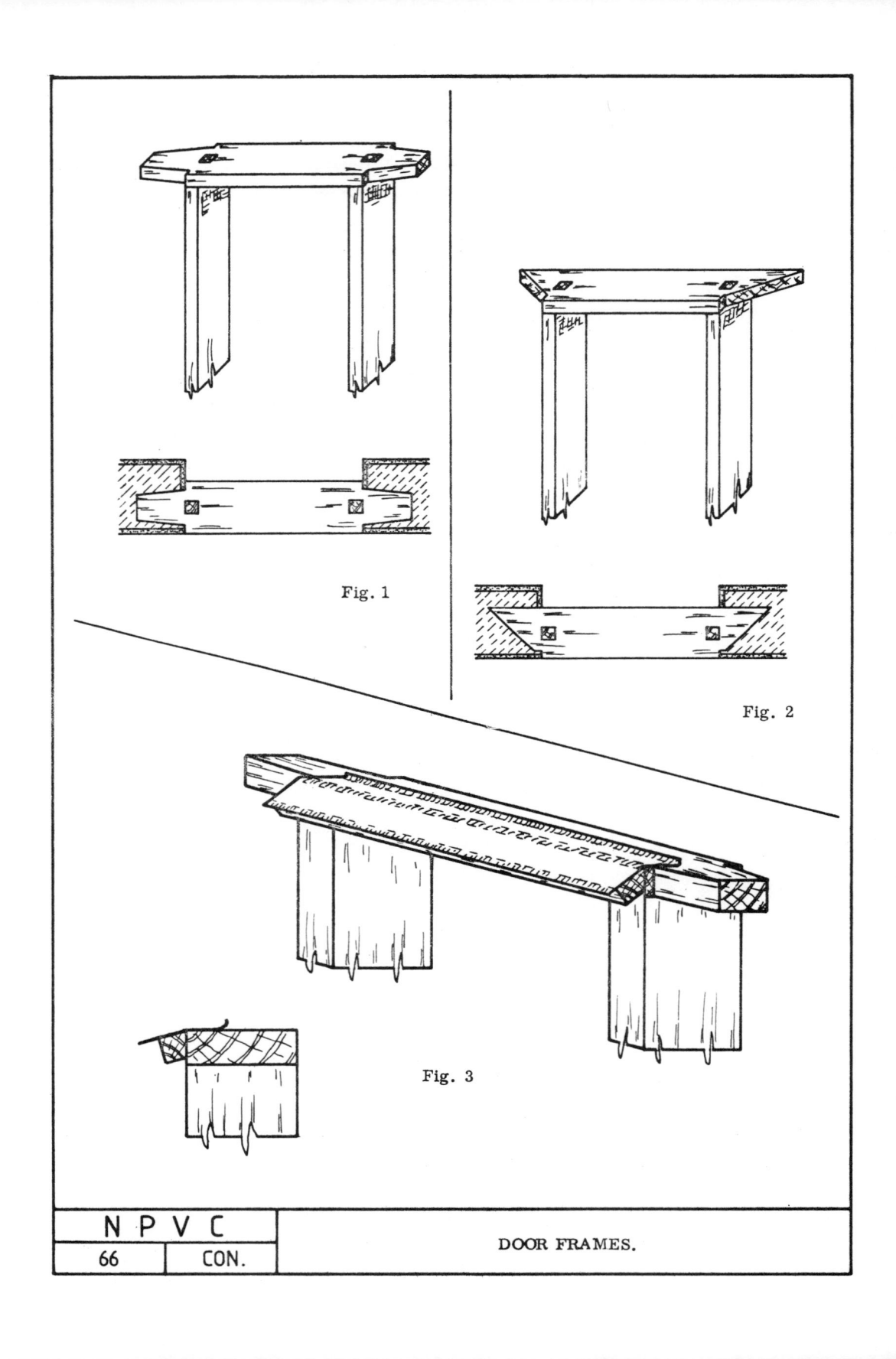

Fig. 1

Fig. 2

Fig. 3

NPVC 66 CON.   DOOR FRAMES.

HEADS OF DOOR FRAMES

- HORNS: In order to provide an additional attachment to the wall, the ends of the head and the threshold can project beyond the posts into the wall. This is especially necessary in mud walls.

  These projecting parts can be shaped like horns (Fig. 1), or else the front corners can be cut off at an angle (Fig. 2). These shapes give a good appearance to the frame head when the frame is built into the wall.

- WEATHER STRIPS: If a door frame is to be installed in an outside wall, you need to take some extra precautions to prevent water from entering between the lintel and the frame head, and between the door itself and the door frame.

  Weather strips (Fig. 3) can be fixed on the door head to keep water out. We will learn more about weather strips later in the book, in the section on window frame heads.

- STEEL CRAMPS: These are pieces of mild steel rod which are fixed into the door posts (Fig. 1, next page). They are fixed about 60 cm apart; near the head, middle and foot of the frame.

  The function of these cramps is to secure the frame rigidly to the blockwork. They are set in the posts in positions which correspond with the bedjoints of the blocks, so that they can be built into the bedjoints.

  To fix the cramps in the door frame, drill holes in the posts. The holes should have a diameter a little smaller than the rods, and should not go all the way through the post. Bend one end of the rod to form a head (Fig. 2, next page) and knock the head into the post. Bend the other end to a right angle so it fits into the cross joint between the blocks (Fig. 2, next page).

NOTES:

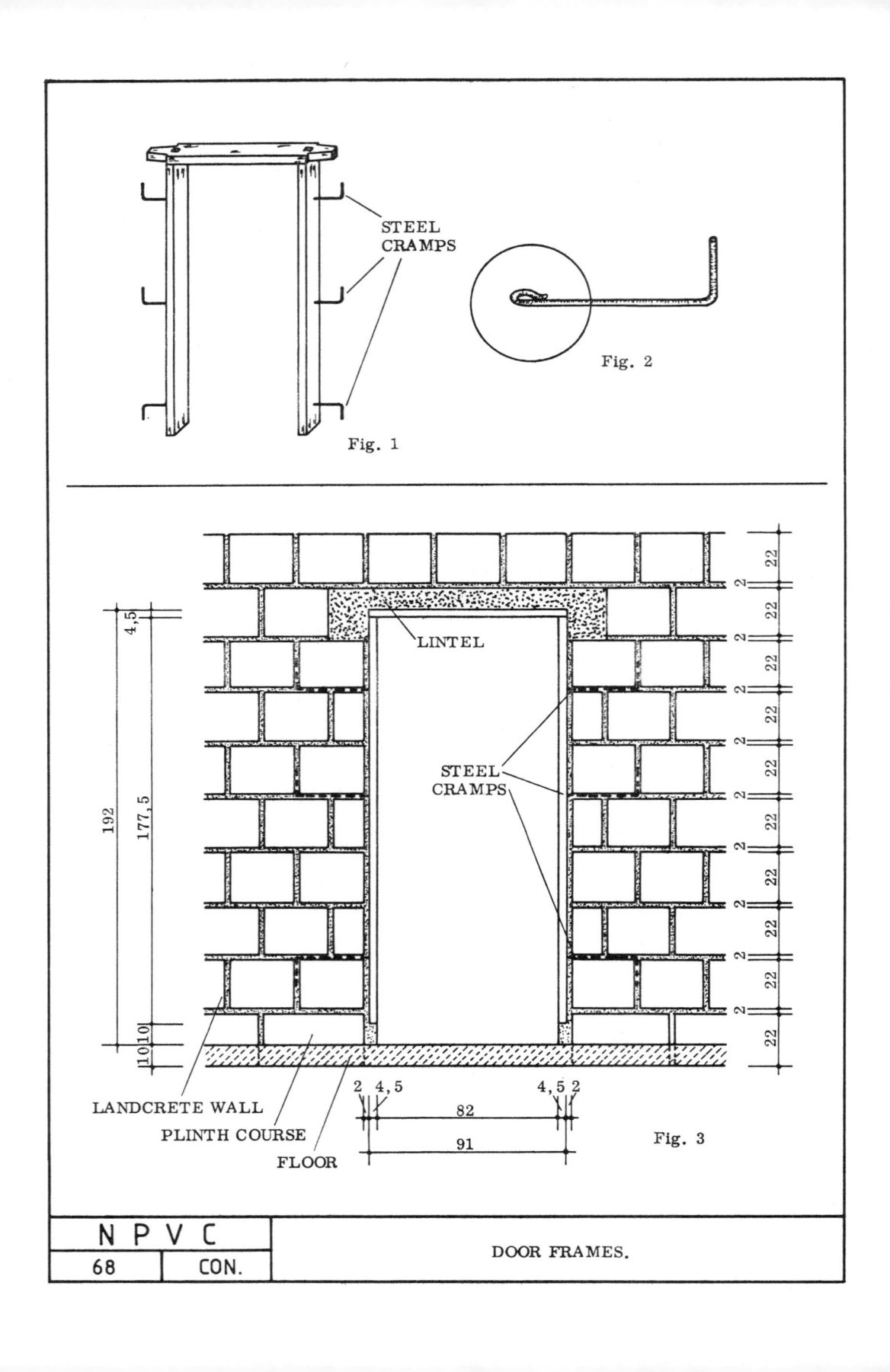

Fig. 1

STEEL CRAMPS

Fig. 2

Fig. 3

LINTEL

STEEL CRAMPS

LANDCRETE WALL
PLINTH COURSE
FLOOR

N P V C		DOOR FRAMES.
68	CON.	

MEASUREMENTS FOR DOOR FRAMES

- MEASUREMENTS OF THE TIMBERS: Usually door frames are made of timbers which are about 10 cm wide and about 5 cm thick. If the posts are held in the wall by steel cramps, it is possible to reduce the size of the timbers to 7,5 by 5 cm. The cramps provide a good grip to the wall so the frame doesn't have a lot of stress on it. This means that the timbers can be made smaller and thus less costly. The cramps are made by cutting iron reinforcement rods.

    The frame itself is always measured from inside face to inside face. Unless it is stated otherwise on the estimate, measurements given for door frames are always inside measurements.

- MEASUREMENTS OF THE FRAME WITH RESPECT TO THE MASONRY: To make the work easy, to avoid extra block-cutting, and to reduce waste, the outside measurements of the frames should correspond to the building unit measurement of the blocks used to construct the wall. For example:

    - The building unit measurement for landcrete blocks is 31 cm wide by 24 cm high.

    The outside height of the door frame will be 8 x 24 = 192 cm.

    The ouside width of the door frame will be 3 x 31, minus 2 cm for one cross joint = 91 cm (Fig. 3).

    The steel cramps are also placed according to the building unit measurement of the blocks, so that they will fit into the bedjoints of the wall (Fig. 3).

NOTES:

DOOR FRAMES.

Fig. 1 SQUARING STRIPS / SPACERS

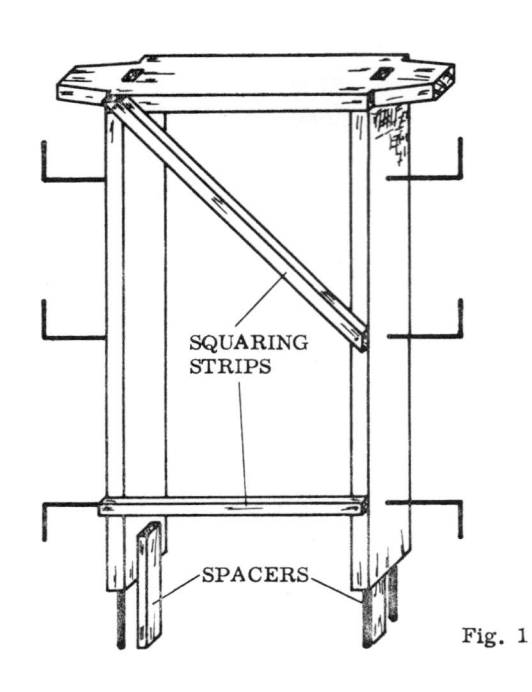

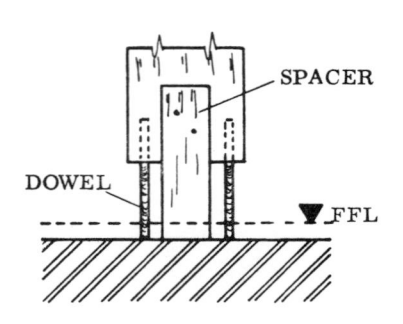

Fig. 2 SPACER / DOWEL / FFL

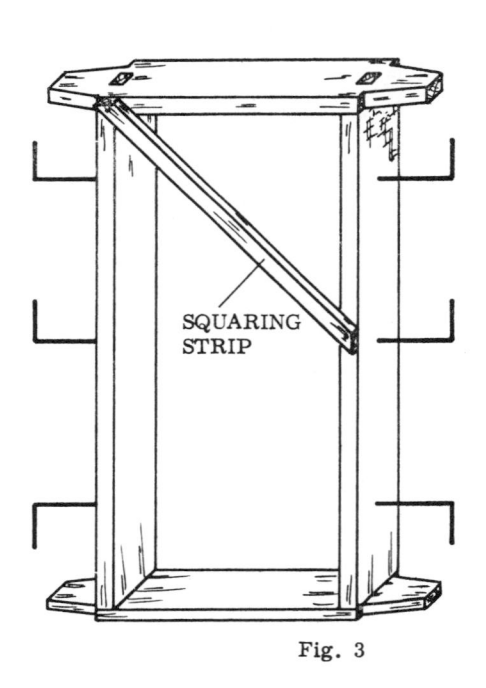

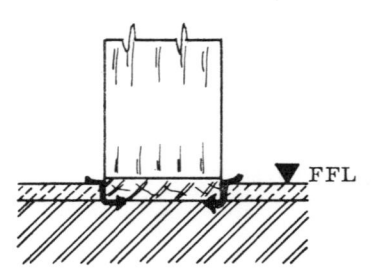

Fig. 3 SQUARING STRIP

Fig. 4a WRONG - WATER CAN PENETRATE BETWEEN THRESHOLD AND FINISHED FLOOR

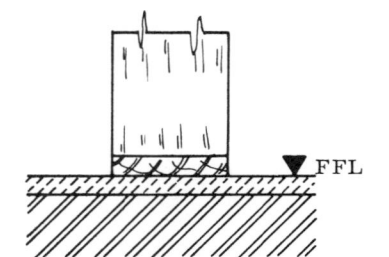

Fig. 4b BETTER - THRESHOLD ON TOP OF FINISHED FLOOR

NPVC 70 CON.

DOOR FRAMES.

CONCRETE SHOE

In order to give a firm base to a door frame with no threshold, concrete "shoes" can be fitted on the posts.

To make the shoes, drive steel dowels into the bottom end of each post before setting the frame in the wall (Fig. 1). The projecting ends of the dowels are enclosed later by the finished floor and the shoe (Fig. 2).

Spacers are used to keep the frame at the correct height during the building operations (Fig. 1). The steel dowels are short pieces cut from reinforcement rods.

- ADVANTAGES AND DISADVANTAGES OF THE THRESHOLD (Fig. 3):

  advantages:
  - The door closes tightly into the framework.
  - If the threshold is thick enough, it can be made with a rebate.

  disadvantages:
  - The threshold is easily attacked by wood diseases or termites.
  - More materials are needed.
  - More joints have to be made, therefore construction is more difficult.
  - Water can penetrate between the threshold and the finished floor (Figs. 4a & 4b).

- ADVANTAGES AND DISADVANTAGES OF THE CONCRETE SHOE

  advantages:
  - The wood is not in direct contact with the floor, so there is less danger of attack by termites, etc.
  - Less materials are needed.
  - The shoe secures the doorframe to the floor.

  disadvantages:
  - The door might not close as tightly as with a threshold.
  - If the frame is not well braced during the building operations, there may be a danger of distorting it while fixing it in place.

- SEQUENCE OF OPERATIONS FOR MAKING A REBATED DOOR FRAME WITH A CONCRETE SHOE: (see Drawing Book, page 67)

  a. Study the drawing.
  b. Prepare a cutting list.
  c. Select the wood.
  d. Plane the wood to the required sizes (as indicated on the cutting list).
  e. Make the face marks.
  f. Mark out the joints (take special care in marking out the shoulders of the

tenons on the posts) and mark the length of the posts.

g. Mortice the head and rip the tenons at the posts.

h. Plane the rebate.

i. Cut the shoulders of the tenons.

j. Drill the holes for the steel dowels.

k. Clean up the inside edges.

l. Paint all the members with a primer, especially the joints.

m. Fix the steel dowels and the wooden spacers.

n. Drill the holes and set anchors at the correct places (steel cramps).

o. Assemble the frame with squaring strips.

p. Cut the horns to the required shape and paint them.

Take care when marking the lengths of the posts. They should be the length of the door; plus the upper joint, and minus the height of the concrete shoe.

- ASSEMBLING THE DOOR FRAME: To keep the door frame rigid and square during the various building operations, it needs to be braced. This bracing is done with squaring strips (Fig. 1, last page). These strips are about 2 x 5 cm, and they are cut into the rebate or nailed to the face of the frame. They hold the frame square and keep the correct distance between the posts.

The proper steps for assembling a door frame are as follows:

a. Knock the frame together.

b. Place the frame, rebate side up, on the workbench.

c. Clamp the posts to the head (use sash clamps or the wedges on the bench).

d. Nail a squaring strip to hold the posts at the correct distance apart.

e. Test for squareness from corner to corner (diagonally).

f. With the clamp still in position, fasten the joints.

g. Cut and nail the other squaring strip diagonally from the head to the post (Fig. 1, last page). This holds the frame square.

h. Cut the posts to the required length.

i. Drill the holes for the steel dowels and fix them. They should be long enough to anchor into the concrete floor.

j. Nail wooden spacers into the inside of the rebate. These keep the frame at the correct height during the building operations (Figs. 1 & 2, page 70).

k. Finish off the frame. The priming coat of paint should be applied to the external woodwork before it is fixed in place.

When timber with a small section, for example 5 x 7 cm or 5 x 10 cm, is used for the posts, an extra squaring strip should be fixed horizontally in the middle of the frame. This prevents the pressure of the blockwork from distorting the frame.

INSTALLING DOOR FRAMES

Frames which are improperly built-in can cause problems later, when the plastering is done or the doors are hung. Therefore we must give special attention to setting the frames properly.

A door frame should be fixed in such a way that the door can open flat to the wall. Otherwise, the door will form a lever to the frame, and the hinges will be forced out when the door swings wide open suddenly.

Door frames can be fixed in position either during the masonry construction or after the walling has been completed.

- SETTING FRAMES DURING MASONRY CONSTRUCTION: The first frames to be set are those of the outside walls. When the masonry work has been built to floor level, the door frame is placed in position according to plan and at the correct height, and held there by means of wedges (Fig. 1, a, next page).

To hold the frame in position during the following building operations, we use struts. These are braced against blocks on the ground and lean up to the head of the frame, where they are secured with two nails (Fig. 1, b, and Fig. 2, next page).

The masonry line is used to align the frame with the face of the wall. The line is fixed at the two corners of the wall and two small wooden blocks are used to hold the line out from the wall by the same amount as the thickness of the future render layer; so that the line shows the position of the face of the render on the finished building (Fig. 1, c, next page).

The line should be free (not touching any frame or block) from corner to corner of the building. It should be separated from the frame faces by about a trowel's thickness (Fig. 3, next page).

Because the posts are long, the plumb bob is the best tool for making sure that the frames are straight and upright. Hold the bob at the inside top of the frame,

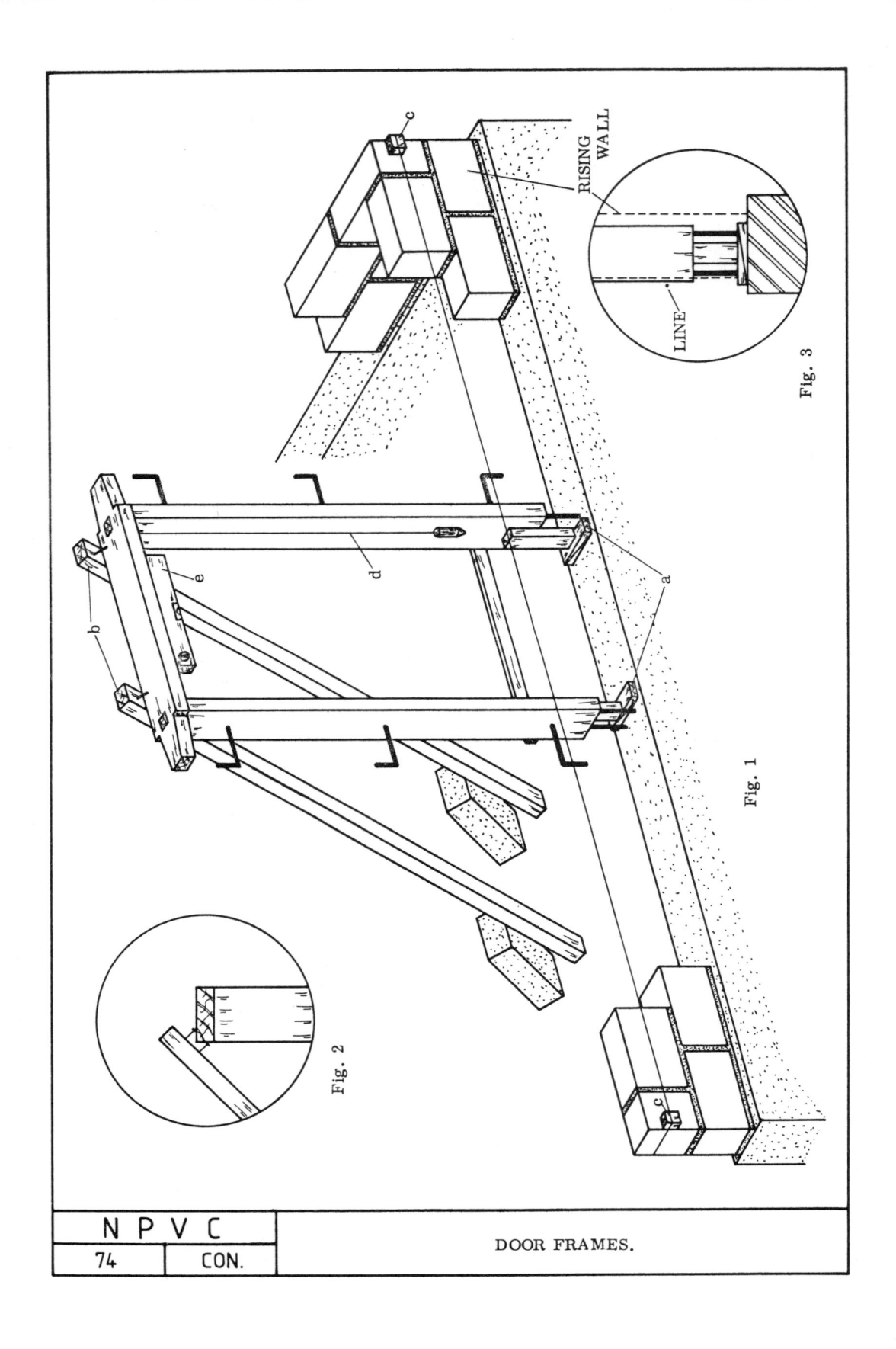

Fig. 1

Fig. 2

Fig. 3

RISING WALL

LINE

николая			
N	P	V	C
74		CON.	

DOOR FRAMES.

so it has room to swing freely (Fig. 1, d). Check both posts (Reference Book, page 5)

To adjust the frame, the wedges are knocked a bit in or out, thus raising or lowering one post at a time until the frame is straight.

The face of the frame can be plumbed at the same time. To do this, step to one side of the frame and sight the edges of the post and plumb line. A helper should stand by the struts and either "give" or "pull" them until the frame is straight. Take care that the space between the face of the frame and the mason line remains correct (Fig. 3).

As a final check, the soffit (underside) of the head can be levelled with a spirit level (Fig. 1, e).

- REMEMBER: Be sure that the face of the frame is in line with the mason line, not flush with the face of the blocks. When the building is finished, the render surface and the face of the frame should be flush.

Make sure that the frame remains at the correct height, so that the steel cramps can fit into the bedjoints and the frame does not get twisted.

After several courses of blocks have been laid, the frame should be rechecked to make sure that it has not become distorted by the blocklaying. Blocks should be laid on both sides of the frame at the same time, so that it does not get pushed out of plumb.

The struts and braces are taken off only when the blockwork has reached the head of the frame.

If more than one frame is erected, walk around the building and look from a distance to compare the frames with each other and make sure that they are all in the correct alignment.

- SETTING FRAMES AFTER THE WALLING HAS BEEN COMPLETED: In this method, the installation of the frames is postponed until the roof construction is complete, in order to reduce the risk of damage to the woodwork from the rain and sun.

Hardwood or plastic plugs (Reference Book, page 215) are driven into the bedjoints to serve as anchors for the frame. The frame is then put into position and nailed or screwed to the plugs.

NOTES:

DOOR FRAMES.

Fig. 1

Fig. 2

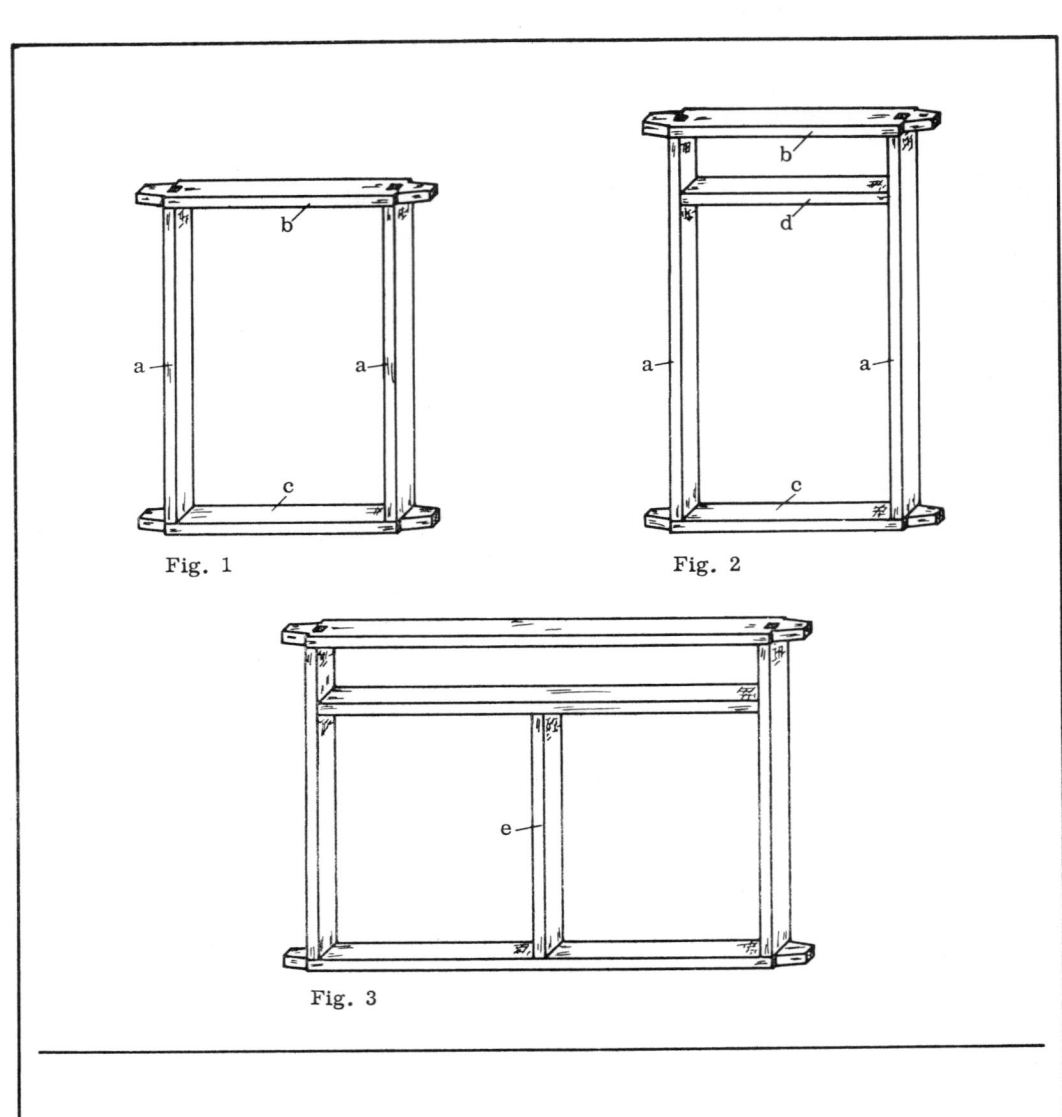

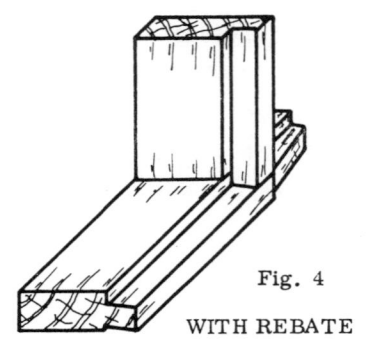

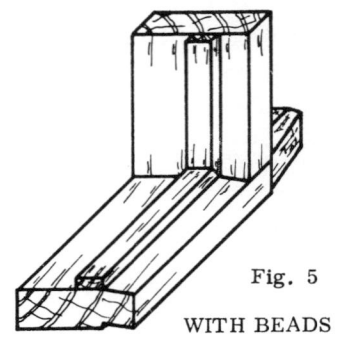

Fig. 3

Fig. 4
WITH REBATE

Fig. 5
WITH BEADS

N	P	V	C
76		CON.	

WINDOW FRAMES.

WINDOW FRAMES

TYPES OF WINDOW FRAMES

The function of a window frame is to admit light and air into the building. The most common frame used in Ghana is the solid window type, consisting of an outer frame into which a casement or louvres are fitted.

The most common window frame consists of four parts (Fig. 1):

- Two posts (a)

- One head (b)

- One cill (c)

When additional light and ventilation are needed, a fanlight can be added. The extra piece is called a transom (Fig. 2, d).

If wider windows are required, a mullion (Fig. 3, e) can be added.

Any of these frames can be made with a rebate or with beads, or with a combination of the two (Figs. 4 & 5).

For a list of the advantages and disadvantages of beads or rebates, see the section on door frames. If beads are used, they are fixed after the casement is installed, so they can be adjusted (see Drawing Book, pages 42 & 72).

MEASUREMENTS OF WINDOW FRAMES

Before you start constructing a window frame, you have to decide:

- whether the window will have shutters, louvres or any other kind of closing;

- if there will be any burglar proofing, and if so what kind;

- if any mosquito proofing is required;

- and where the window will be installed (does it have to be opened frequently, for example).

All these factors can influence the kind and sizes of timber we use, and the size of the frame itself.

- REMEMBER: Window frame measurements which are given with no further explanation are always inside measurements.

NOTES:

- MEASUREMENTS OF TIMBER SIZES FOR WINDOW FRAMES: A very common size of timber used for window frames is 7,5 to 10 cm wide, by about 5 cm thick. This size of timber is readily available and relatively inexpensive. If large frames are required, the timber size should be about 15 cm by 5 cm.

- MEASUREMENTS WITH RESPECT TO THE CASEMENTS: The height and width of the window frame are partly determined by the kind of casements which will be used. If ready-made casements are used, the frame has to be the right size to fit them. If the casements are to be made, consider the timber sizes which are available and choose the size of the casements according to the size of the boards, to avoid waste.

- MEASUREMENTS WITH RESPECT TO LOUVRES: If glass louvres will be installed, their size will determine the height and width of the frame. Find out the size of the louvre first, then start making the frame, not the other way around. Mistakes here can be very expensive!

Also make sure that there is room enough to open and close the louvres and that the mosquito proofing, if present, does not interfere with them.

Louvre windows will be discussed thoroughly in one of the next lessons.

The type of burglar proofing which is chosen will also affect the construction of the window frame, and we will consider this in one of the later lessons.

STEEL CRAMPS

If steel cramps are used to hold the frames, they should be positioned to fit into the horizontal joints (bedjoints) of the walls (see Measurements of Door Frames section, page 69).

JOINTS FOR WINDOW FRAMES

When you are marking out the joints, first decide whether the frame is to have a rebate or beads.

In general, the joints for window frames are the same as those for door frames (see the section on Joints for Door Frames, page 65).

NOTES:

- SEQUENCE OF OPERATIONS FOR MAKING A WINDOW FRAME WITH TRANSOM AND MULLION: The joint for the transom-mullion connection should be a stub tenon, to prevent water from entering the mortice at the top.

  Take care when marking out and cutting the shoulder of the tenon where it meets the slope of the cill, so that they fit together exactly (Drawing Book, pages 46, 47, 70, 71 and 72).

  a. Study the drawing.

  b. Prepare the cutting list.

  c. Select the timber.

  d. Cut the timber to size.

  e. Plane the timber to size and make the face marks.

  f. Mark out the joints. Take special care in marking out the shoulders of the joints. The head, transom, and cill must be marked out together, as the two posts and the mullion must also be marked out together.

  g. Mortice the head and cill, and make stopped mortices in posts and transom.

  h. Rip the tenons for the transom, posts and mullion.

  i. Rip the horns on the head and cill.

  j. Prepare the rebate and throating if required.

  k. Cut the shoulders of the tenons.

  l. Smooth and clean up the inside edges.

  m. Paint the frame with a primer coat, especially at the joints.

- ASSEMBLING THE WINDOW FRAME:

  a. Place the frame on the workbench, with the rebated side up.

  b. Fix the transom to the posts.

  c. Fix the mullion to the transom.

  d. Clamp the head and the cill to the posts and mullion.

  e. Test for squareness at the four corners, and check the diagonals.

  f. Cut and nail the squaring strips.

  g. With the clamps still in position, fasten the joints.

  h. Finish off the frame.

WINDOW FRAMES.

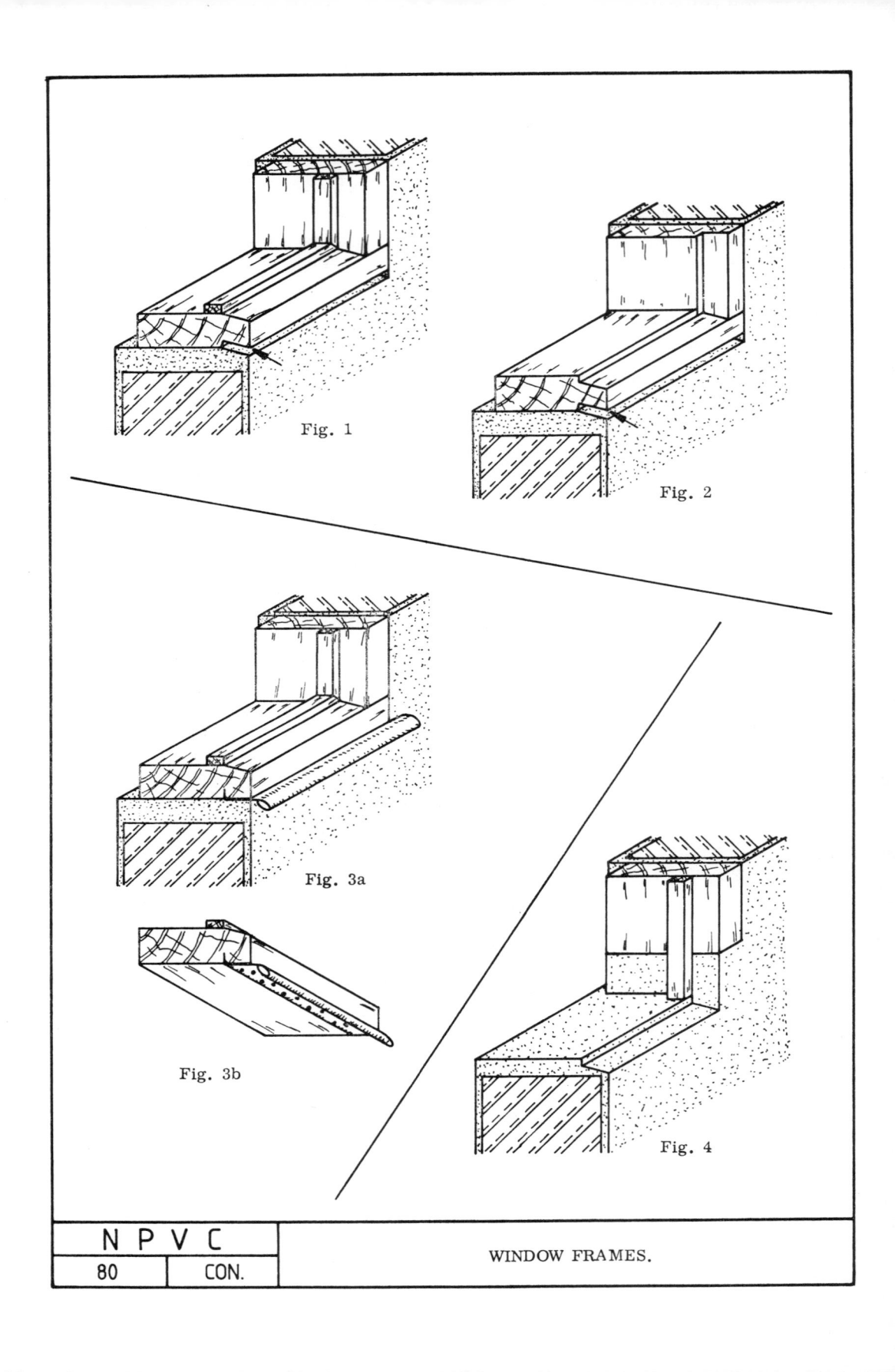

Fig. 1

Fig. 2

Fig. 3a

Fig. 3b

Fig. 4

N P V C		WINDOW FRAMES.
80	CON.	

CILLS OF WINDOW FRAMES

The cill should have a slope on the outside, to keep rain out of the building. This slope can be constructed by chamfering the cill (Fig. 1) or by a combination of chamfering and rebating (Fig. 2).

It is important to prevent rainwater from entering the space between the cill and the wall, where it could cause the cill to rot. A throating at the soffit of the cill (Figs. 1 & 2, arrows) allows water to drip down instead of running back under the cill. The angle of the throating should be parallel to the slope of the cill.

Another way to keep water out is to fix a metal strip, called a drip, under the cill (Figs. 3a & 3b). The drip is fixed in a sawcut in the underside of the cill, so that the water which drips onto the metal will not run inwards. The slope of the metal comes to just inside the face line of the cill (Fig. 3b).

The metal strip should extend past the ends of the cill. If it does not, the water which runs down the posts of the frame will damage the render around this part at the corners of the frame.

It is also possible to combine the two solutions, by fixing a metal strip onto the throating.

Every cill should have some kind of protection to keep water from entering between the cill and wall. All cills must come out at least as far as the finish line of the render.

As an alternative to the wooden cill, the window can be constructed with a plaster cill. The posts are fitted with steel dowels and a concrete shoe. The construction is similar to that of a door frame with a concrete shoe (Fig. 4, also see Concrete Shoe, page 71).

NOTES:

WINDOW FRAMES.	N P V C	
	CON.	81

Fig. 1

Fig. 1a

Fig. 2

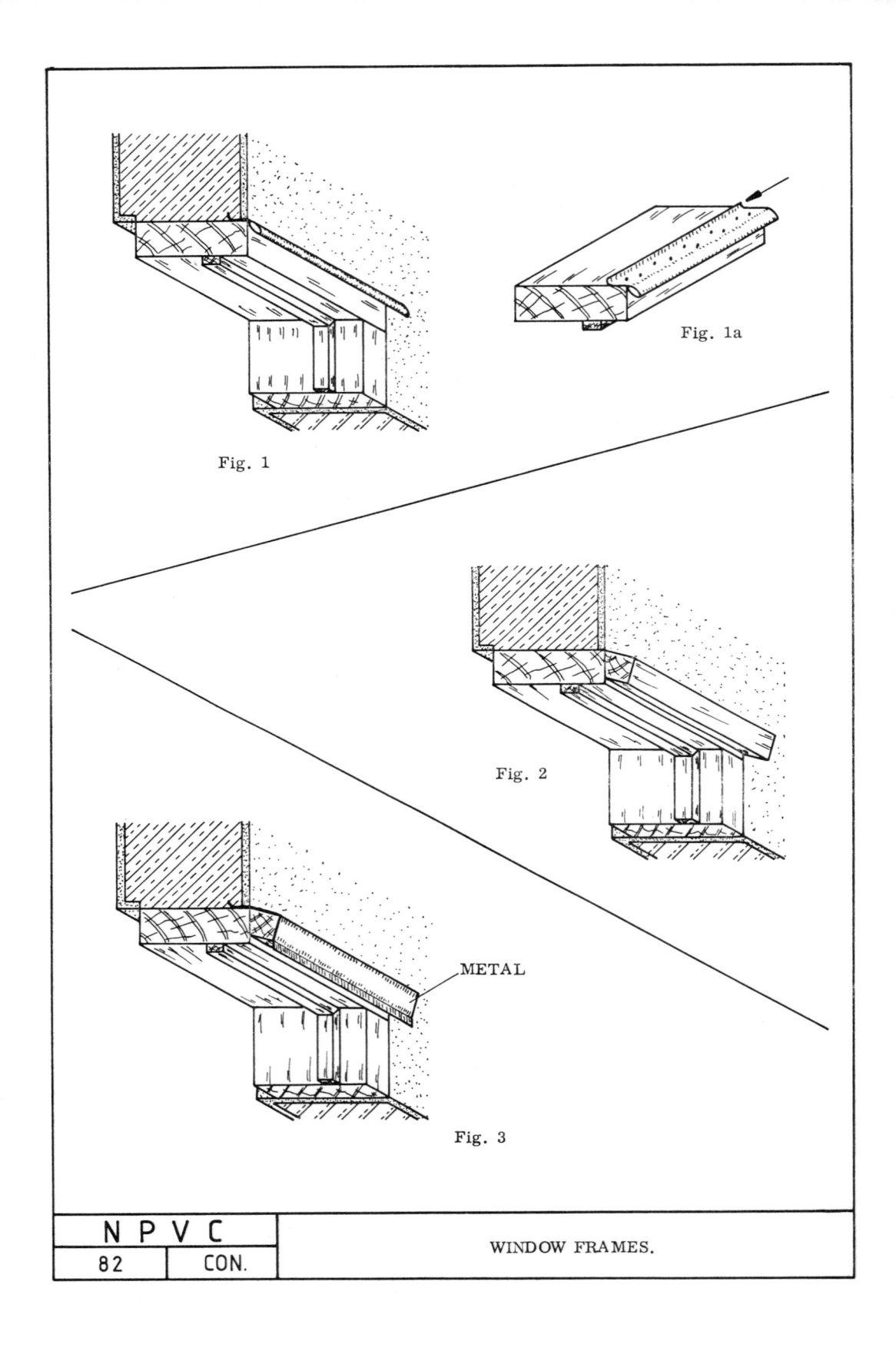

METAL

Fig. 3

N P V C		WINDOW FRAMES.
82	CON.	

HEADS OF WINDOW FRAMES

Some protection must also be added at the head of a window frame to prevent any water from entering the space between the head and the lintel and causing damage to the wood. The simplest method is to fix a metal strip similar to the one on the cill. The part of the strip which is plastered over should be bent slightly upwards to prevent water from entering (Fig. 1a, arrow).

Note that the line of nails is staggered so that the drip (or weather strip) is firmly secured to the head (Fig. 1a). Otherwise, movement of the strip during heavy rains or storms would probably loosen the render.

These metal drips may be put on when the frame is set up, or left until just before the rendering is done.

An alternative method would be to fix wooden battens, shaped as shown in Fig. 2, on the face of the window head. This method gives a more satisfactory appearance to the window frame.

A combination of both these methods is also possible (Fig. 3).

The importance of fixing weather strips on both the cill and the head of the frame cannot be over-emphasized. They are easily overlooked during the construction of the building and one might be tempted to do a quick job and leave them out. But in the long run the strips will save money by preventing water damages to the wood and expensive repairs.

The same kind of protection can be applied to the heads of door frames.

The heads and cills of window frames can be provided with horns to give an additional secure fixing to the wall (see Heads of Door Frames).

NOTES:

Ⓐ – FRAME SHOULD REST ON A BED OF
CONCRETE, NOT DIRECTLY ON THE
LANDCRETE.

Fig. 1

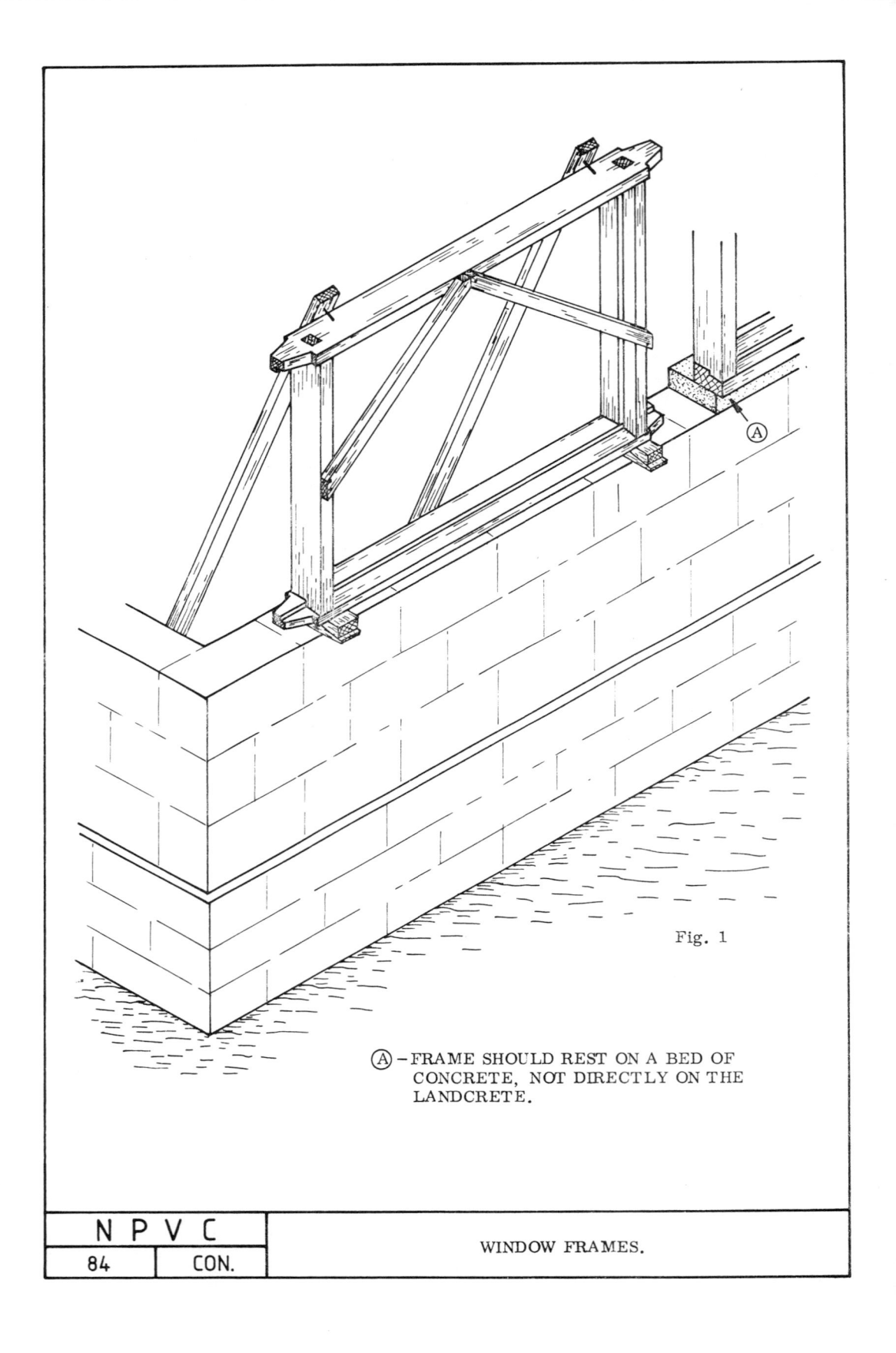

N P V C		
84	CON.	WINDOW FRAMES.

INSTALLING WINDOW FRAMES

Installing window frames is similar to installing door frames (see Door Frames section). A window frame should be fixed in such a way that the casement can open flat to the wall, otherwise the casement will form a lever with the edge of the wall. This can cause the hinges to be forced out when the window is blown open by the wind.

Window frames can be installed either during the construction of the walls, or after the walling has been completed.

- INSTALLING THE FRAMES DURING THE WALL CONSTRUCTION: When the blockwork has reached window cill level, the frames can be set and aligned in the same way as a door frame. The chief difference is that in the window frame the horizontal members are usually longer than the vertical members, and for that reason more attention is given to levelling the head and cill of a window frame than to levelling the head of the door frame (Fig. 1).

  If more than one frame is installed, check them by comparing from a distance to see if they are aligned, as with door frames.

- INSTALLING THE FRAMES AFTER WALLING HAS BEEN COMPLETED: This second method is not so commonly used, but it also has its advantages. In order to keep the frames clean, square and dry, they are kept in a store until the building is roofed. Openings are left in the blockwork to receive the frames, horns, and steel cramps. The frame is set into this opening and the steel cramps and horns are secured with mortar.

  Sometimes hard wood or plastic plugs are used to secure the frame in the wall.

  - NOTE: The window frame does not rest directly on the landcrete wall; this would provide a path for moisture to get into the landcrete and weaken or damage it. The window frame is set upon a bed of concrete or cement mortar (Fig. 1, A).

NOTES:

WINDOW FRAMES.

b

a

THICKNESS OF RENDER

UP TO 10 cm CONCRETE BELOW THE CILL

Fig. 1

HOLE FOR STEEL CRAMP (CRAMP NOT SHOWN)

GROOVE

THICKNESS OF RENDER

FACE OF FRAME

Fig. 2

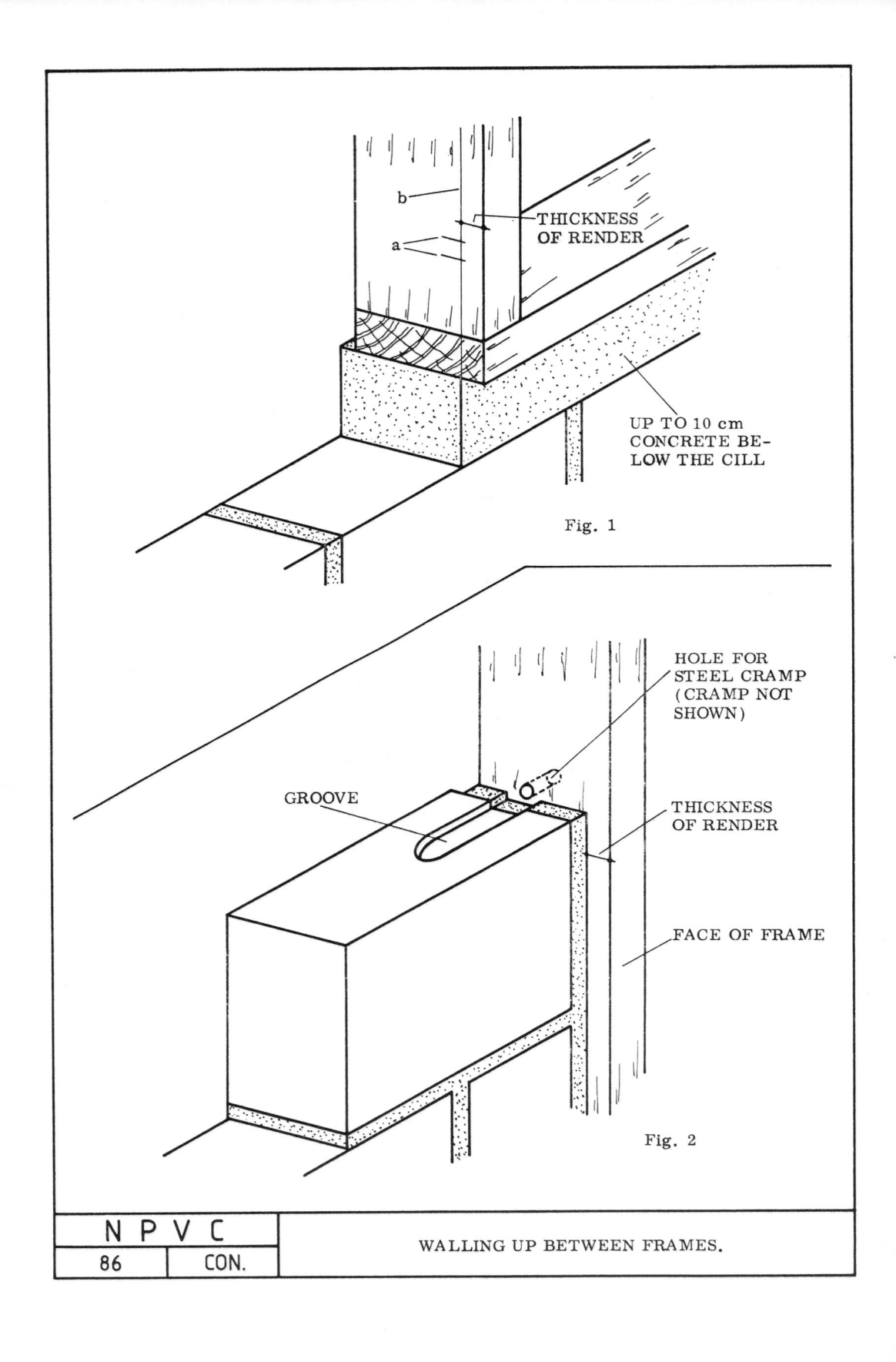

N P V C	
86	CON.

WALLING UP BETWEEN FRAMES.

WALLING UP BETWEEN FRAMES

First mark the positions of the blocks on the outsides of the frames. Mark both the height of the blocks (Fig. 1, a) and the face line of the blockwork (Fig. 1, b). The face line of the blockwork will be inside the face line of the frames, because the faces of the frames must be flush with the finish line of the render.

- NOTE: The window frame does not rest directly on the landcrete blocks. This is because moisture could enter between the cill and the landcrete and get into the blocks, where it would cause them to swell and fall apart. To protect the landcrete blocks, the cill must rest upon a bed of concrete which can be up to 10 cm thick. By adjusting the height of this concrete bed, the anchorage irons and the heads of the frames can be brought into alignment with the bed joints of the blocks.

The first blocks laid are those which touch the frames. The courses in between are completed and aligned with the aid of either a straight edge, spirit level, or mason line, depending on the distance between the frames.

Pay special attention to the anchorage irons of the frames. Make sure that they are not just embedded in the bed joint; instead make grooves in the bed faces of both blocks, which can be filled with mortar to anchor the iron and to protect it from rust (Fig. 2).

MOSQUITO PROOFING

It is often necessary to install mosquito wire on doors and windows. The wire screen is normally attached with wooden beads directly onto the window frame. For louvred windows it is fixed on the outside. For casement windows which open outwards, the wire screen has to be attached on the inside or onto a separate framework which is hinged onto the inside of the frame.

A separate framework for the mosquito wire is commonly constructed on doors. Make this framework like a panelled door (this door and its construction are described in one of the following sections, under Doors). Use one or two braces in the framework to keep the door rigid. Often a spring is attached to the wire door to keep it in the closed position.

Fig. 1

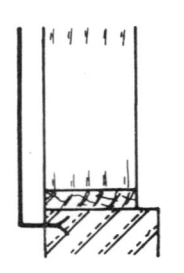

Fig. 2a

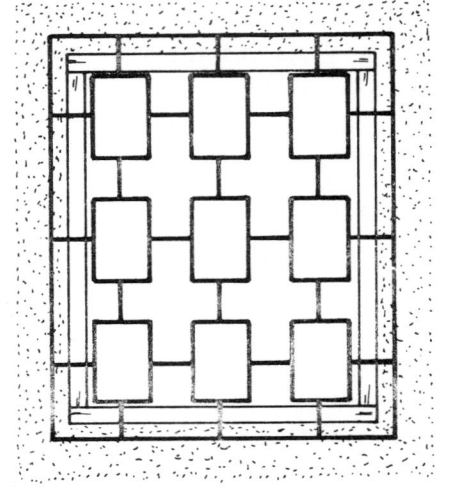

Fig. 2

Fig. 3

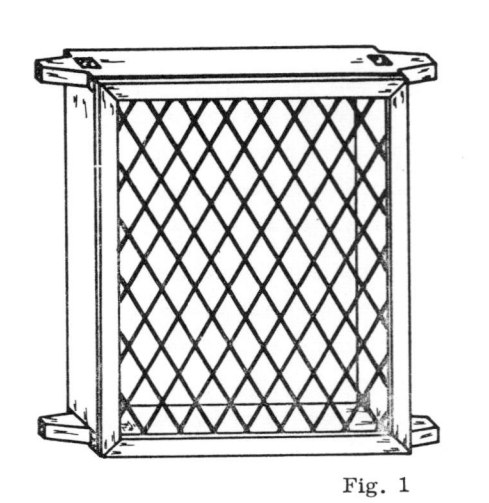

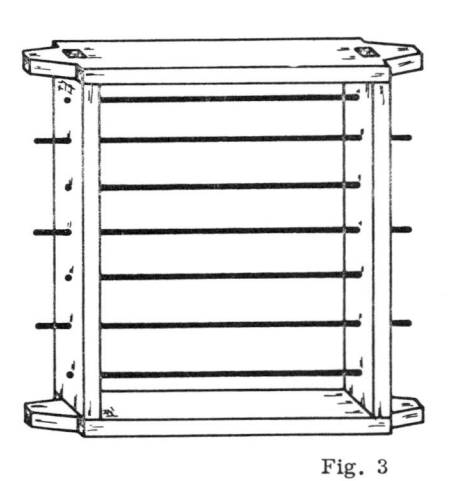

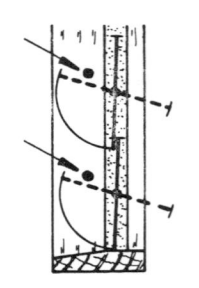

Fig. 3a

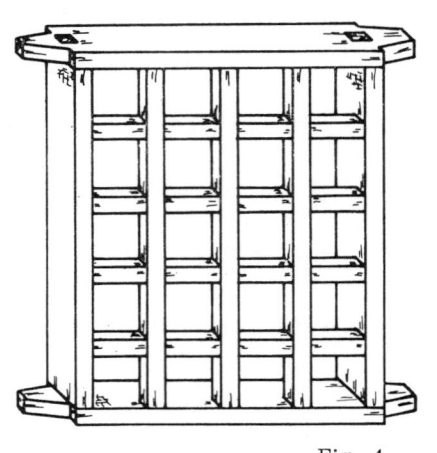

Fig. 4

NPVC 88 CON.

BURGLAR PROOFING.

BURGLAR PROOFING

It is practically impossible to keep burglars out of a house when the occupants are away, but it is at least possible to make the house burglar proof to the extent that burglars cannot enter without making considerable noise and so alerting people in the area.

TYPES OF BURGLAR PROOFING FOR WINDOWS

- One of the most common methods is to fix mesh wire or expanded metal directly onto the frame, in a way that it cannot be torn off without making considerable noise or using a special cutting tool. Use strong beads to attach the wire to the frame and plenty of screws or nails (Fig. 1).

- Sometimes burglar proofing is made from welded iron rods in different patterns and fixed directly into the wall around the window. It should be anchored well and deeply enough so that it cannot be pulled out (Figs. 2 & 2a).

- Another way to make a window safer is to fix iron bars horizontally into the frame. This must be done during the assembly of the frame. If some of the bars stick out at the sides of the frame, they can also be used as an additional anchor for the frame in the wall (Fig. 3). If the span between the posts of the window is large, the horizontal bars may need to be reinforced by having vertical bars welded onto them, so that they cannot be bent up or down to allow the burglar to enter.
  The bars can also be applied to windows fitted with louvres. In that case, the iron bars are placed between the louvre glasses (Fig. 3a). They must be placed in such a way that they will not obstruct the louvres when they are opened.

- A r latively inexpensive and reliable burglar proofing is the wooden type. During assembly of the frame, this burglar proofing is installed into mortices made in the head, cill and posts of the frame. The separate pieces are fitted together with halving joints in a crosswise pattern (Fig. 4).

If outward-opening casements are installed in a frame, install the burglar proofing on the inside face of the frame. Leave a small opening in the burglar proofing for the hand to pass through in order to open and close the window.

For louvred windows the burglar proofing is always installed on the outside.

Fig. 1

Fig. 2

Fig. 3

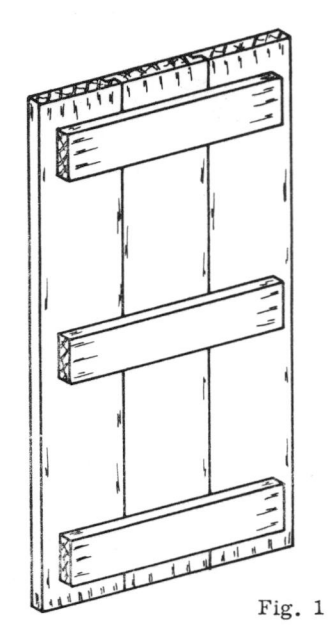

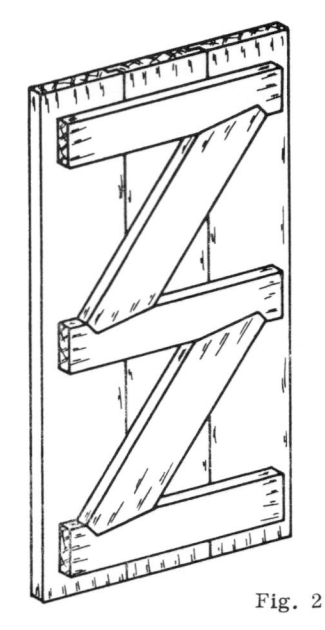

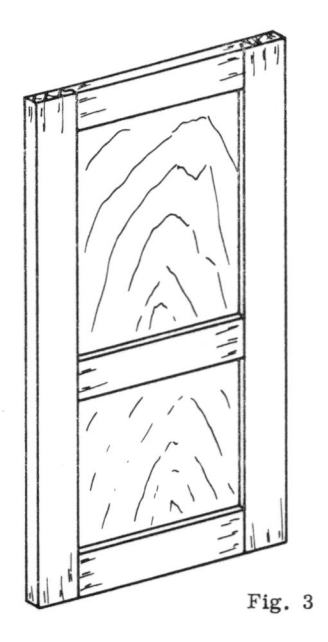

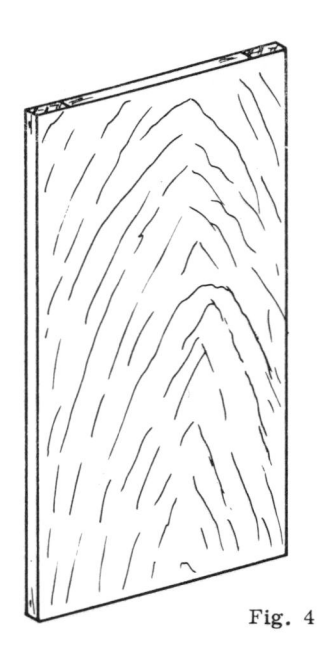

Fig. 4

DOORS.

DOORS

The door is an essential part of the building structure. It must have the right proportions: not so large that it weakens the wall of the building, and not too small either. People, goods, and equipment, according to the uses of the building, have to be able to pass through the door easily.

TYPES OF DOORS

In Rural Building, doors are classified as follows:

- Ledged and battened doors (Fig. 1)

- Ledged, braced and battened doors (Fig. 2)

- Panelled doors (Fig. 3)

- Flush doors (Fig. 4)

For an outside door which is exposed to rain and sun, it is best to choose a weather resistant door like a battened or panelled door. If the door is protected inside the building then a flush door made of plywood will be satisfactory.

SIZES OF DOORS

The size of the door depends on:

- Where it will be used (entrance doors will naturally be slightly larger than other doors, and bathroom doors will be a bit smaller)

- The materials which are available (plan the door according to the sizes of wood and plywood which are available).

A common door size is 198 cm by 76 cm. This is wide enough to allow most sizes of furniture to pass through. Nearly all modern furniture has one dimension which is 76 cm or less: a standard table, for example, is 76 cm high.

Living room and bedroom doors are sometimes larger for reasons of ventilation.

POSITION OF THE DOOR

Outside doors should always open to the outside. This makes them more weatherproof and easier to open in case of an emergency such as a fire, when you want to get out quickly.

Try to position the door so that you cannot look directly into the room from the outside.

| DOORS. | N P V C |
| | CON. 91 |

TOPS OF LEDGES CAN BE BEVELLED

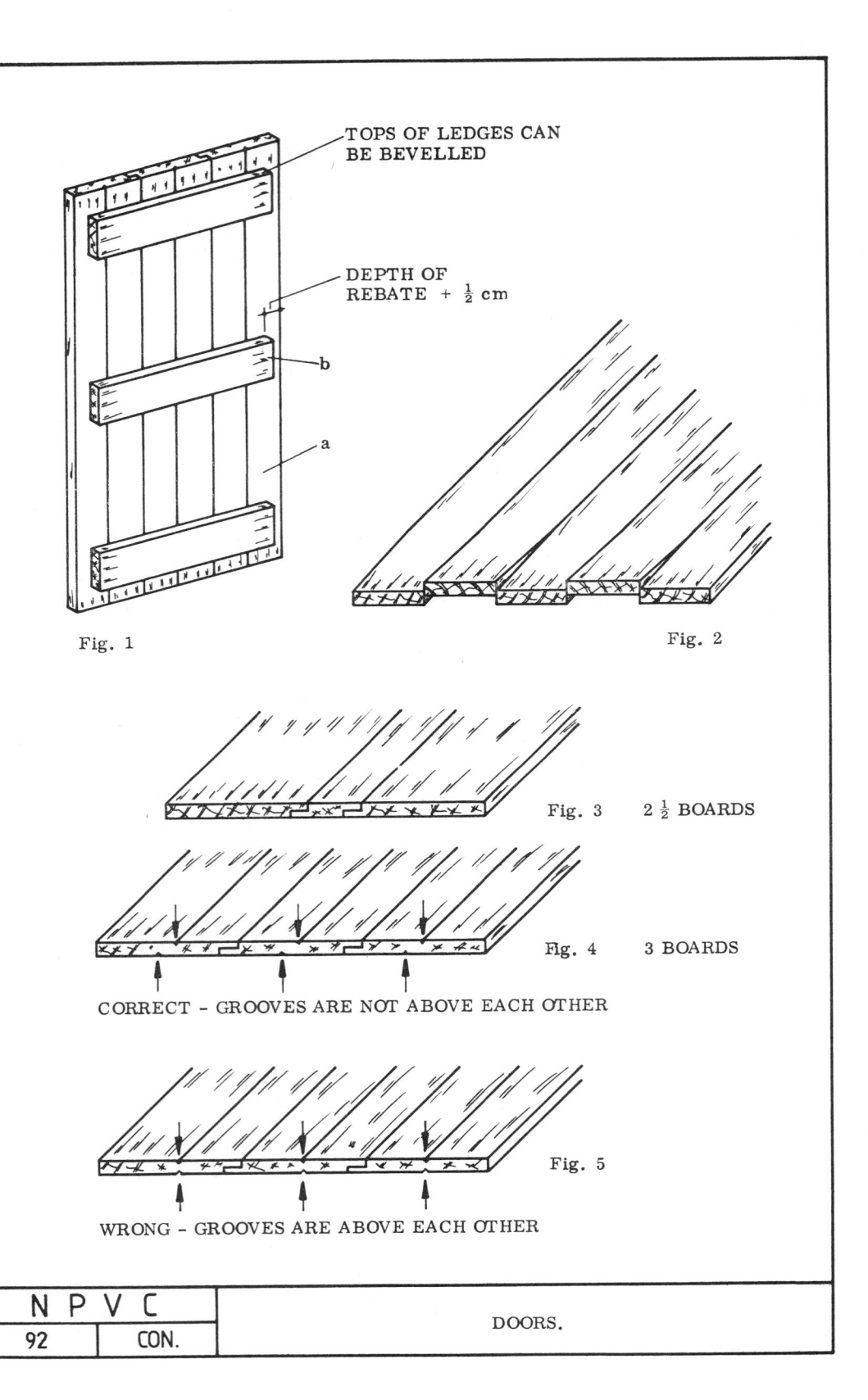

DEPTH OF REBATE + $\frac{1}{2}$ cm

b

a

Fig. 1

Fig. 2

Fig. 3   2 $\frac{1}{2}$ BOARDS

Fig. 4   3 BOARDS

CORRECT - GROOVES ARE NOT ABOVE EACH OTHER

Fig. 5

WRONG - GROOVES ARE ABOVE EACH OTHER

N P V C	
92	CON.

DOORS.

FINISHING TREATMENT OF THE DOOR

All doors should be painted or varnished, or preserved in some way. A good paint job preserves the door from decay and reduces shrinkage and swelling, thereby lengthening the life of the piece (Reference Book, pages 200 & 201).

Paint is especially important for flush doors. They should be well painted and the edges of the door should be soaked in paint.

LEDGED AND BATTENED DOORS

These doors consist of vertical boards called battens (Fig. 1, a) which are nailed or screwed to the horizontal members, called ledges (b). Often the battens are about 15 to 18 cm wide and 2 to 3 cm thick. Doors made with narrow battens like these have a better appearance.

Because of the climate in the northern part of Ghana, here it is better to use boards which are up to 30 cm wide. These wider boards are less likely to twist and warp at the ends (Fig. 2). The boards should be well seasoned so that they won't crack at the ends.

The width of this door is usually the width of $2\frac{1}{2}$ or 3 boards (Figs. 3 & 4).

Grooves can be made in the wide battens, to produce a pleasing appearance as if the door were made with narrow battens (Fig. 4). The grooves should not be made from both sides of the door at the same spot, above each other, because this will make the door weak (Fig. 5).

The ledges should be as long as possible to prevent cupping of the battens, and to provide a solid attachment. The length of the ledges will be the width of the door minus the depth of the two rebates or beads in the door frame, and minus a $\frac{1}{2}$ cm allowance on each side (Fig. 1). The ledges are usually 2 to 3 cm thick. When this construction is used for an external door, the tops of the ledges should be bevelled (Fig. 1).

The ledged and battened door is the simplest type of door. It is often used for narrow openings. It is relatively cheap to construct, but unfortunately it tends to sag because of its weight.

NOTES:

DOORS.

Fig. 1 FEATHER

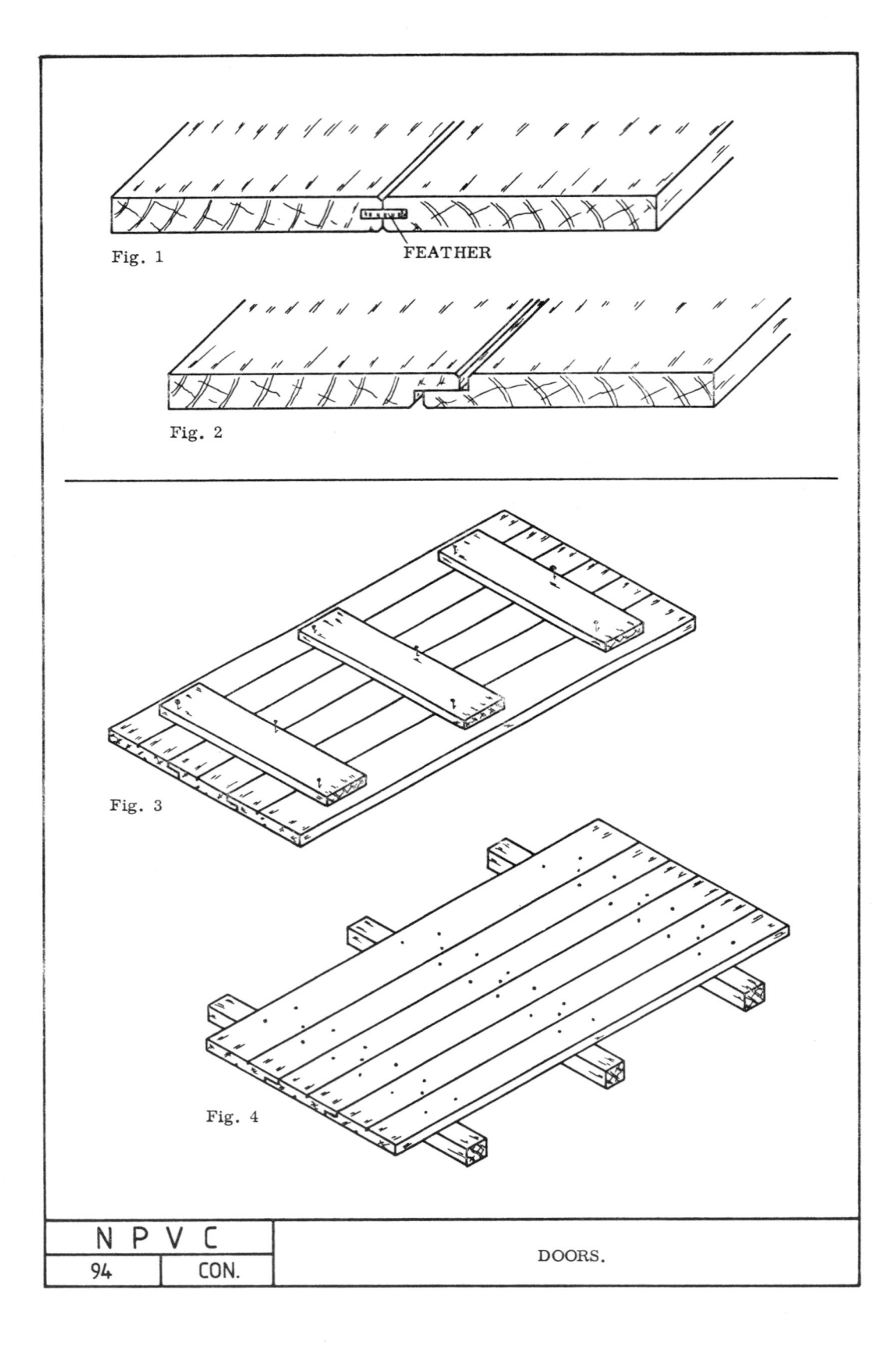

Fig. 2

Fig. 3

Fig. 4

N P V C		
94	CON.	DOORS.

- JOINTS FOR LEDGED AND BATTENED DOORS: We use special types of joints to connect the battens of this door. These joints are known as matchboarding joints.

A matchboarded door will have a good appearance even when the boards have shrunken and the joints are opening. Matchboarding joints are not glued like ordinary widening joints (Basic Knowledge, pages 130 to 136). They are left loose so that shrinkage can take place, and the battens are held together by the ledges.

In Rural Building we use two different matchboarding joints: the loose tongued joint (Fig. 1), and the rebated joint (Fig. 2).

The loose tongued joint has a tongue made of plywood. It is glued into only one side of the groove so that shrinkage can take place (Basic Knowledge, page 134).

There are several possible ways to make the rebated joints used in battened doors (Fig. 2).

When you assemble this joint, put a strip of wood in between the battens with the same thickness as the gap you want. This keeps the gap between the battens open as the door is assembled.

Any batten-type door made during the rainy season should be assembled with the matchboarding joints closed. If the door is assembled during the dry season, then the joints should be kept open (Fig. 2, also see Reference Book, pages 129 to 130). There should be enough allowance between the battens to allow them to swell up to 5% of the width of the board (for a 30 cm board size the gap will be up to $1\frac{1}{2}$ cm). If this is not done the door will start to warp when it swells in the rainy season.

- SEQUENCE OF OPERATIONS FOR MAKING A (NAILED) LEDGED AND BATTENED DOOR:

    a. Study the drawing.

    b. Prepare the cutting list.

    c. Select the wood.

    d. Cut the timber to the right size, leaving a 1 cm allowance in the length of the battens for finishing off.

    e. Plane the timbers to size. Mark the boards (Basic Knowledge, page 88).

    f. Plane the matchboarding joints.

    g. Paint or varnish the edges of the battens.

DOORS.

Fig. 1   NO BRACES
         DOOR MAY SAG

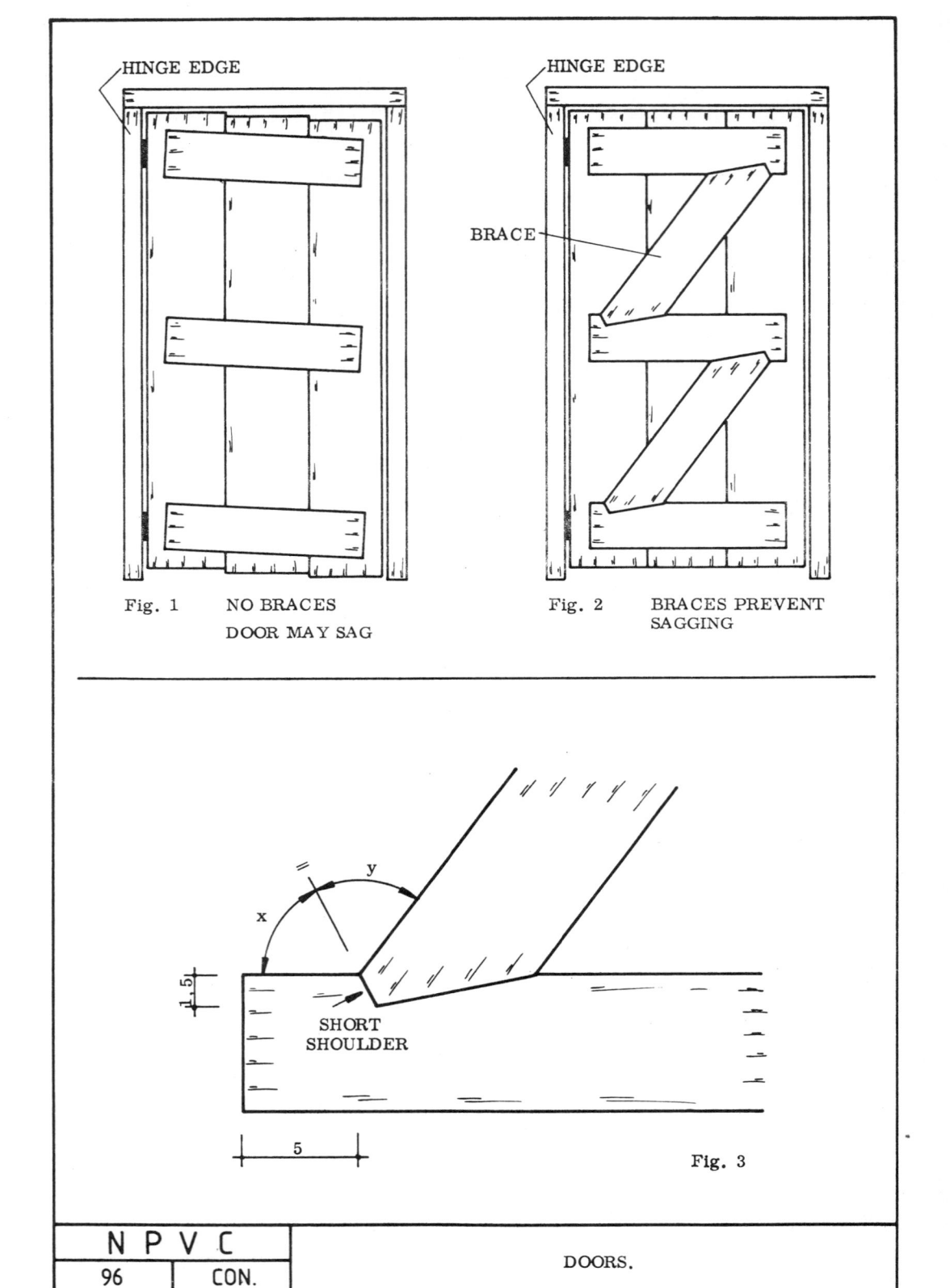

Fig. 2   BRACES PREVENT
         SAGGING

HINGE EDGE

BRACE

x, y

1.5

5

SHORT SHOULDER

Fig. 3

N P V C		
96	CON.	DOORS.

h. Fit the battens together. Use a wooden or metal sash clamp to hold them.

i. Mark the positions of the ledges.

j. If the door is to be an external one, bevel the tops of the ledges.

k. Paint or varnish the inside faces of the ledges and battens.

l. Nail the ledges lightly to the battens (Fig. 3, previous page).

m. Turn the door over put three pieces of wood under the ledges for support during nailing (Fig. 4, previous page).

n. Drive the nails through the battens and ledges and punch the heads under the surface of the wood.

o. Turn the door over and clench the nails. Remove the nails used in step "l" to nail the ledges lightly to the battens.

p. Cut off and level the battens at the top and bottom.

- REMEMBER: To avoid splitting the wood, always blunt the nails and drive them in a staggered pattern (Fig. 4, previous page). Clench all the nails to prevent them from being pulled out if the wood warps or swells. For clenching, the nails have to be long enough to go through the wood and out the other side by about 5 mm (Basic Knowledge, pages 92 to 94).

LEDGED, BRACED AND BATTENED DOORS

This is a ledged and battened door to which braces have been added to prevent sagging (Figs. 1 & 2). These braces must slope upwards from the hinge edge of the door, and they are housed with a skew notch into the ledges (Drawing Book, page 73).

The skew notch helps to distribute the force from the weight of the door, so that the ledges have an even pressure on them. To achieve an equal distribution of the force, the angle of the short shoulder of the skew notch has to bisect (divide in two) the angle between the ledge and the brace (see Fig. 3, angle "x" = angle "y").

The skew notch should start from a point 5 cm from the end of the ledge, to prevent shearing off. The depth of the skew notch will be 1,5 cm (Fig. 3).

This type of door construction may be used for large openings because of its greater strength.

The sequence of operations to make this door is the same as for the ledged and battened door. The braces are fitted to the door after the ledges are nailed lightly to the battens (step "l" in the construction).

Fig. 1 (m)

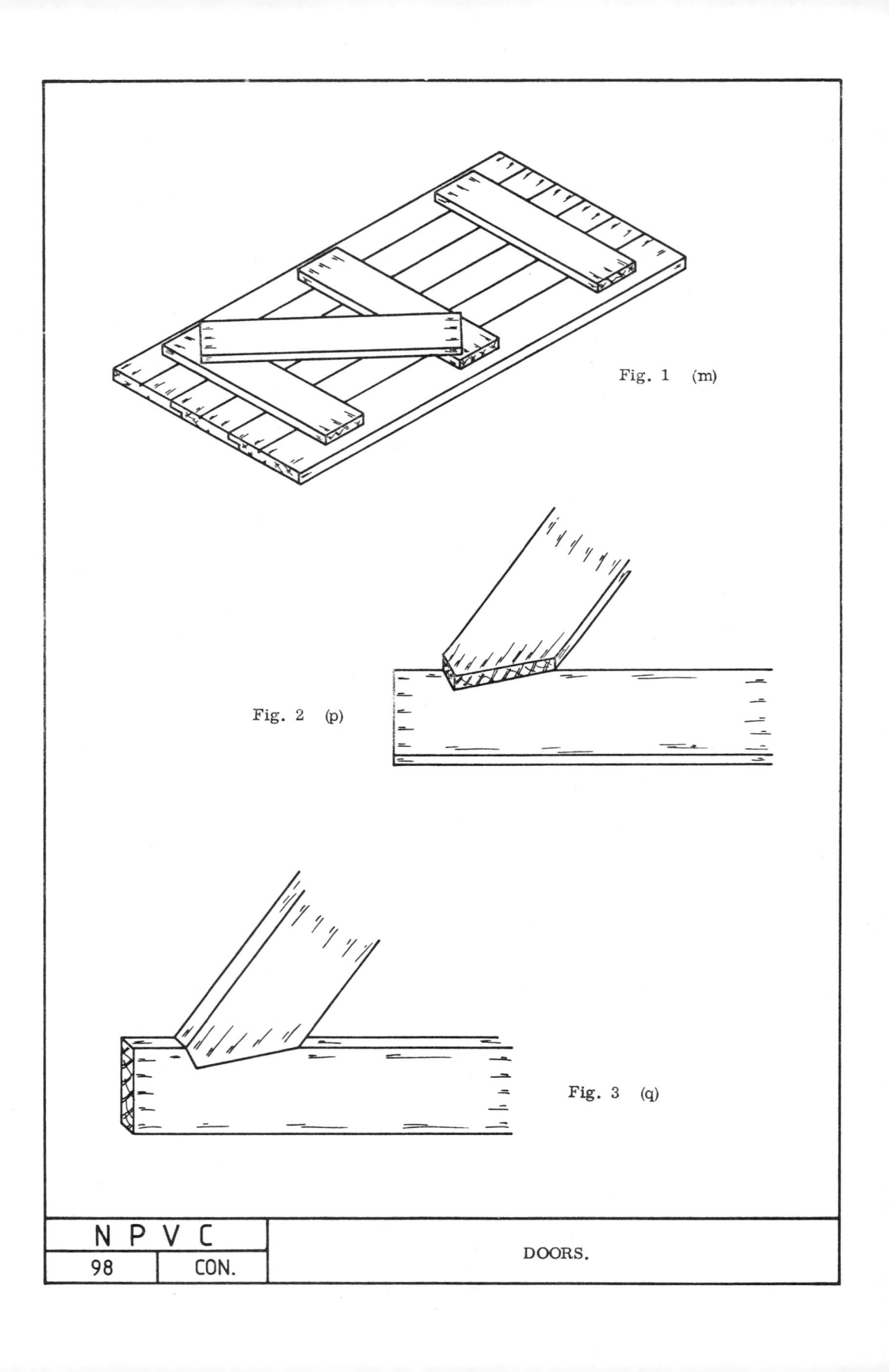

Fig. 2 (p)

Fig. 3 (q)

| N P V C | DOORS. |
| 98 CON. | |

- SEQUENCE OF OPERATIONS FOR MAKING A (NAILED) LEDGED, BRACED AND BATTENED DOOR: Proceed in the same manner as for the ledged and battened door, until you reach step "l". Continue as follows:

  m. Place the braces on top of the ledges (Fig. 1).

  n. Mark out the skew notches on each end of the braces.

  o. Cut the braces according to the marks.

  p. Place each brace in position and mark out the positions on the ledges (Fig. 2).

  q. Chisel the notches in the ledges (Fig. 3).

  r. Paint the skew notches.

  s. Nail the braces lightly to the battens.

  t. Turn the door over and place three pieces of wood under the ledges for support during nailing.

  u. Drive nails through the battens, ledges and braces, and punch the nail heads under the surface.

  v. Turn the door over and clench the nails. Remove the nails which were used to hold the braces and ledges lightly to the battens in step "s".

  w. Cut off and level the battens at the top and bottom.

NOTES:

DOORS.

NPVC
CON. 99

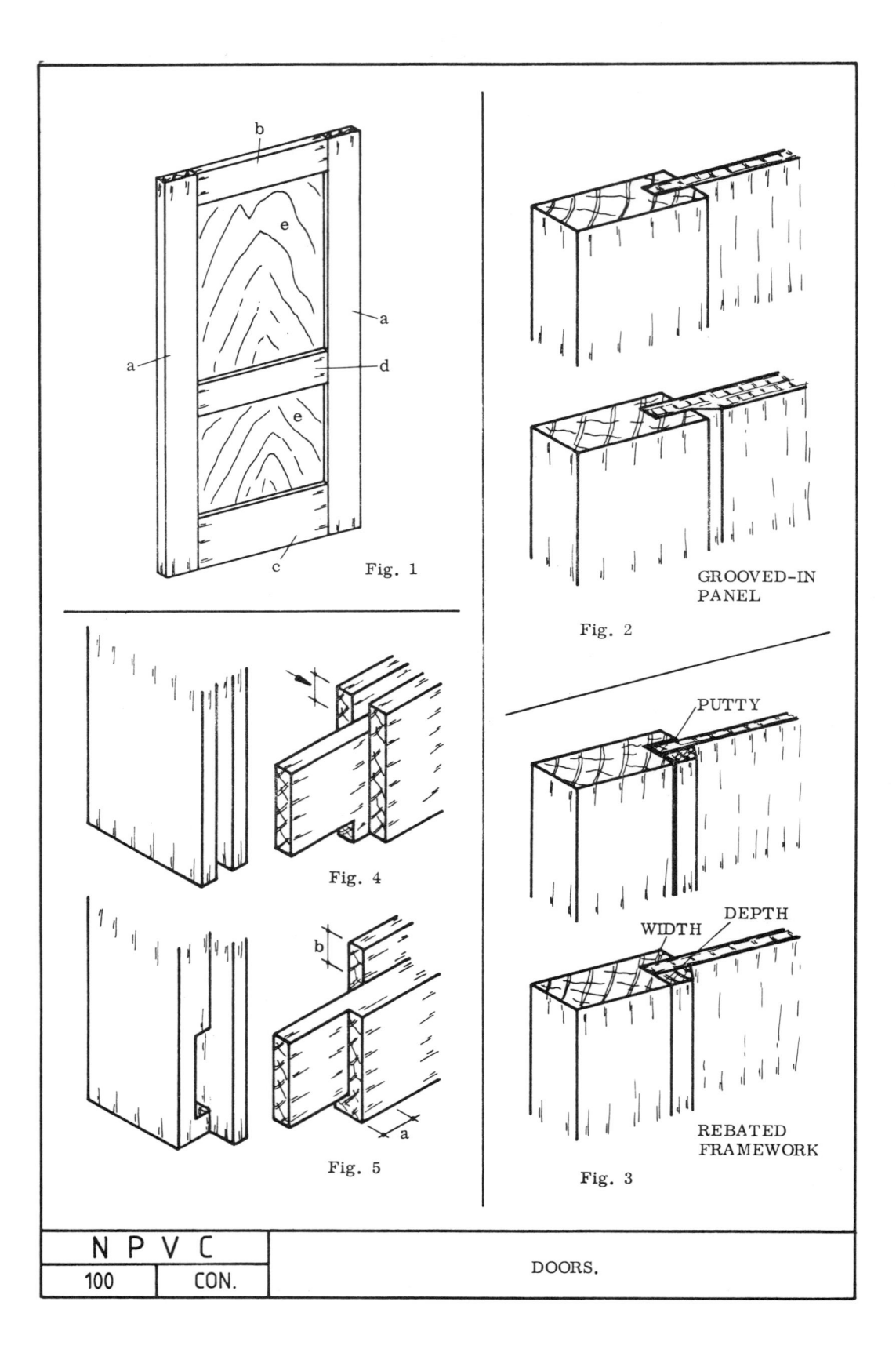

Fig. 1

Fig. 2
GROOVED-IN PANEL

Fig. 3
PUTTY
WIDTH
DEPTH
REBATED FRAMEWORK

Fig. 4

Fig. 5

N P V C		
100	CON.	DOORS.

PANELLED DOORS

These doors (Fig. 1) consist of a frame made up of stiles (a), a top rail (b), a bottom rail (c) and sometimes an intermediate rail (d). Into this framework a plywood panel (e) is fitted. This panel may fit into a groove or a rebate (Drawing Book, pages 75 to 78).

- GROOVED-IN PANEL: This construction can be used only on doors which will be protected from the rain. If water enters the groove, it will not be able to dry out and can cause the plywood to rot (Fig. 2).

If the door will be painted or varnished, the edges of the plywood panel and the inside of the grooves should be treated before the door is assembled.

The depth of the grooves is usually 1/3rd of the thickness of the members. The grooves are made as explained in the Basic Knowledge book, page 136.

The joints for this kind of door will be a haunched stub tenon for the top and bottom rails (Fig. 4), and a stub tenon for the intermediate rail (Basic Knowledge, pages 118 to 122). Note that the width of the tenon has to be reduced by the depth of the groove (Fig. 4, arrow) and the mortice is reduced accordingly.

- REBATED FRAMEWORK: When the panel will be fixed after the frame is assembled, a rebated framework is used. A rebate is made in the frame and the panel is secured with beads (Fig. 3).

Unlike the grooved-in panel door, this door can be used in places where it is exposed to rain. In that case, the plywood is set in putty so that water cannot penetrate inside, and the tops of the intermediate and bottom rails should be bevelled. The beads are also set in a thin layer of putty, then secured with nails.

If the door will be painted or varnished, the joints, edges of the plywood, rebates and beads all have to be painted before assembly.

The width of the rebate is usually 1/3rd of the thickness of the frame member, and the depth is usually 2/3rds of the thickness of the member (Fig. 3).

The joints for this kind of door will be haunched stub tenons for the top and bottom rails and stub tenons for the intermediate rail. Note that one shoulder of the tenon has to be longer to fill the rebate (Fig. 5, a). The widths of the tenon and mortice have to be reduced by the size of the rebate (Fig. 5, b; also Drawing Book, page 78, and Basic Knowledge, pages 118 to 120).

NOTES:

DOORS.

Fig. 1a

Fig. 1b

HEAD OF NAIL PINCHED OFF

STRIP OF SHEET METAL

TOP RAIL

STILE

STILE

BOTTOM RAIL

Fig. 2

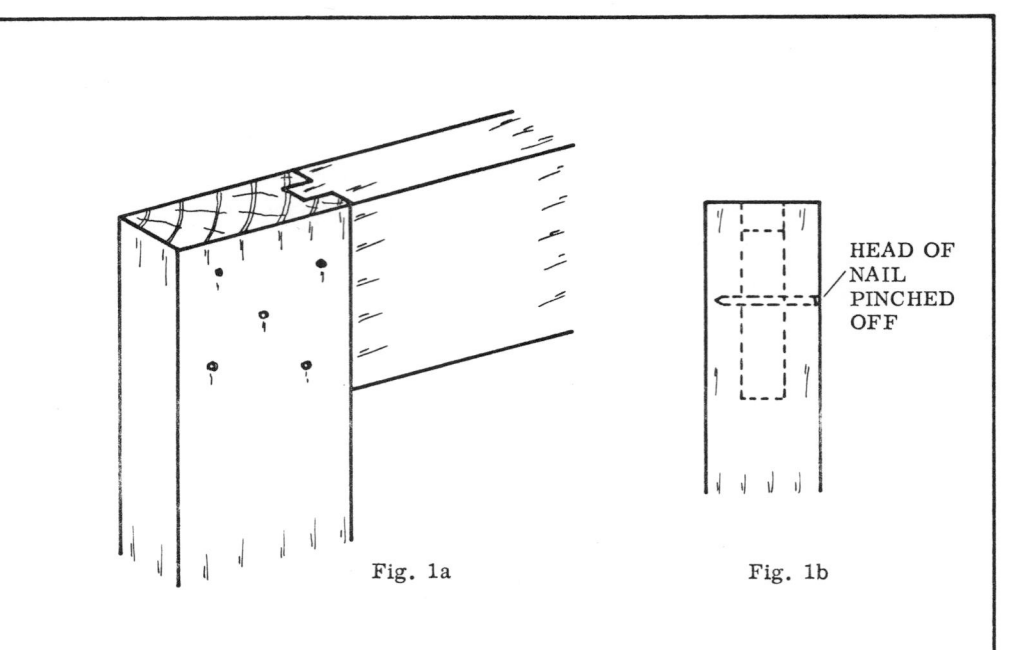

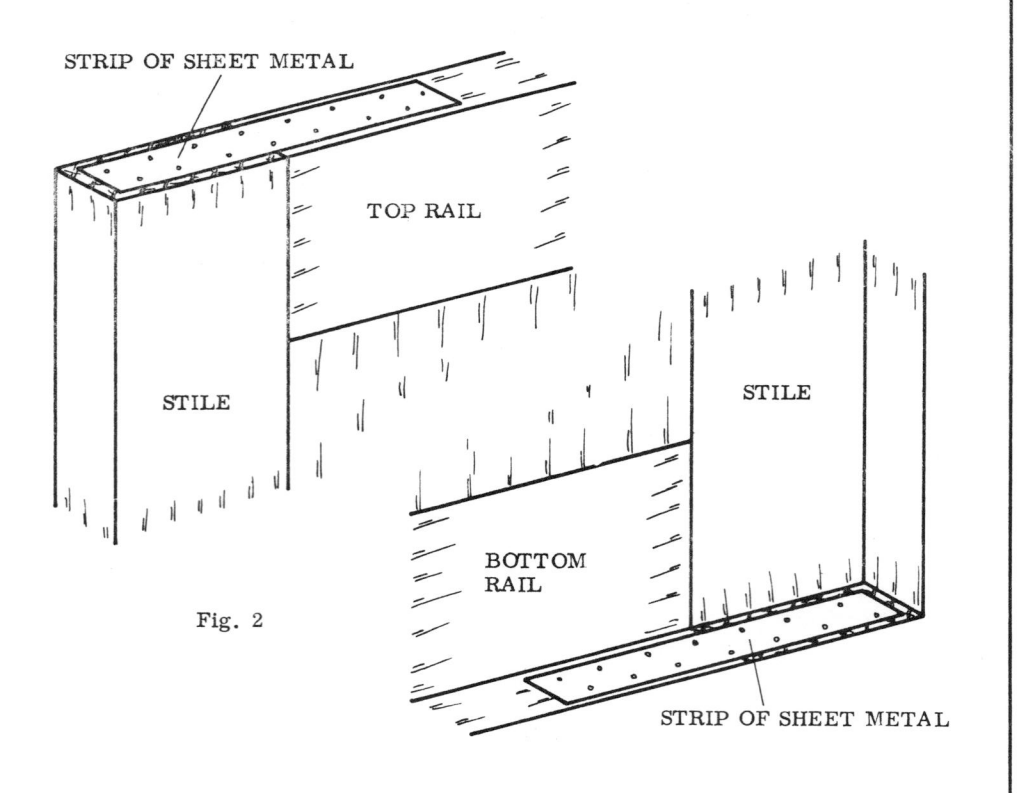

STRIP OF SHEET METAL

N P V C	
102	CON.

DOORS.

- TIMBER MEASUREMENTS FOR PANELLED DOORS:

    Stiles and top rail   -   3 to 4 cm thick, 10 cm wide
    Intermediate rail   -   3 to 4 cm thick, 10 cm wide
    Bottom rail   -   3 to 4 cm thick, 15 to 20 cm wide

- HOW TO PREVENT A PANELLED DOOR FROM SAGGING: Very often panelled doors will sag after a certain length of time. This is due to the joints drying out and loosening, and to the weight of the door.

    The key to the door's ability to resist sagging is the plywood panel. For this reason the panel must be properly secured and fit tightly into the frame.

    The tenons must also be well secured. If we can prevent the joints from opening we prevent the door from sagging. A good method of securing the joints is to nail through the joint and pinch off the heads of the nails. They should only be long enough so that they can be punched under the surface but do not come through the other side of the frame (Fig. 1b).

    Note the staggered positions of the nails in Fig. 1a on the left. This prevents the nails from splitting the wood.

    On the hinge stile of the door, the joints tend to pull apart at the lower edge. To prevent this, fix strips of sheet metal at the top hinge corner and at the lower corner opposite from the hinge stile. Use small nails to fix the metal (Fig. 2).

NOTES:

DOORS.

VENT HOLES

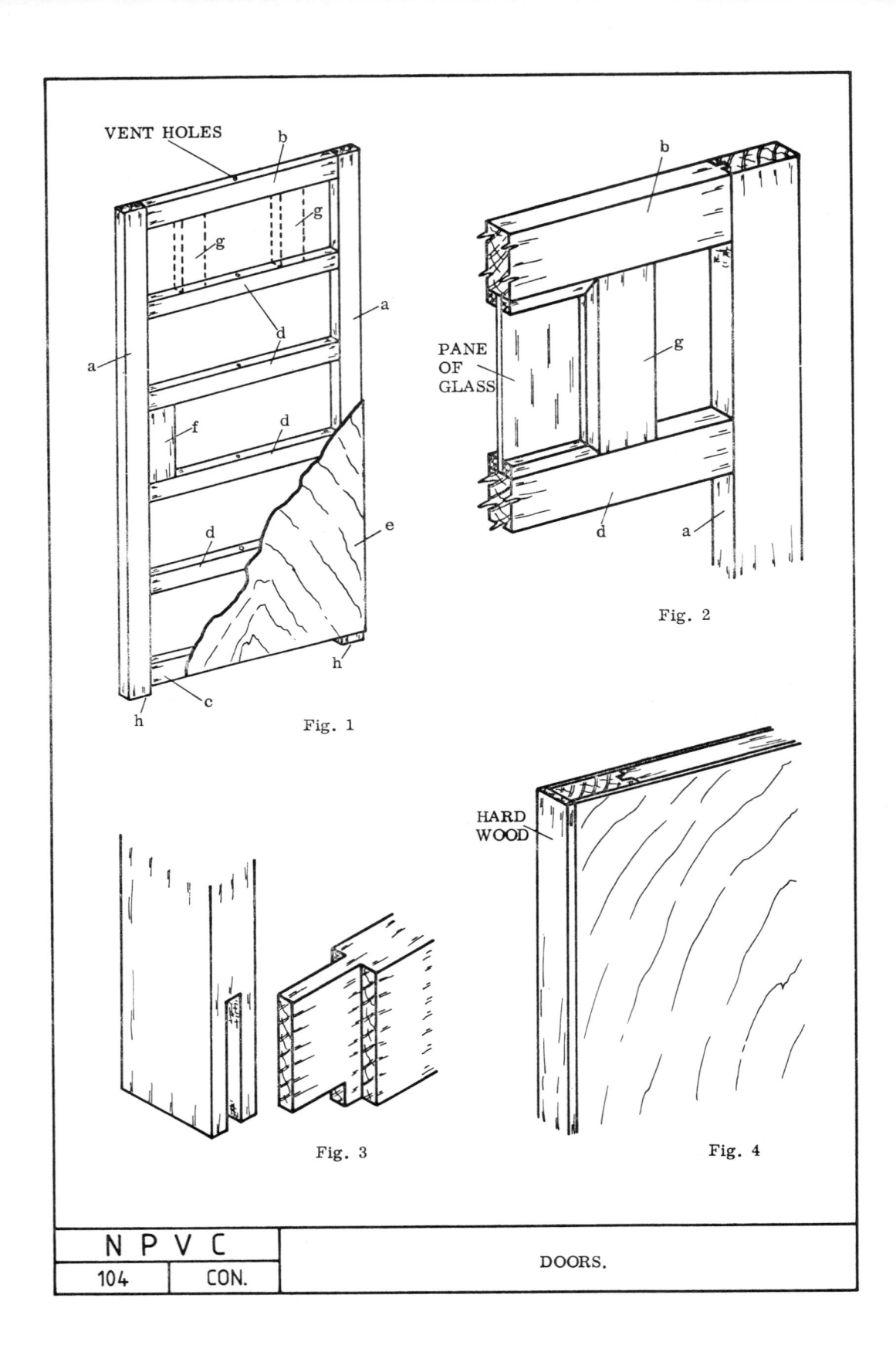

Fig. 1

PANE OF GLASS

Fig. 2

Fig. 3

HARD WOOD

Fig. 4

N P V C		DOORS.
104	CON.	

FLUSH DOORS

The flush door with a framed core (Fig. 1) is a type of door that we frequently make in Rural Building. This door consists of a frame which has stiles (a), top and bottom rails (b & c), and narrow intermediate rails (d). It is covered on each side by a sheet of plywood (e). Sometimes flush doors for the outside of the building are covered on one or both sides by sheets of thin metal, usually aluminium or galvanized iron. Plywood-covered flush doors cannot be used where they will be exposed to rain and sun.

Well seasoned Wawa is used for the frame. If the stiles are not wide enough to provide room for the mortice lock (Reference Book, page 223), then a lock block (f) is added. This block should be long enough so that does not restrict the lock position to only one spot. The location of the lock block should be indicated on the outside edge of the frame so it can be found later after the plywood or metal sheets are fixed.

When two additional members are placed between the rails (g), a pane of glass can be set in them to provide additional light for the room (Figs. 1 & 2).

To prevent damage to the top and bottom rails when the door is stored and moved around, the stiles can be left projecting past the rails (h), thus forming a rest for the door.

Ventholes can be drilled through each of the rails to ensure good air circulation within the frame. Take care that the ventholes in the top and bottom rails are not blocked by paint.

The joints for the flush door can be haunched stub tenons for the top and bottom rails (Fig. 3) and stub tenons for the intermediate rails.

- LIPPING: A strip of hard wood (Fig. 4) can be fixed to the striking stile to prevent damage to the plywood edges. Lipping is always done after the plywood has been fixed. A thoroughly painted lipping can be added to the top of the door if necessary to keep water out. The lipping joint should be a mitre joint.

NOTES:

- SEQUENCE OF OPERATIONS FOR MAKING A FLUSH DOOR: (also see Drawing Book, page 74)

   a. Study the drawing.

   b. Prepare a cutting list.

   c. Select the wood.

   d. Cut the wood to the right size. The plywood should be cut slightly larger than the door, so it can be planed to size later.

   e. Plane the wood to size, and make the face marks.

   f. Mark out the joints.

   g. Prepare the joints.

   h. Drill the ventholes.

   i. Glue and clamp the frame together.

   j. Fix the lock block. Indicate the position of the block on the outside edge of the door.

   k. Plane the frame so the surfaces are flush.

   l. Glue the plywood on both sides of the door and secure it with wire nails to the rails and stiles. If more than one door is being made, the doors can be set on top of one another and weighted down until the glue hardens.

   m. Fix the lipping and finish off the door.

- TIMBER MEASUREMENTS FOR FLUSH DOORS:

Top and bottom rails	– 3 to 4 cm thick, 7 cm wide
Stiles	– 3 to 4 cm thick, 7 cm wide
Intermediate rails	– 3 to 4 cm thick, 5 cm wide
Plywood	– 6 mm thick
Distance between rails	– 30 cm

NOTES:

LARGE DOORS

Sometimes a large door is needed for a store or a similiar building.

Large doors can be constructed like panelled doors. In addition to this framework, braces are used. The braces slope up from the bottom of the hinge side to the top of the opposite side.

The frame can be covered with plywood on the inside, and with either sheet metal (Fig. 1) or horizontally fixed overlapping boards (Fig. 2) on the outside. If corrugated sheet metal is used, the top edge has to be covered to prevent water from entering between the sheet metal and the door frame (Fig. 1, arrow).

The frame used for this kind of door should be heavier than the one used for an ordinary panelled door. The timber for the frame should be no less than 5 cm thick.

Each large door should have 3 band-and-hook hinges. If T-hinges are used, the strap has to be long enough to overlap a part of the rails. If the hinges are fixed to the stile only, the door will tend to sag.

In the Reference Book, page 229, there is a description of a simple locking device which can be made for this type of door.

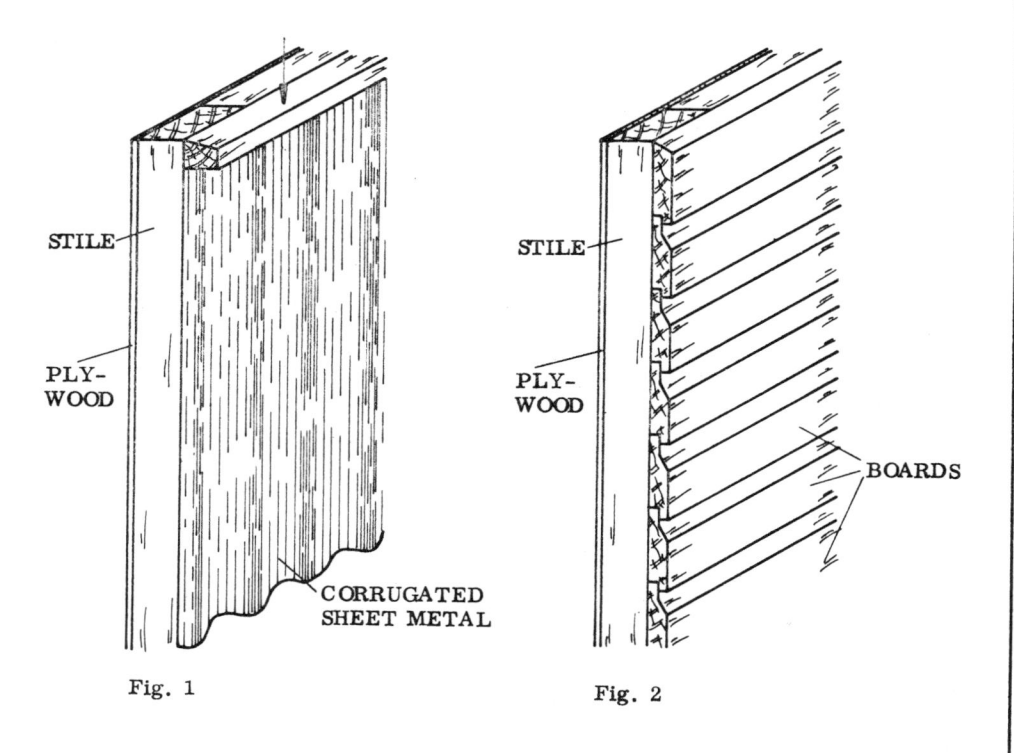

Fig. 1

Fig. 2

DOORS.

NPVC
CON.   107

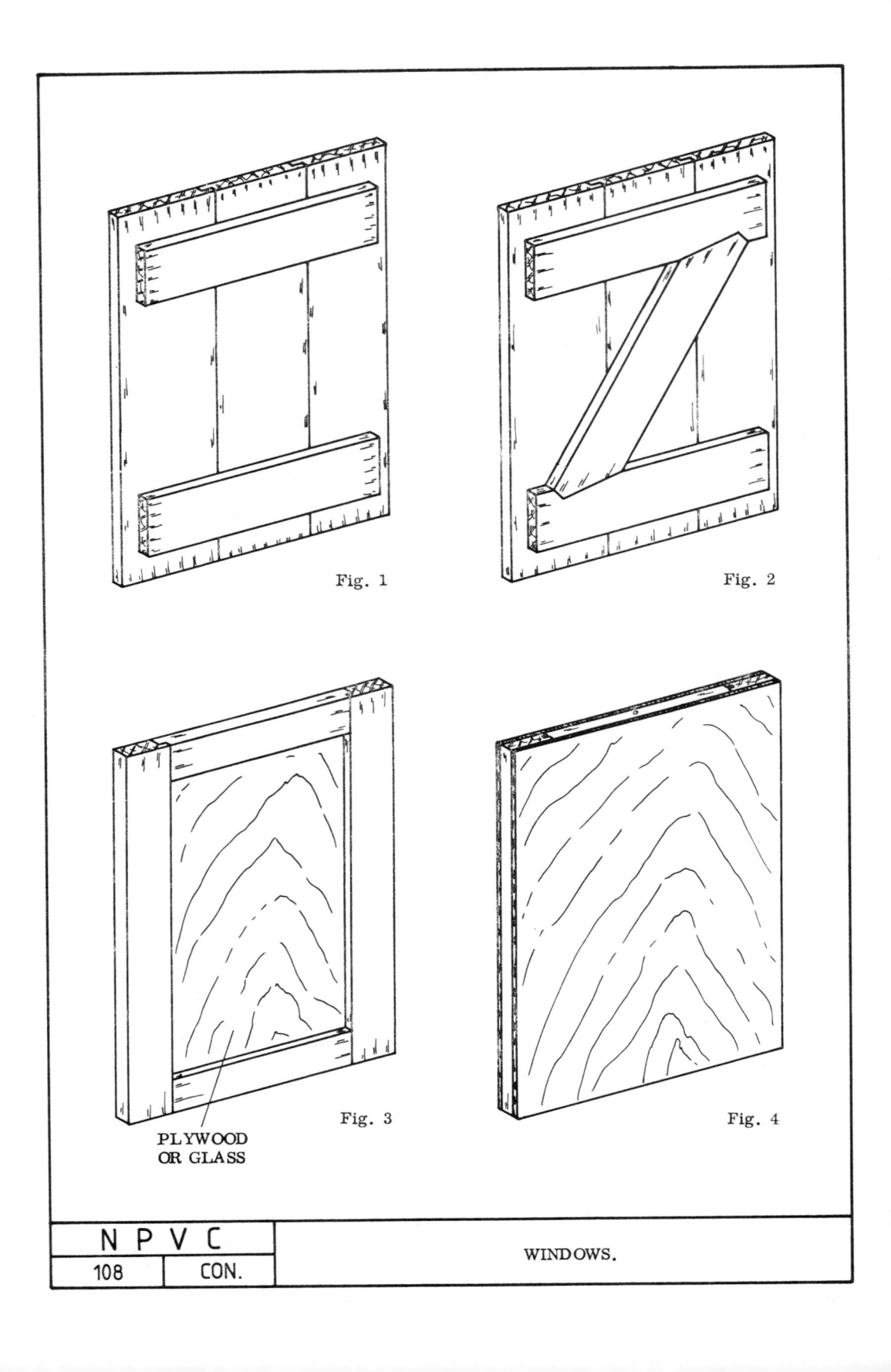

Fig. 1

Fig. 2

Fig. 3

PLYWOOD
OR GLASS

Fig. 4

N	P	V	C	
108	CON.			WINDOWS.

WINDOWS

The function of a window is to let light and air into the building. When closed, the window should be draught-proof, weather-proof and burglar-proof. The size of the window should be appropriate for the room.

In Rural Building, windows consist of a window frame into which louvres or casements are fitted.

Casements are hinged to open to the outside. This makes them easier to waterproof than if they opened to the inside. The hinges should be fixed so that the casement opens flat to the wall. The frame of the window is set flush to the outside surface of the wall (see Installing Window Frames section, page 85).

Sometimes the hinges are fixed at the head of the window frame. This allows the casement to open upwards, thus providing shade for the opening and the wall. Refer to the Drawing Book, page 50, for ways to show the positions of hinges in the drawing.

TYPES OF CASEMENTS

In Rural Building, we deal with 5 different kinds of casements:

- Ledged and battened casements (Fig. 1)

- Ledged, braced and battened casements (Fig. 2)

- Panelled casements (Fig. 3)

- Glazed casements (Fig. 3)

- Flush casements (Fig. 4)

It is necessary to leave at least 3 mm of play on all sides between the casement and frame (more if the casement is made during the dry season) so that the casement can open freely. This allows for expansion and shrinkage in the frame and casement.

BATTENED CASEMENTS

Both types of battened casements are constructed in the same way as the corresponding type of door.

Matchboarding is also done in the same manner as for a door. Depending on the size of the casement, usually only two ledges are needed (Drawing Book, page 73).

GLASS

Fig. 1a
BEADS

Fig. 1b
PUTTY

Fig. 1

a
b

Fig. 2

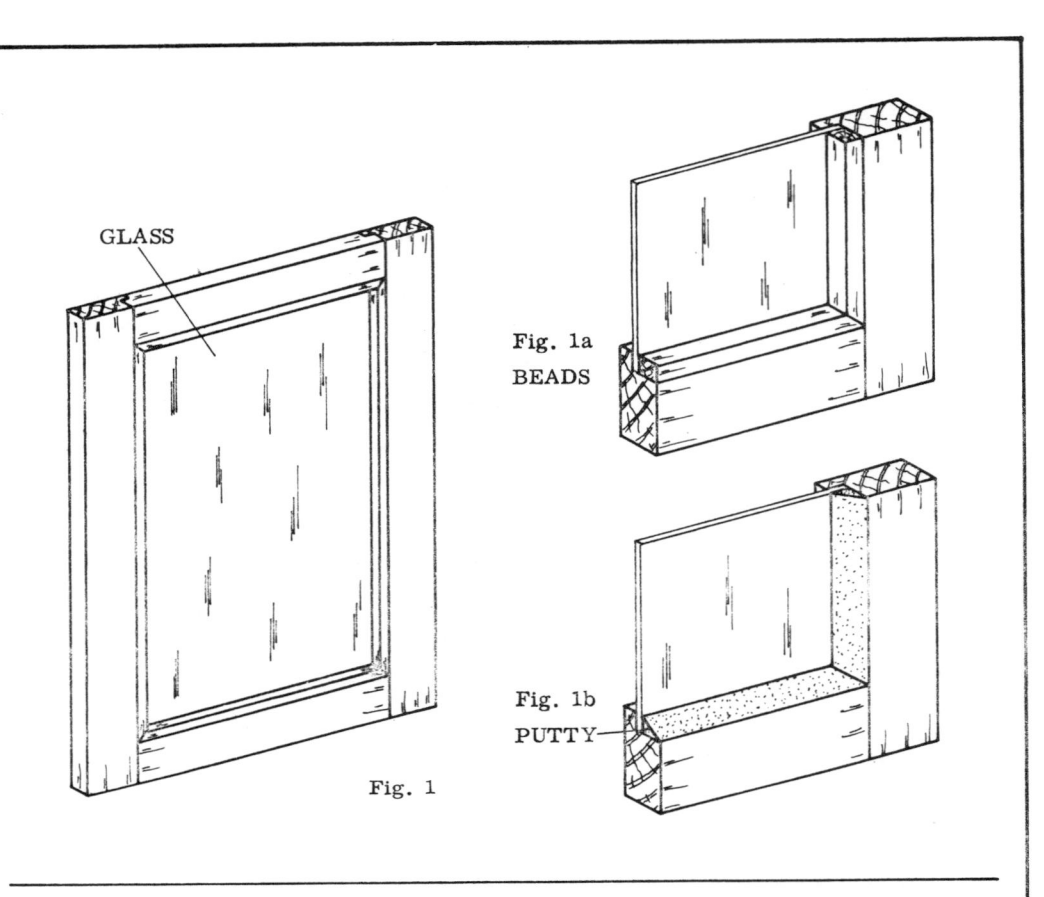

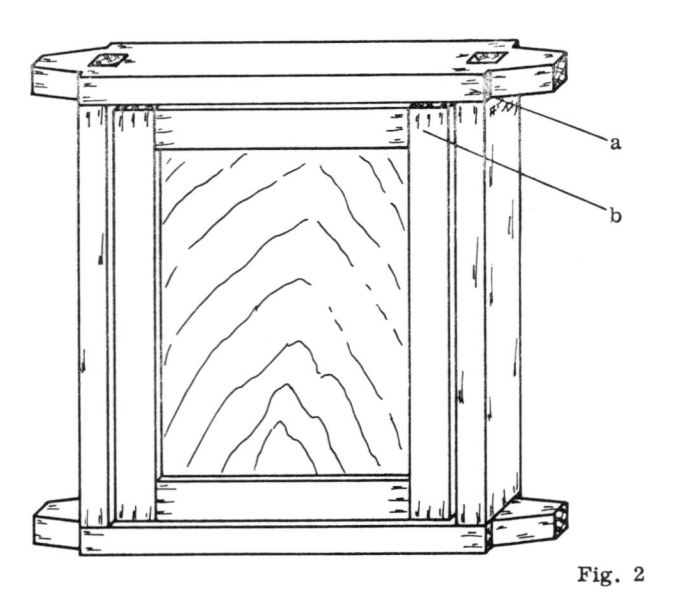

N P V C		WINDOWS.
110	CON.	

PANELLED CASEMENTS

These can be made with grooved-in or rebated frameworks, in the same way as panelled doors (see Doors, page 101).

Usually the panels are made from plywood. Be sure to cut the bottom rail of the casement according to the slope of the cill.

The measurements of the members can be 5 to 7 cm wide and 3 to 4 cm thick.

The overall size of a panelled casement equals the inside measurement of the window frame, minus the allowance of at least 3 mm all around (Drawing Book, page 75).

GLAZED CASEMENTS

These are made with a rebated framework. The construction is the same as for a panelled door or casement (see Panelled Doors, page 101). Instead of a plywood panel, a pane of glass is installed in the framework (Fig. 1).

Select the wood for this casement very carefully. It should be well seasoned and straight grained, so that the frame will not twist and cause the glass to break.

The joints are usually haunched stub tenons. Paint the joints and rebates before assembling the frame. The glass pane is always fitted after the frame has been assembled. The rebated framework is used because it permits the replacement of broken glass.

There are different ways of securing the glass in the rebated framework:

- By using beads, where the casement does not have to be waterproof (Fig. 1a)

- By using putty, to make a waterproof seal (Fig. 1b)

- By a combination of a thin layer of putty and beads.

The glass should always be cut 2 mm shorter in the length and the width, to make it easier to fit in the frame.

The sections of the frame members will be: 5 to 7 cm wide and 3 to 4 cm thick.

FLUSH CASEMENTS

These are constructed in the same way as flush doors. Flush casements can be covered on one or two sides with plywood or sheet metal (aluminium or galvanized iron).

Flush casements should only be used where they will be protected from the rain and sun, for instance on a verandah.

Plywood should be glued and nailed to both sides of the casement, not one side only, to keep the casement rigid, straight and strong. For outside work, use waterproof glue.

- TIMBER MEASUREMENTS:

    Top and bottom rails - 3 to 4 cm thick, 5 to 6 cm wide

    Stiles - 3 to 4 cm thick, 5 to 6 cm wide

    Intermediate rails - 3 to 4 cm thick, 4 to 5 cm wide

    Plywood - 6 mm thick

The lower part of the bottom rail must be planed to fit the slope of the cill. The distance between the rails is approximately 30 cm (Drawing Book, page 74).

JOINTS

Note that in the construction of window frames and casements, different methods of framing are used.

For window frames, the horizontal members (heads and cills) have mortices, and the vertical members (posts and mullions) have tenons (Fig. 2, a; previous page).

For the casements, the vertical members (stiles) have mortices and the horizontal members (rails) have tenons (Fig. 2, b; previous page).

The same system applies for door frames and doors.

NOTES:

LOUVRE WINDOWS

- SEQUENCE OF OPERATIONS FOR INSTALLING LOUVRE WINDOWS: (see Reference Book, page 217)

  a. If there will be mosquito wire on the window, you will have to be careful to set the louvre channels so that the louvres will be able to open without touching the wire.

  b. Mark the position of the channel on the wooden frame, at the side where the operative channel (the one with the handle) will be fixed.

  c. Grease the moving parts on the inside of the channel (use grease, not oil).

  d. Fix the operative channel with one screw near the top.

  e. Plumb this side. Secure the channel firmly with screws (the screw size will be 3 x 30, round head screws).

  f. Install the non-operative channel, parallel to the first channel.

  g. Secure this side loosely near the top, leaving the bottom free so it can be adjusted later.

  h. Put in the louvre glasses, starting from the top and working down. Close the clip ends to secure the glasses in place.

  i. When all the glasses are installed, close the louvres.

  j. Move the bottom of the non-operative channel until the louvres fit tightly together and are parallel.

  k. Now secure this side firmly with screws.

  l. Check the installation (screws, clip ends, etc.) and operate the handle several times to make sure that the louvres move freely.

  m. Fix the waterbar (either a wooden bead or a ready-made aluminium bar) at the head (outside the glass) and the cill (inside the glass), so that it is just touching the glass when the louvres are locked.

NOTES:

OUTSIDE

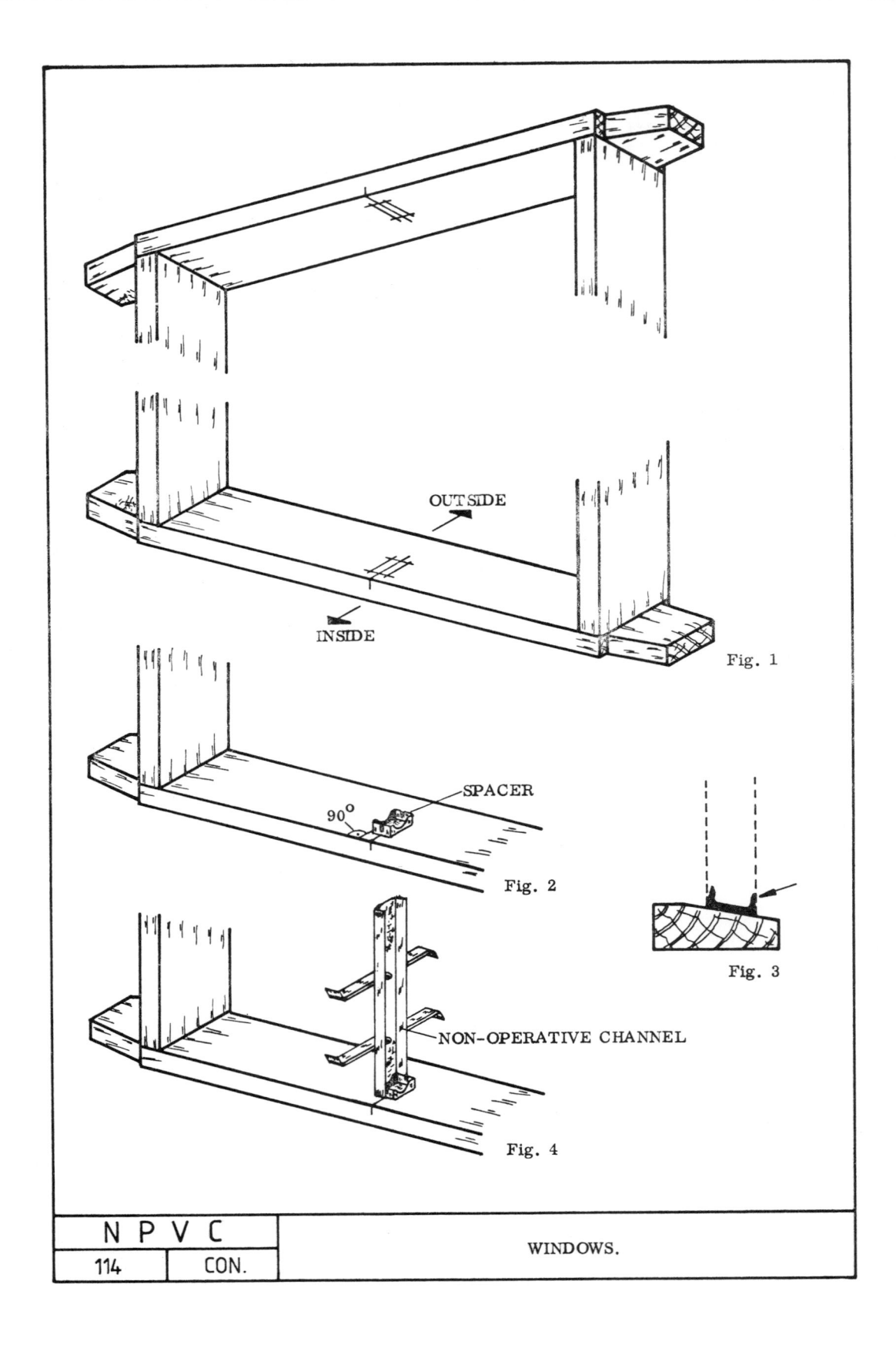

INSIDE

Fig. 1

SPACER

90°

Fig. 2

Fig. 3

NON-OPERATIVE CHANNEL

Fig. 4

N P V C	
114	CON.

WINDOWS.

- SEQUENCE OF OPERATIONS FOR INSTALLING A SELF-MULLIONING LOUVRE WINDOW: If two or more louvres are to be set across the window frame, the metal channels can be fixed together with screws, to form a metal mullion. The metal mullion is installed before the other channels.

   a. Mark the centres of the head and cill. The width of the window frame has to correspond to twice the length of a louvre glass, plus 3,8 cm for each set of channels, making 7,6 cm for both sets.

   b. Mark out the position of the metal mullion on the head and cill. Keep in mind the position of the mosquito wire (Fig. 1).

   c. Fix the bottom spacer on the cill (Fig. 2) at right angles to the face of the frame. The frame must be painted before the spacers are fitted.

   d. If the cill is sloping, bend the lugs of the spacer so that they are upright. The bottom end of the louvre channel should be cut to fit the slope of the cill, so that it fits tightly (Fig. 3).

   e. To position the top spacer, place a louvre channel against the bottom spacer and plumb it with a spirit level. When it is straight, fix the top spacer exactly above the bottom spacer.

   f. Take the non-operative channel and fit spacers into it at the places indicated by holes; these serve to stiffen the mullion.

   g. Attach the non-operative channel to the spacers at the head and cill of the frame (Fig. 4).

   h. Attach the operative channel to the non-operative channel with the special bolts and nuts.

   i. Check the installation and operate the handle several times.

   j. Install the other louvre channels to the window frame as in the previous page on installing louvre windows.

   k. Make sure that all the channels are the same distance from the inside face of the window.

NOTES:

Fig. 1

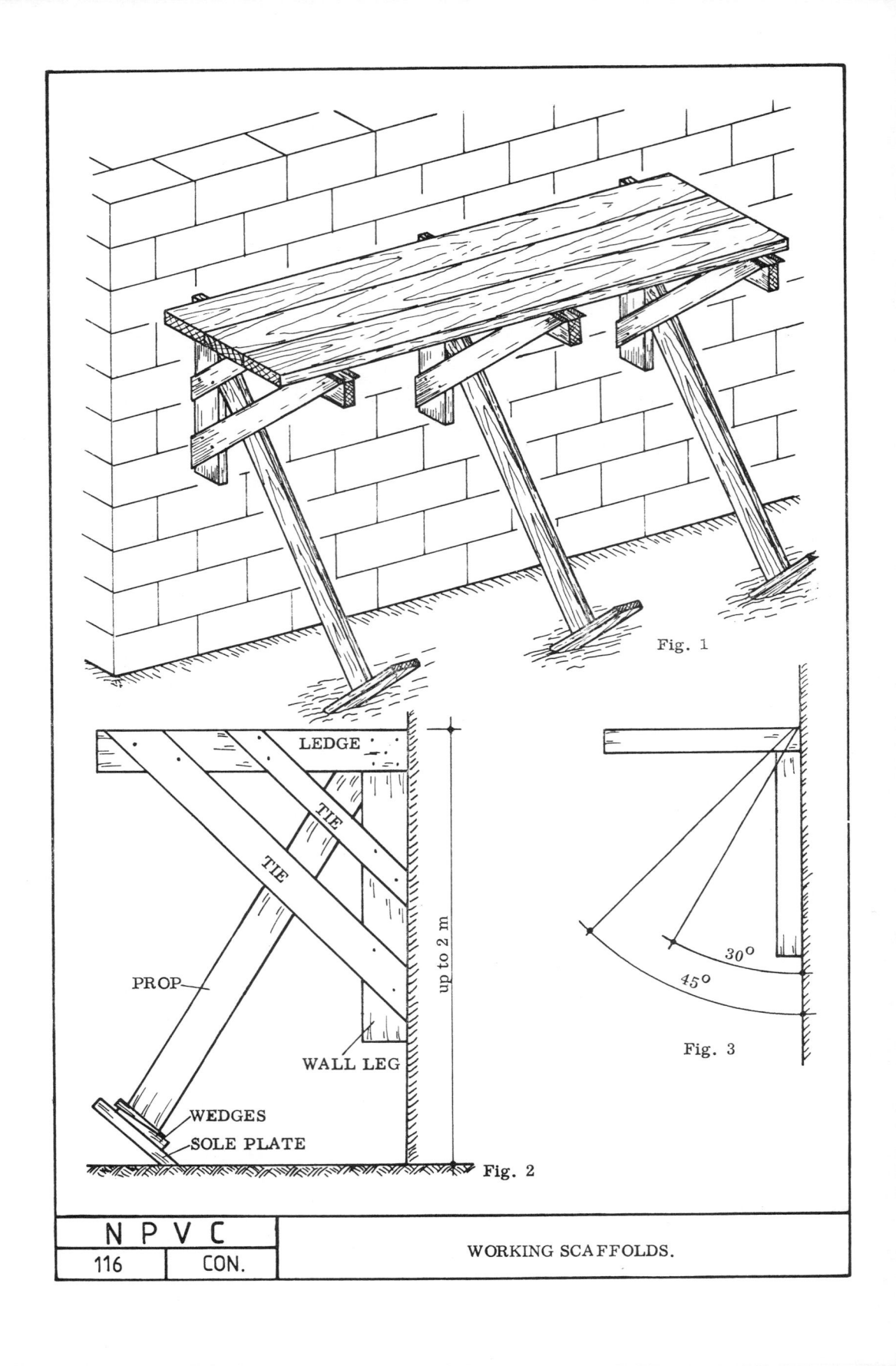

LEDGE
TIE
TIE
PROP
WALL LEG
WEDGES
SOLE PLATE
up to 2 m
Fig. 2

30°
45°
Fig. 3

NPVC	
116	CON.

WORKING SCAFFOLDS.

WORKING SCAFFOLDS

These are scaffolds which the workers stand upon to reach the higher parts of the construction. The tools and a supply of materials are also on the platform with the worker.

JACK SCAFFOLD

Although it can be no more than 2 m high, the jack scaffold is a satisfactory working scaffold for most jobs in Rural Building (Fig. 1). The block and trestle scaffolds which may be used for low level work are described in Basic Knowledge, page 158.

The jack scaffold is a dependent scaffold, which means it is propped against the wall for support instead of standing independently like other scaffolds. The feet of the props have to be properly set in place so that they will not move and settle when the scaffold is in use. Never try to put any support under the wall leg of the jack, as this would actually make the scaffold unstable.

Each jack scaffold consists of three jacks, three props and three (approximately 4,5 m long) planks. If the scaffold is to be only 2 m long or less, two jacks are enough.

- CONSTRUCTING THE SCAFFOLD PIECES: The jacks consist of a wall leg and a ledge which is nailed edgewise to the top of the wall leg. Both boards must be at least 5 cm thick by 10 cm wide, and about 100 cm long. The right angle is braced with two ties on each side, in a way similar to the construction of a large square (Fig. 2, also see Reference Book, page 13).

  The shorter tie should be set so that it does not cover the corner of the jack, where the prop meets the angle. An opening is left, through which the position of the prop can be checked. The prop has to fit securely into the corner of the jack, without any "play" in the connection. The ties should be at least 2,5 cm thick, and the prop will be 5 by 10 cm by whatever length is required (remember that this scaffold should not be more than 2 m in height).

- ERECTING THE SCAFFOLD: Hold the jack with its wall leg against the wall, and fit the prop into the corner of the jack. Set the foot of the prop securely in the ground. The other jacks are set up in the same way. Check that all three are at the same height and the distance between them is not more than 2 m. The angle between the prop and the wall should be no less than 30 degrees and no more than 45 degrees (Fig. 3). In soft soil, sole plates must be set under the feet of the props.

SOLE PLATE

PROP

WEDGES

Fig. 1

WRONG !

Fig. 2

CORRECT

Fig. 3

FOOT-STOP

PLATFORM BOARDS

BEADS

25 cm

Fig. 4

| N P V C | WORKING SCAFFOLDS. |
| 118 CON. | |

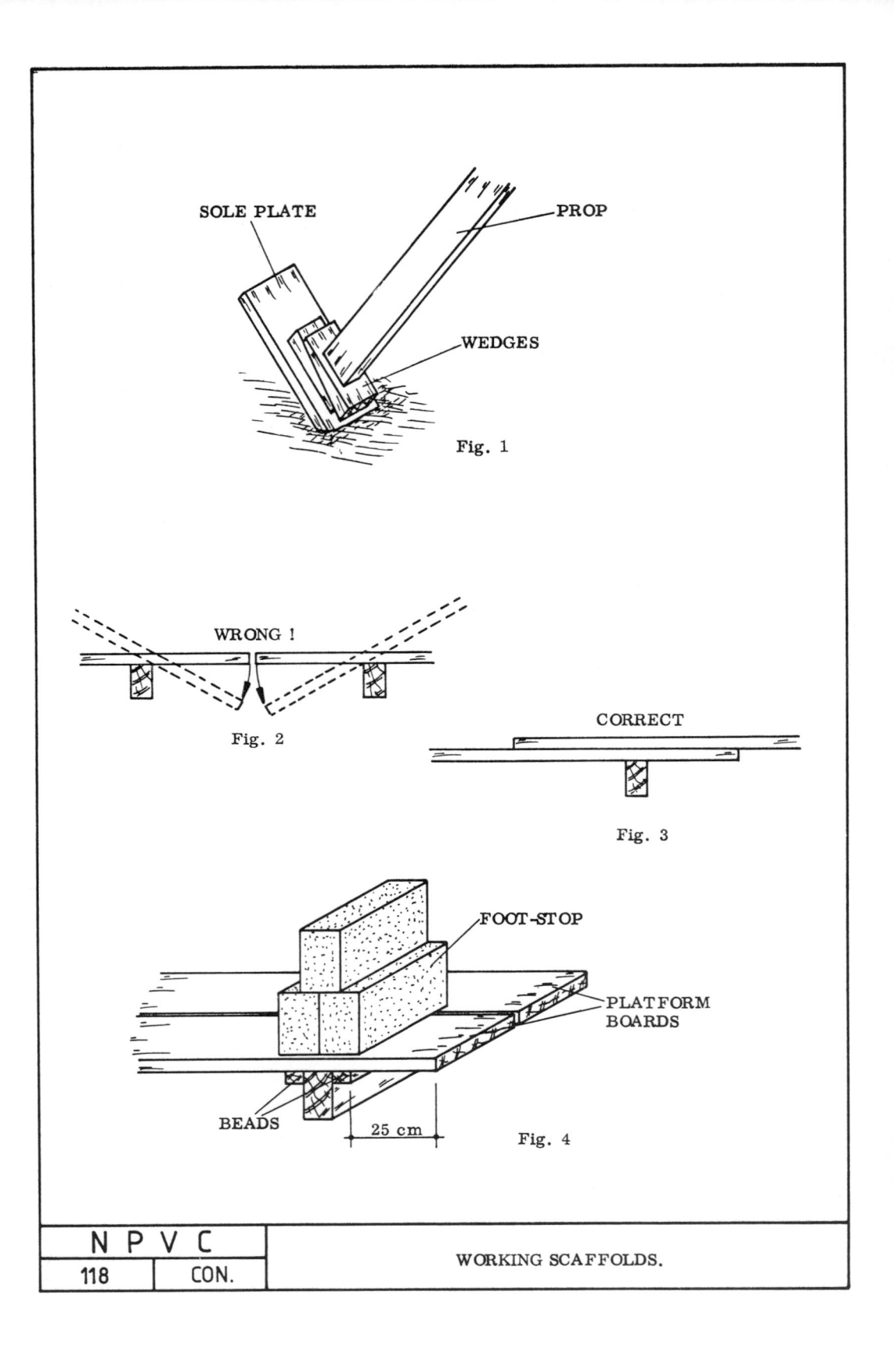

- SAFETY PRECAUTIONS:
    - Check that all the members of the scaffold are sound and without cracks.
    - Check whether the upper ends of the props fit well and without play into the corners of the jacks
    - Make sure that the feet of the props will not slip. In soft soil, they should be supported by sole plates set in the ground, at right angles to the props (Fig. 1).
    - Pay attention when you lay the platform boards, so you don't create a scaffold trap. This is a place which looks safe but is not (Figs. 2 & 3. The free ends of the platform must not project by more than 25 cm past the ledge, and they should be blocked off by a foot-stop (Fig. 4).
    - Wedges are used between the foot of the prop and the sole plate. By adjusting these, the prop can be tightened against the angle.
    - Beads can be attached to the bottom of the platform boards to prevent them from shifting along the ledges (Fig. 4), or the platform boards can be nailed directly to the ledges.
- NOTE: Always pay attention to the safety of the scaffold. A carelessly erected scaffold is a danger to the lives of workers and passers-by.

NOTES:

WORKING SCAFFOLDS.

NPVC
CON. 119

GUARD RAIL
STANDARD
STRUT
TOE BOARD
CLEAT
PUTLOGS
BRACES
LEDGER
150 cm
150 cm
150 cm
Fig. 1
SOLE PLATE
STANDARD **BEADS**
Fig. 2

N P V C		WORKING SCAFFOLDS.
120	CON.	

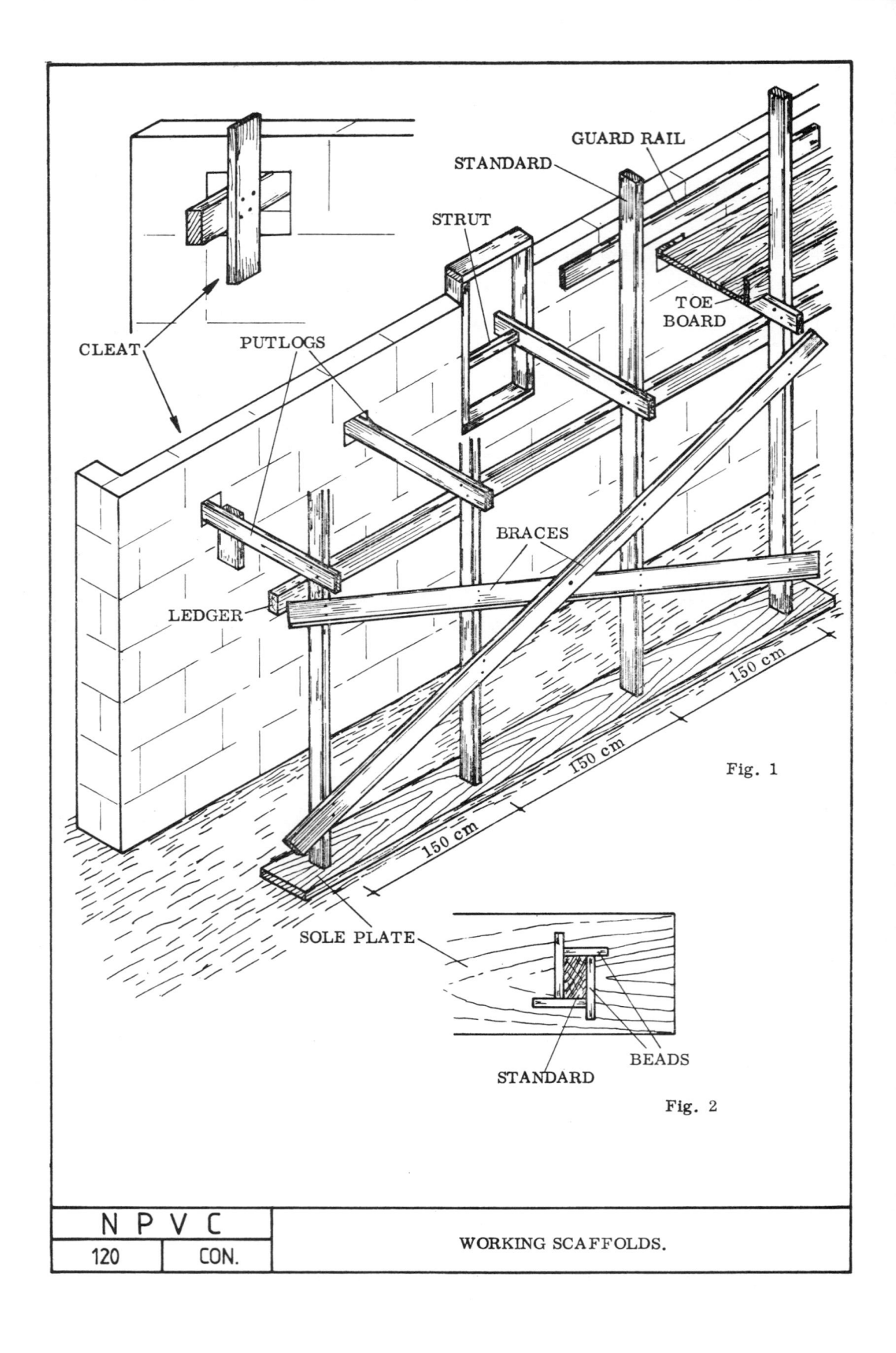

BLOCKLAYERS SCAFFOLD

Like the jack scaffold, this scaffold is dependent on the wall for support. The construction is fixed with "putlogs" into the wall, and secured on both sides of the wall with cleats which are nailed to the putlog (Fig. 1).

Erect the standards on top of the soleplate (Fig. 2) and nail the ledger at the required height. Each putlog will have one end in a wall opening which was left during the blocklaying; and the other end is supported by the ledger and nailed to the standard. Secure the putlogs into the wall with two cleats, one on either side of the wall. Make sure that the cleat on the outside of the wall does not project above the putlog, where it will interfere with the platform boards.

Once all the putlogs have been secured, nail braces across all the standards to make the whole construction rigid.

Check the construction, then lay the platform boards. Fix the toe board, and nail the guard rail at a height of 90 cm above the platform.

If a putlog is secured inside of a window frame, add an additional strut inside the frame (Fig. 1).

The advantage of this type of scaffold is that it requires less wood to make the frame. The holes left in the wall are easily filled with small blocks before the plastering is done.

- MEMBERS OF THE SCAFFOLD: (see Fig. 1)

  - Sole plate
  - Standards
  - Putlogs
  - Cleats
  - Ledger
  - Braces

  - Platform boards
  - Toe board
  - Guard rail
  - Struts (if needed)
  - Pieces to secure the standards to the sole plate (Beads)

NOTES:

WORKING SCAFFOLDS.

Fig. 1

N P V C		WORKING SCAFFOLDS.
122	CON.	

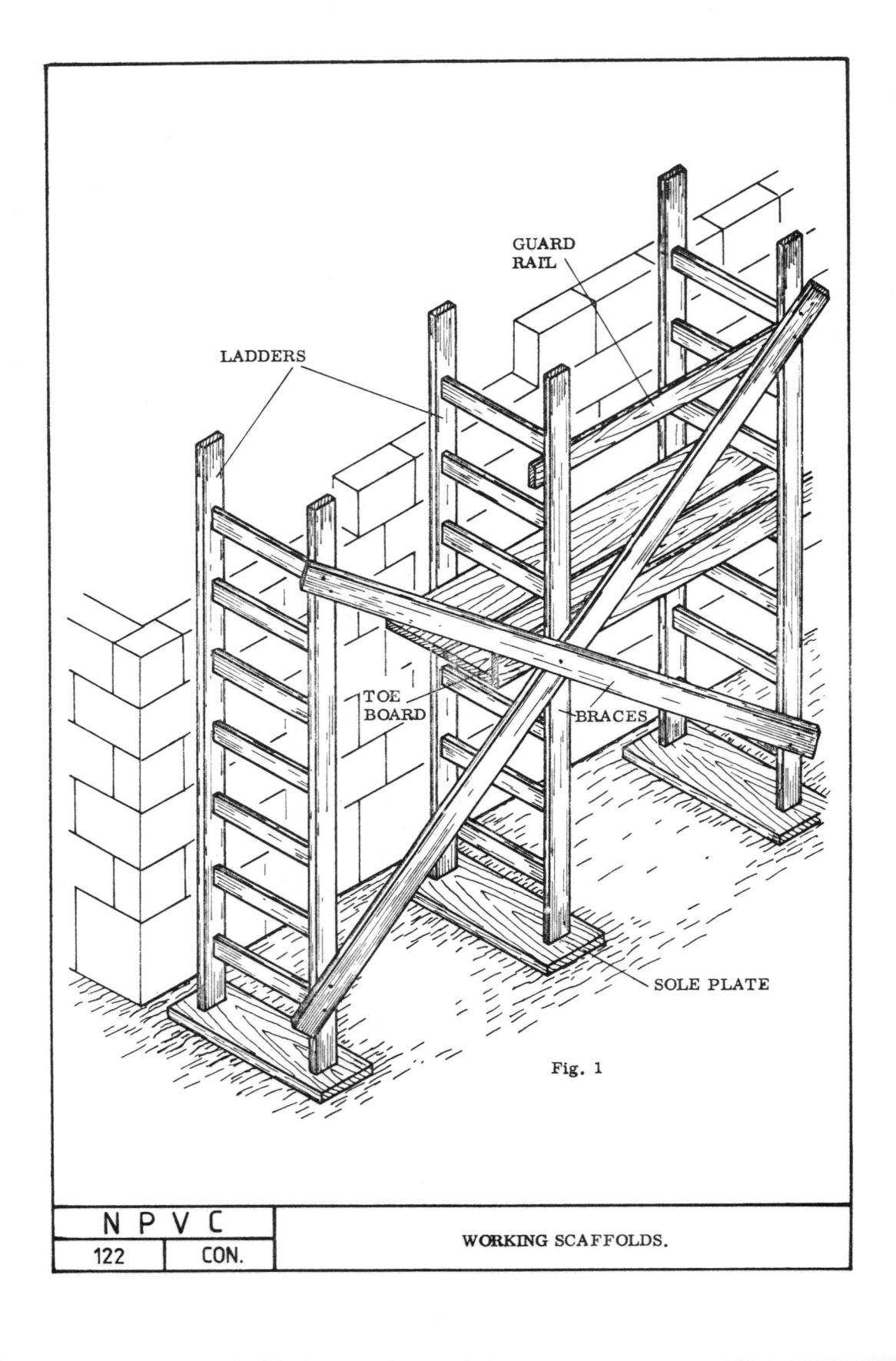

LADDER SCAFFOLD

This is an independent working scaffold, that is, it is not supported on the building but stands free. A ladder scaffold can be used where the height of the building is no more than 3,5 m.

Three or four ladders are usually used. Erect the ladders on top of the sole plates so that they do not sink into the ground under use. Keep the ladders plumb and straight, and nail two braces diagonally across the side.

If necessary use some wedges under the ladders to keep the structure steady.

Lay the platform boards and fix the toe board and the guard rail. The guard rail is fixed 90 cm above the platform.

Before any workers use the newly erected scaffold, make sure that all the parts are well secured and fixed according to the safest manner. Be aware that the scaffold has to carry not only the workmen but also a load of building materials.

NOTES:

GUARD RAIL

STANDARDS

PUTLOG

LEDGER

Fig. 1

Fig. 2

N P V C	WORKING SCAFFOLDS.
124 CON.	

INDEPENDENT SCAFFOLD

This is similar to a ladder scaffold with the difference that no ladders are used. Timbers are used instead; these should be 5 x 10 cm by approximately 4,5 m long.

Sort out the timber, keeping the straight pieces for standards and the bent pieces for shorter members like putlogs.

Mark the height of the working platform on the standards, and nail the putlogs to the standards. Erect the standards on top of the sole plates, and nail the necessary braces to the outside of the structure.

Nail two additional boards (ledgers) under the putlogs. These give support to the putlogs.

Lay the platform boards and secure them with nails or battens. Fix the toe board. Nail the guard rail 90 cm above the platform.

If necessary, place some wedges under the standards to make sure that the whole structure is steady. Extra braces can be added to the outside of the structure; secure them to the ground with pegs (Fig. 1, broken lines).

Fig. 2 shows an alternative way to support the putlogs, with a wedge-shaped piece of wood nailed to the standard underneath the putlog.

SUPPORTING SCAFFOLD

As the term indicates, a supporting scaffold is used to support members of the structure until the mortar or concrete has hardened or cured; or until the parts are assembled, when the member will be able to support itself and the scaffold can be taken away.

The supporting scaffold is seldom used in Rural Building because in most cases a lighter and simpler strutting will serve for the purpose of supporting the usual reinforced concrete members as they are cast-in-situ. See page 167 of the Basic Knowledge book for an illustration of a typical strutting for a reinforced concrete lintel.

The heavier supporting scaffold is used only in cases where the member to be supported is very large and heavy. The supporting scaffold arrangement is shown in the illustration on the next page.

Fig. 1 SIDE VIEW

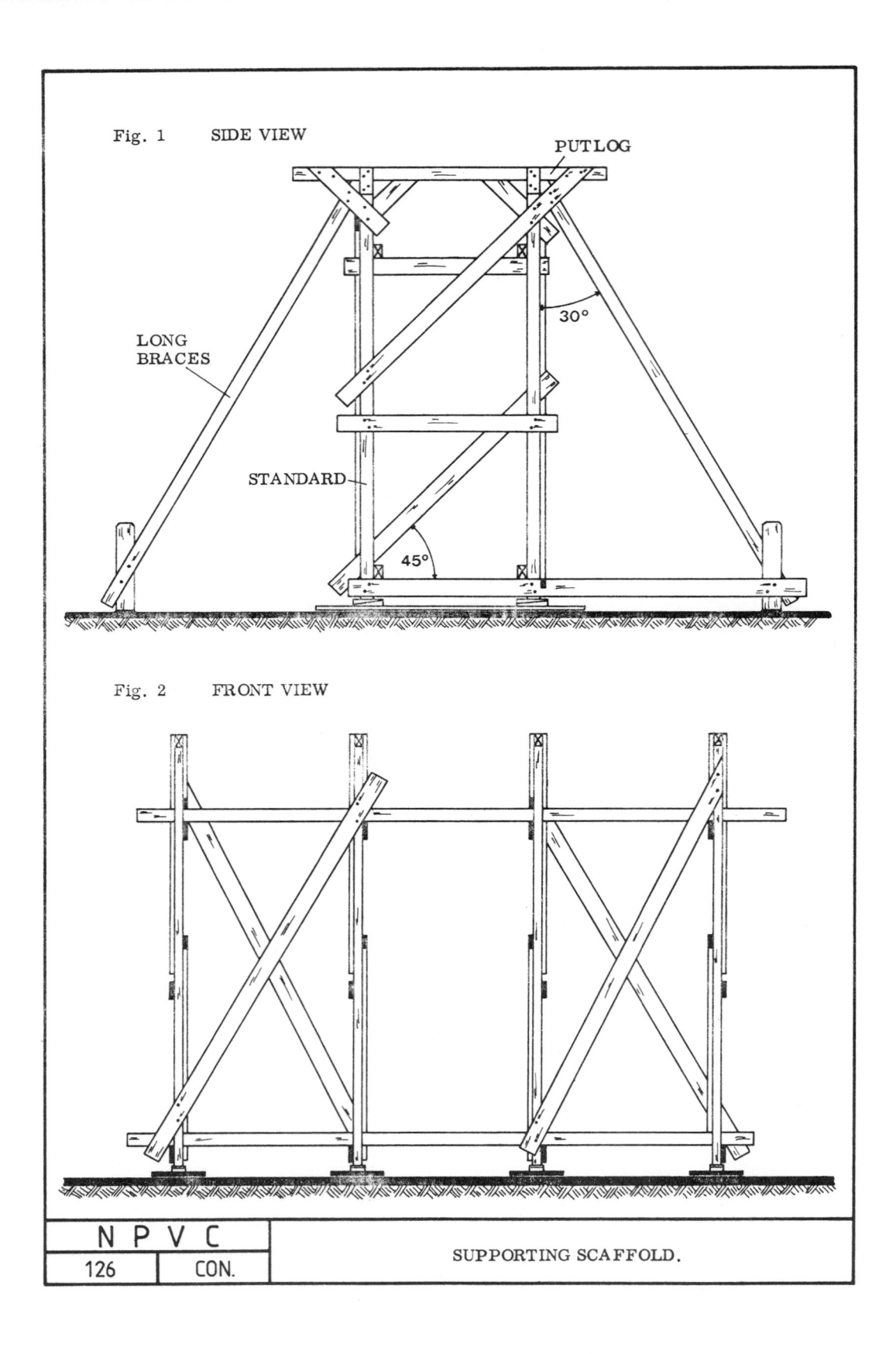

PUTLOG

LONG BRACES

STANDARD

30°

45°

Fig. 2 FRONT VIEW

N P V C		SUPPORTING SCAFFOLD.
126	CON.	

When you construct a supporting scaffold, pay particular attention to the stability of the structure, since it usually has to support very large and heavy members. The distance between the standards (also called struts) must be no more than 1 m (Figs. 1 & 2). All the shorter braces, which reinforce the connections between the putlogs and the standards and between the standards themselves, are fixed at 45 degree angles. The long braces, which go from the top of the structure down to pegs in the ground, are fixed at an angle of no less than 30 degrees. These help to keep the scaffold from tilting or overturning.

PROTECTIVE SCAFFOLDING

On most Rural Building construction sites, protective scaffolding is not necessary. This is because the structures are relatively low in height. However, when a higher structure such as a water tower is constructed, protective scaffolding becomes necessary to protect the people underneath from falling objects. On jobs where walls are being broken down, or where large and heavy objects are handled, it is advisable to erect a protective scaffold of some sort.

A second platform is made below the working platform, and a side projection is fixed at an angle to this (Fig. 3).

The projecting piece is attached to the standards by wooden braces.

Any tools or materials which fall from the working platform will hit the projecting piece and roll onto the second platform, instead of falling to the ground.

NOTES:

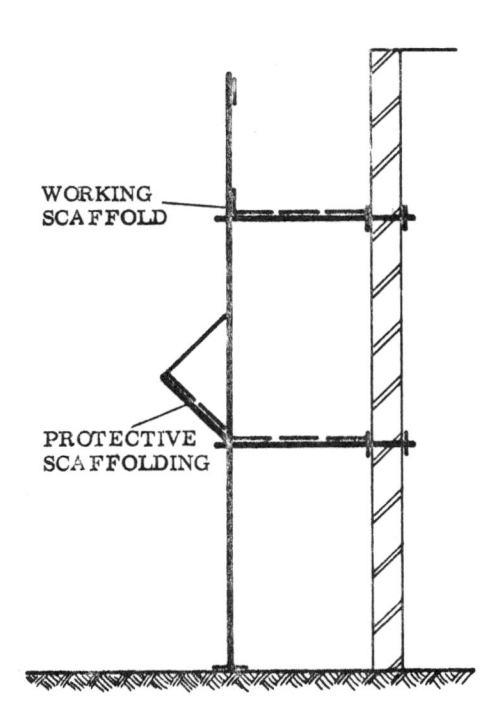

Fig. 3

SUPPORTING SCAFFOLD & PROTECTIVE SCAFFOLDING.

ARCHES

Openings in walls such as doors and windows have to be closed at the top in such a way that the structure above them is supported. Before the use of reinforced concrete became common, smaller openings were bridged with wooden lintels, solid stone lintels and arches (see Basic Knowledge book, pages 162 to 164).

Wider openings were generally spanned by arches, or by a series of arches carried on piers or columns: the so-called arcade. In this book we discuss only one type of arch, the "segmental arch" where the arch is formed like a segment of a circle. Many other forms are possible with arches, but they are not important for Rural Building.

Before we continue with this chapter, it is essential to understand the technical terms that are used in connection with arches. The terms below are indicated in Fig. 1.

TECHNICAL TERMS

- ABUTMENT: The word "abut" means to meet at one end or to border on something. An abutment (Fig. 1, a) is that portion of solid masonry that carries the weight of the structure of an arch. In Rural Building, the minimum width of an abutment at the end of a wall should be 3/4 of the span of the arch. This means that the distance from the arch to the end of the wall should be at least 3/4 of the span of the arch.

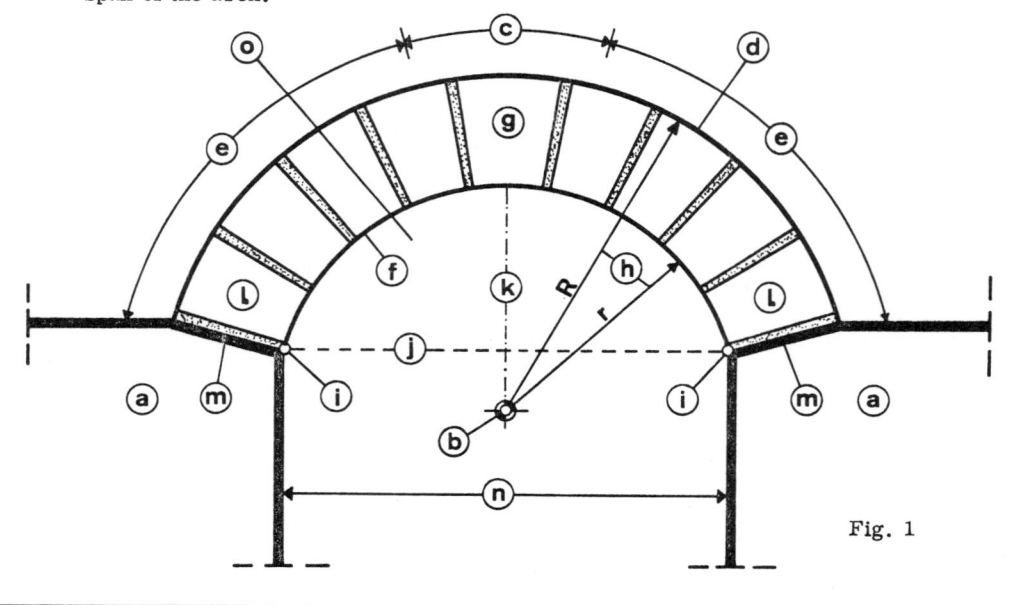

Fig. 1

ARCHES.

- CENTRE: This is the midpoint of the circle which describes the curve of the arch (b).
- CROWN: This is the portion of the arch which forms the top of the curve (c).
- EXTRADOS: The outer curved line of an arch, or the upper surface of the archstones (d).
- HAUNCH: The flanks of an arch, the sides of the curve (e).
- INTRADOS: The under surface or soffit of an arch (f).
- KEYSTONE: The central wedge-shaped archstone at the crown of an arch (g), which is the last stone to be put in place.
- RADIUS: The straight line (h) from the centre of an arch to any point on its intrados (the shorter radius, r); or to any point on its extrados (the longer radius, R).
- SPRINGING POINT: The point of intersection (i) between the intrados and the faces of the wall or pier blocks below. From there the arch "springs".
- SPRINGING LINE: The line across the arch which would connect the springing points (j).
- RISE: The height of an arch measured perpendicularly from the springing line to the highest point of the intrados (k).
- SPRINGER: The first stone laid in an arch on either side (l).
- SKEWBACK: That portion of the abutment which directly supports the springers (m). It is so called because the surface slopes towards the opening.
- SPAN: The horizontal distance between the springing points (n); the length of the springing line.
- CAMBER: This is the space between the springing line and the intrados (o).
- ARCHSTONES: These are any of the stones or blocks which form the arch itself.

NOTES:

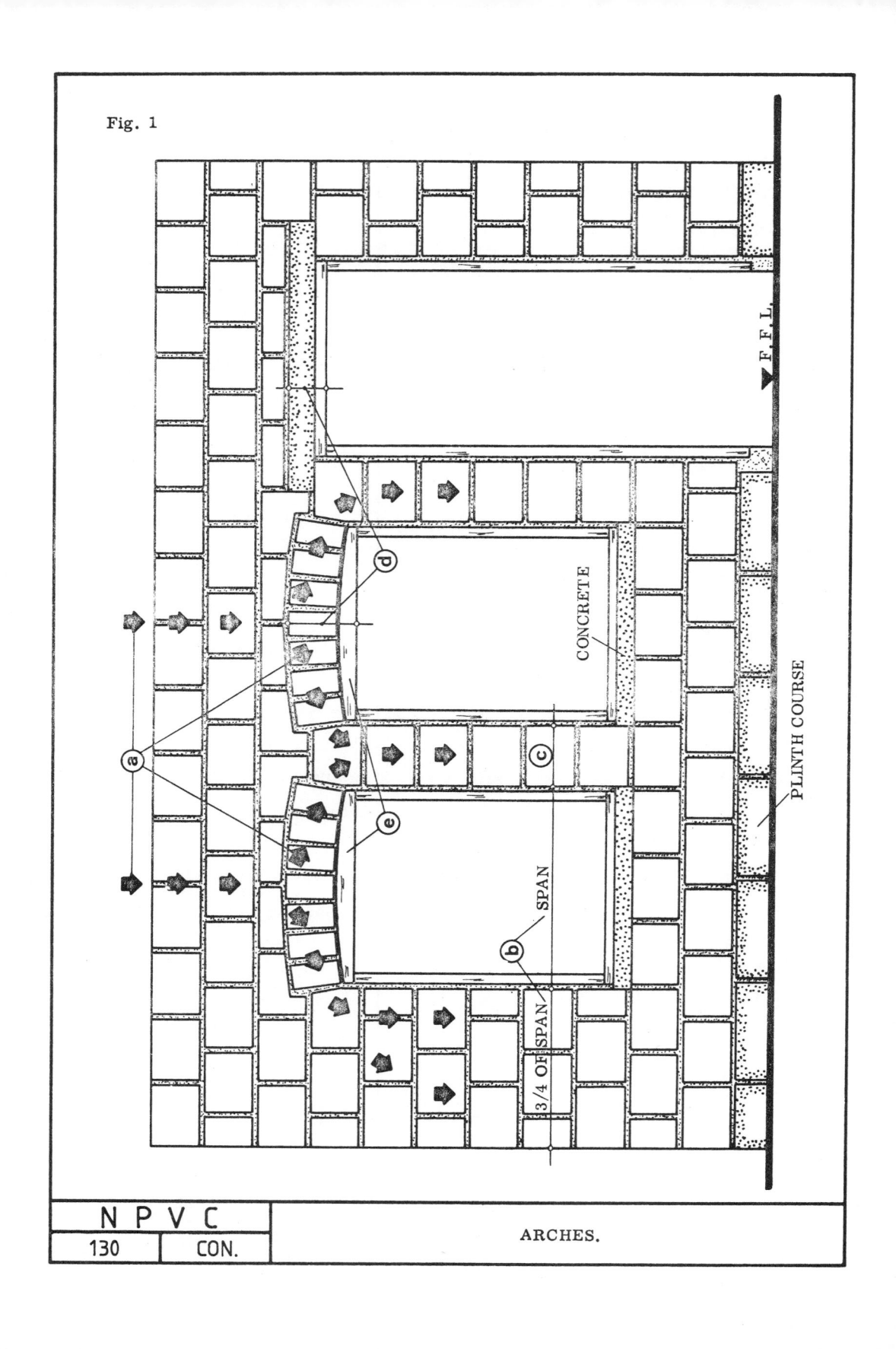

Fig. 1

F.F.L.

CONCRETE

SPAN

3/4 OF SPAN

PLINTH COURSE

N P V C		ARCHES.
130	CON.	

STABILITY OF ARCHES

All arches, regardless of their shape or dimensions, function as wedges. An arch distributes the load imposed on it to the abutments on both sides; these have to be strong enough to withstand the pressures (Fig. 1, a). This is the reason why the abutment which is at the corner or end of a wall must be at least as wide as 3/4 of the span of the arch (Fig. 1, b). Between two arches, thinner supports such as columns or piers can be used (Fig. 1, c).

It is always better to build two or three courses of blocks above an arch because this weight will improve its stability. The preferred material for constructing an arch is sandcrete blocks, because they can be cast to any shape required and can withstand high pressures. Of course landcrete or mud blocks can also be used, but these must be well protected against moisture penetration.

- DISADVANTAGES OF ARCHES:

  - All arches, especially those which have unusual shapes, require more space than a lintel (Fig. 1, d).

  - Frames must be shaped on their top according to the curve of the intrados (Fig. 1, e); or if a square frame is to be fitted, the remaining space above it must be closed with plywood etc.

  - A temporary wooden support, the centring or turning piece, is needed during the construction; this piece has to be exactly the correct shape.

  - Masonry arches cannot be prefabricated.

  - Specially made archstones are often required.

- ADVANTAGES:

  - Arches are more economical because scarce and expensive building materials like reinforcement iron are not needed.

  - Locally available building materials such as mud blocks can be used to build arches.

  - In contrast to cast-in-situ lintels, the walling above an arch can continue immediately after it is completed, because the arch does not have to set.

  - Traditional building styles can be maintained and developed in a simple and cheap way.

  - Arches, if well constructed, give the building a more attractive appearance and demonstrate the skill of the builder.

ARCHES.	N P V C
	CON. 131

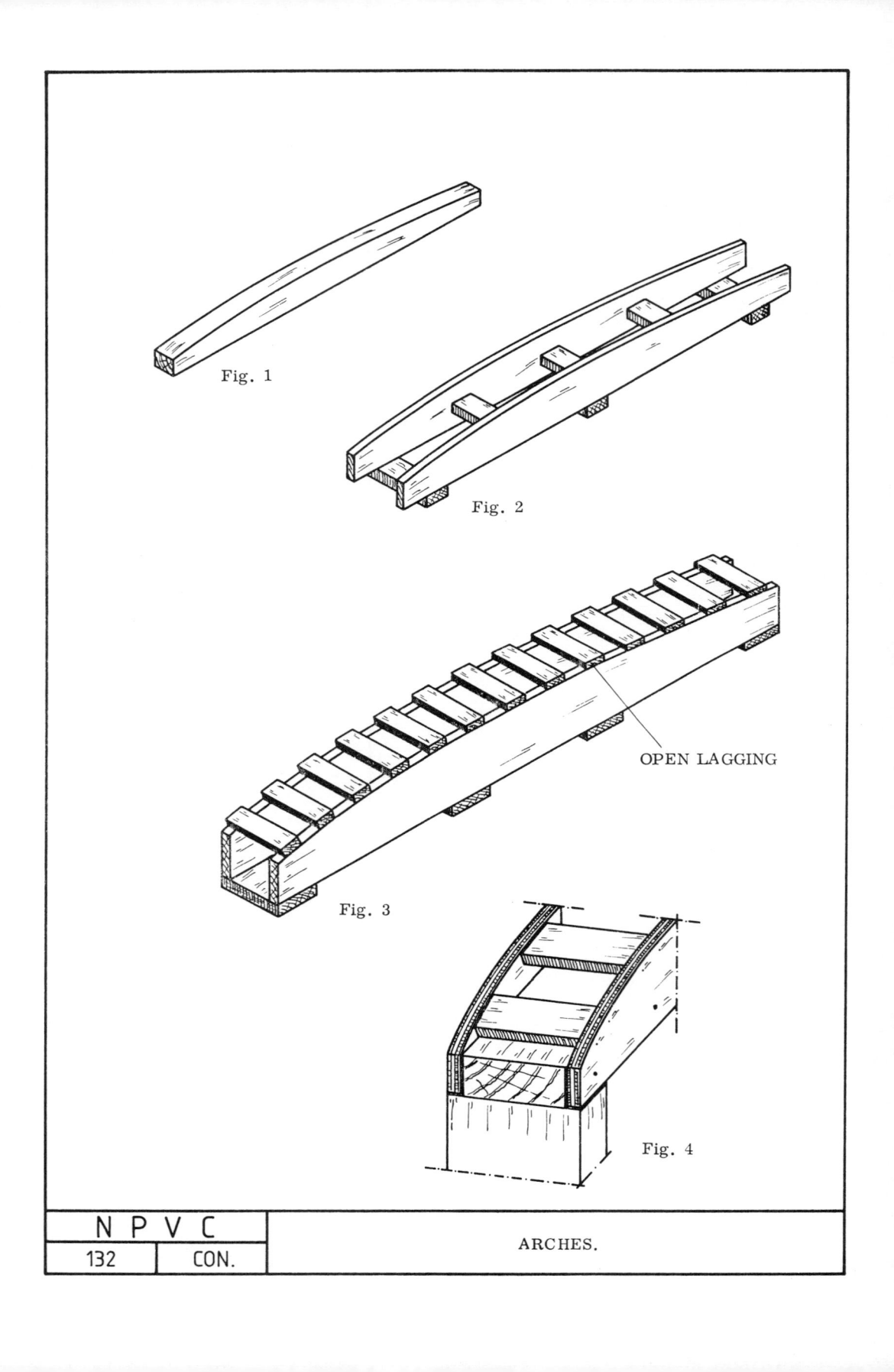

Fig. 1

Fig. 2

OPEN LAGGING

Fig. 3

Fig. 4

N P V C		ARCHES.
132	CON.	

THE SEGMENTAL ARCH

The segmental arch is the most important one in Rural Building. This is mainly because:

- The construction depth measured from the springing line to the crown is comparatively small; with the result that the building costs are low.

- The construction is not too difficult and specially made tapered blocks are not necessarily required.

- The time spent on the construction of this kind of arch can be shorter than the time spent for constructing a reinforced concrete lintel.

TYPES OF CENTRING

"Centring" is the term generally applied for the curved wooden piece which temporarily supports arches or domes during their construction. It is comparable to the strutting (not the shuttering) for concrete members.

The actual shaping part of the centring is called the "turning piece". There are three main types of centring:

- A solid wooden board, about 5 cm by 7,5 cm, placed edgewise. Its upper surface is carefully and evenly shaped according to the curve of the intrados (Fig. 1). Its use is restricted to short-span openings; not exceeding about 100 cm wide. The disadvantage of this type is that the archstones rest on only a small area, which might lead to the blocks tilting during the laying.

- Two wooden planks or sheets of plywood are connected by spacers, so that the archstones are supported on their outer edges (Fig. 2). Both turning pieces must be precisely cut and planed to the same shape, and must match with the curve of the intrados.

- Arches spanning wider openings and those which have a rise of more than about 1/10th of the span need to be supported by a stronger centring, with either open or closed lagging (Fig. 3).

The three types of centring may be combined with the frame or may form part of the frame; the head. In this case no strutting is needed, and the centring piece simultaneously acts as a permanent seal of the space between the springing line and the intrados (Fig. 4). The lagging must then be fixed in between the turning pieces so it is flush with the upper curved edge and is not seen (Fig. 4).

A B
— NOT LESS THAN 2 cm

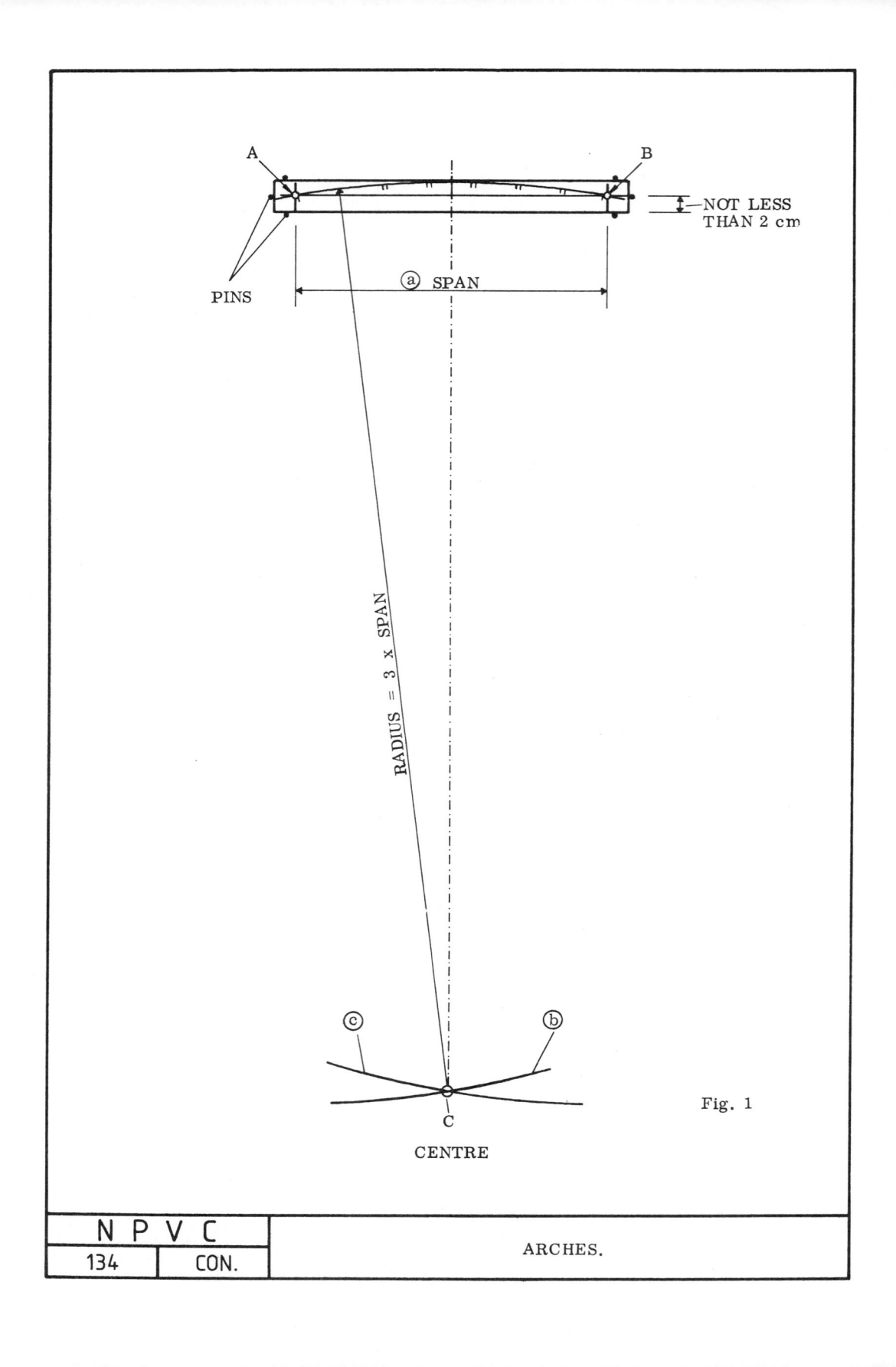

PINS

ⓐ SPAN

RADIUS = 3 × SPAN

ⓒ   ⓑ
C
CENTRE

Fig. 1

N P V C	
134	CON.

ARCHES.

SETTING OUT THE TURNING PIECE

To set out the turning piece means to mark the curve of the intrados on a suitable piece of board or plywood, so that after it is cut and planed it can serve as a temporary support and guide for the arch construction. The positions of the archstones are also marked on the turning piece so that no mistakes can be made during laying.

As far as Rural Building is concerned, the radius of the circle which describes the curve of the intrados will always be three times the span of the opening. For example: if the span of the opening is 80 cm (the clear width), the radius must be 3 x 80 cm = 240 cm; if the span is 180 cm, the radius is 540 cm, etc.

The curve obtained from this rule will result in an arch which is flat enough so that not much space is wasted, but also curved enough to ensure stability.

- SEQUENCE OF OPERATIONS:

  - You need a large, flat and level space for the setting out.

  - Mark the span of the opening on a suitable board (Fig. 1, a).

  - Determine the springing points (at least 2 cm from the edge of the board) and drive in short nails to mark these points. Leave the nails projecting, and make sure that they are both the same distance from the edge of the board (Fig. 1, A & B).

  - Place the board on the ground and secure it with nails or pegs as shown, so that it cannot shift to any side.

  - Fix your mason line on nail A, and measure off from there three times the span of the opening along the line. Use this length to describe a short section of a circle on the ground (Fig. 1, b).

  - Repeat this procedure with the line fixed on nail B (Fig. 1, c). Both circle sections will intersect at point C, which is the centre of the arch. Secure the centre point with a nail or peg.

  - Fasten the mason line on peg C and tie a pencil at the distance of three times the span of the opening.

  - Keep the line taut and describe a circle section, while marking the section on the board. This curved line should intersect with the springing points, A and B, and the curve resembles the curve of the intrados.

  - To check the accuracy of the setting out, measure the rise of the arch perpendicularly from the springing line to the highest point of the curve. This distance must be 4,2% of the span (0,042 x span (in cm) equals the rise (in cm).

ARCHES.

NPVC
CON. 135

SPRINGER   KEYSTONE   SPRINGER

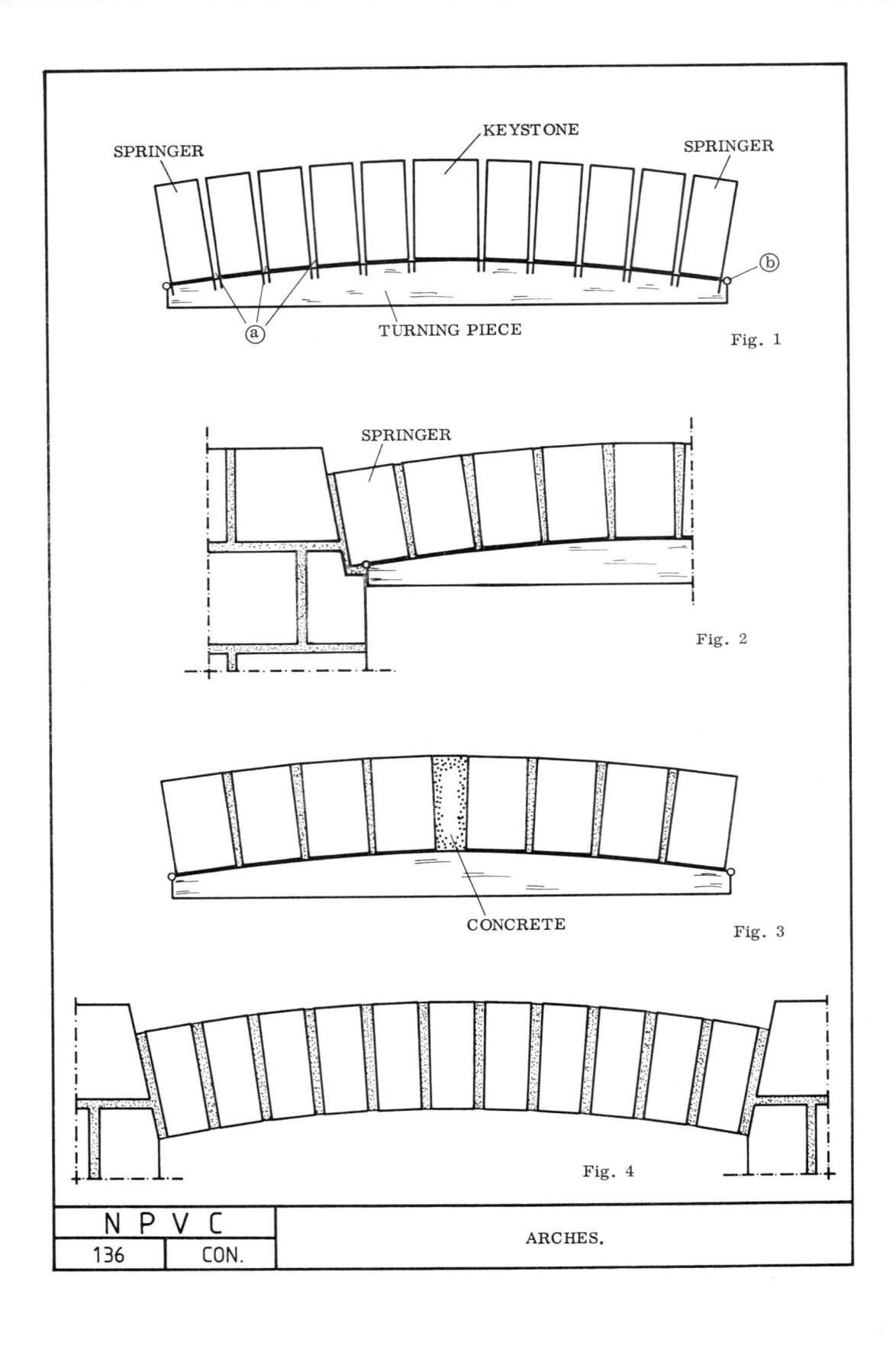

ⓐ   TURNING PIECE   ⓑ

Fig. 1

SPRINGER

Fig. 2

CONCRETE

Fig. 3

Fig. 4

N P V C	ARCHES.
136 CON.	

POSITIONS OF THE ARCHSTONES

Regardless of which type of arch is constructed and what kind of archstones are used, the total number of blocks or bricks must add up to an odd number: the keystone in the middle is flanked by an even number of archstones (Fig. 1).

Therefore the position of the keystone is usually marked first, followed by the archstones to the left and right. Cross joints are 1 cm wide at the intrados (Fig. 1, a), and wider at the extrados, because common blocks are used which results in wedge-shaped joints.

The marking continues until the springing points are either just reached or a little overlapped (Fig. 1, b). If it happens that there is a distance of more than 1 cm between the lower outside corner of the springers and the springing points, then one more block must be added to each side (Fig. 2). These two blocks will then become the springers, although they overlap by far the springing points.

There are two ways to avoid this:

- One can reverse the marking by starting from the springing points and marking by turns from the left and right up towards the crown. The remaining opening is then filled with either a specially made keystone, or with concrete (Fig. 3). This is preferred if the springers would overlap the springing points by too much.
- The most perfect solution would be to use specially made archstones which are wedge-shaped and fit exactly in the arch (Fig. 4). This requires some detailed and accurate planning, and more time and materials.

NOTES:

CAMBER
(FILLED WITH SAND AND ASH)

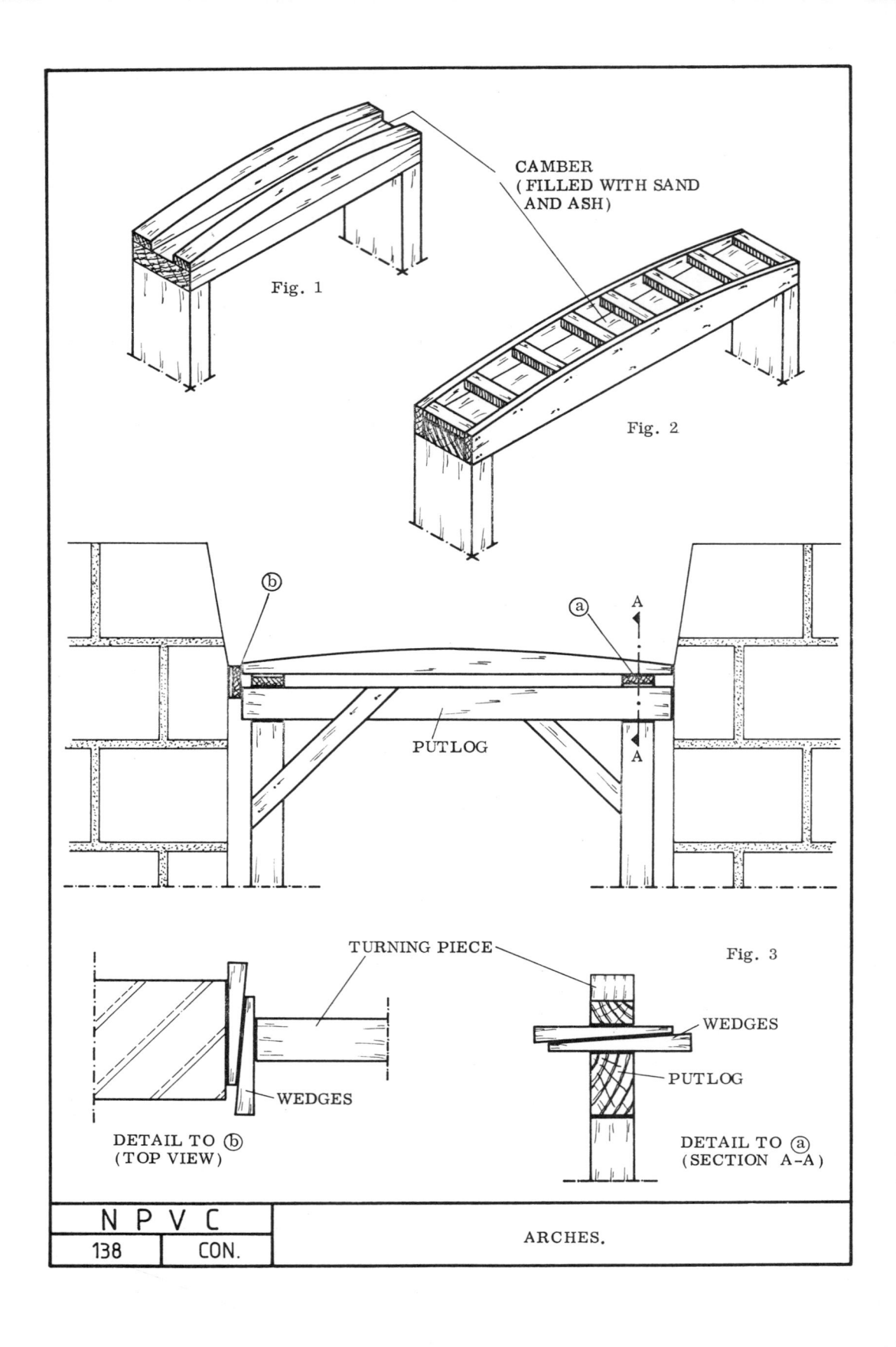

Fig. 1

Fig. 2

PUTLOG

Fig. 3

TURNING PIECE

WEDGES

WEDGES

PUTLOG

DETAIL TO ⓑ
(TOP VIEW)

DETAIL TO ⓐ
(SECTION A-A)

N P V C		ARCHES.
138	CON.	

SETTING UP THE CENTRING

The board on which we marked the curve of the intrados and the positions of the archstones is now prepared for assembly and erection.

The first step is to cut off both ends of the board squarely at the marks which indicate the span. These cuts must pass through the springing points. Then the intrados is roughly shaped with the saw and then exactly planed to the mark, to make a smooth and evenly curved edge.

If the centring will be a part of the frame, it is now fitted to the frame head as in Figs. 1 and 2, and set into its place together with the frame. In this case it is best to make two identical turning pieces to seal the camber on both sides of the wall. They are fixed outside and inside flush with the frame. The hollow part in between the pieces is filled with a mixture of sand and ash; both to keep insects out and -- if there is no lagging -- to give more support to the archstones which otherwise would have to rest on the turning pieces only.

Note that the marks of the springing points are cut off when the turning piece is combined with the frame. This is because it is cut flush with the outside of the frame. This means that the thickness of the joint between the frame and the wall has to be taken into account when marking the slope of the skewback.

If the centring piece is to be removed after the arch is finished, it must be set in a way so that it can easily be taken out. This is done by placing it on wedges as in Fig. 3, (a). It might be necessary to cut the turning pieces slightly shorter on one end so that the centring fits easily into the opening. In that case the end which has not been cut rests against the wall while the gap on the other side is closed tight with wedges to prevent shifting (Fig. 3, b).

Try to avoid making the springing points in line with a bed joint. The sideways force from the arch can possibly cause cracks along the bed joint. The best solution will be to distribute the pressure over two courses (Fig. 3).

When you set the centring piece in position, do not forget to level the soffit, which must be exactly horizontal and parallel to the springing line.

NOTES:

b = 50 cm

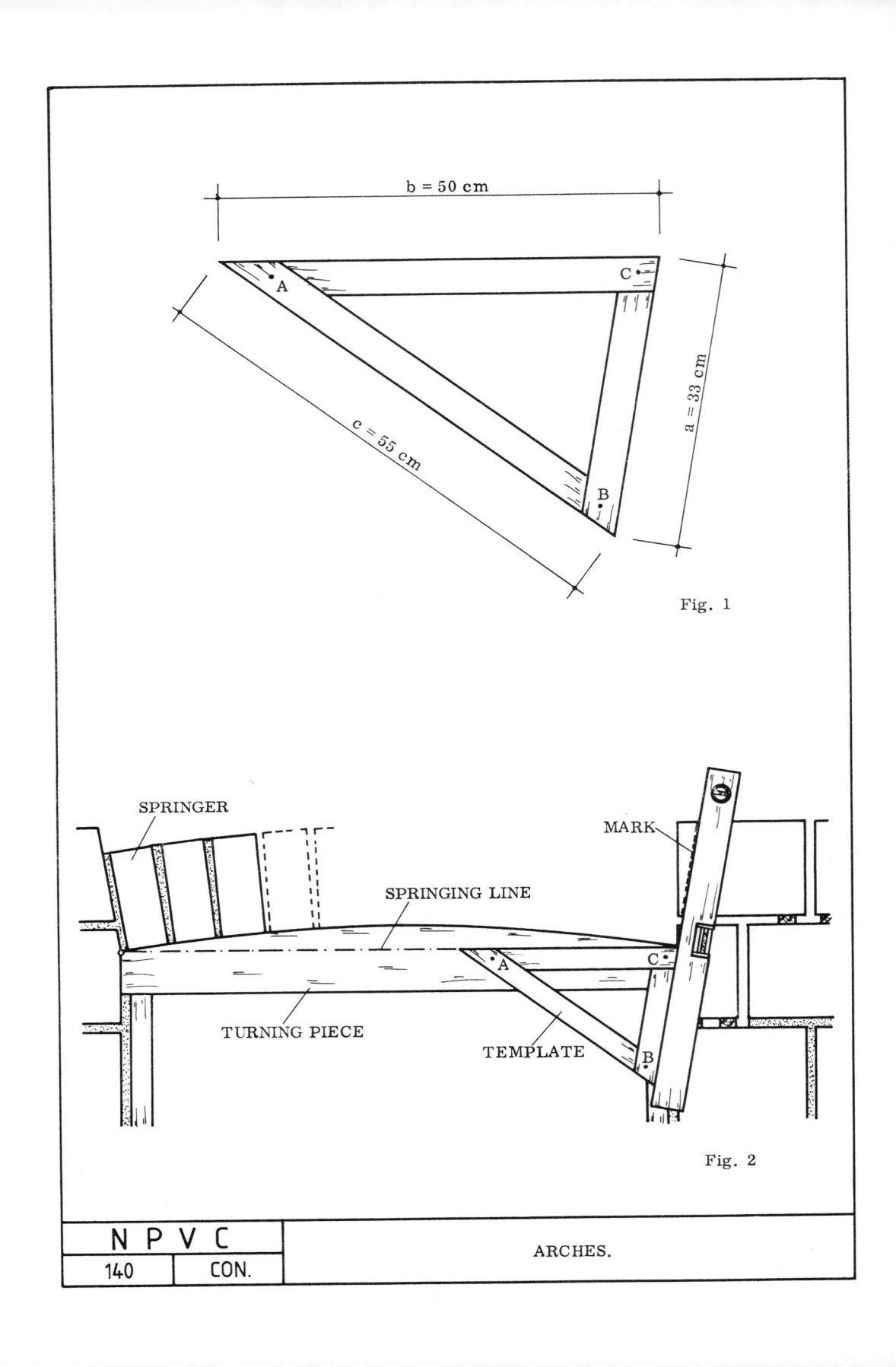

a = 33 cm

c = 55 cm

A

B

C

Fig. 1

SPRINGER

MARK

SPRINGING LINE

TURNING PIECE

TEMPLATE

A

B

C

Fig. 2

N P V C	
140	CON.

ARCHES.

SKEWBACK TEMPLATE

Before you can start laying the archstones, the abutments with their skewbacks must be prepared so that the springers can be laid at the correct angle. To be sure that the angle of the skewback is correct, the Rural Builder can use the template described below (Fig. 1).

This skewback template can only be used for a segmental arch where the radius of the arch is 3 times the span of the opening. Provided the rule of "r = 3 x span" is observed strictly, the same template can be used for any segmental arch, regardless of the length of the span. In other words the width of the opening and the size of the arch make no difference in the angle of the skewback for arches made according to this rule; the angle will always be the same.

- MAKING THE TEMPLATE: The template is made from wooden battens. It is simply a triangle with the outside dimensions as follows: side "a" = 33 cm, side "b" = 50 cm, and side "c" = 55 cm (Fig. 1). The angle at C is the angle we need to mark the skewback.

  Use three battens that are a few centimetres longer than the measurements required. This allowance can be cut off when the template is assembled. Plane one edge of each until it is exactly straight and mark the length on this edge. Nail the battens together as in Fig. 1; or make accurate halving joints and join the pieces together with wing nuts instead of nails, so that they can be taken apart and put back together when they are needed in the future. The corners should be marked on both sides with the letters A, B and C, so that the ends are not accidentally mixed up when it is reassembled.

- USE OF THE TEMPLATE: The outside edge measuring 50 cm (side b) is held along the springing line, while the tip of the corner C is at the springing point (Fig. 2). The direction of the shortest side (side a) extended upwards with a straight edge gives the slope of the skewback. To mark the skewback on the other side, the template is simply turned around and the procedure is repeated.

  - Note that corner B must always point downwards during the marking.

NOTES:

ARCHES.

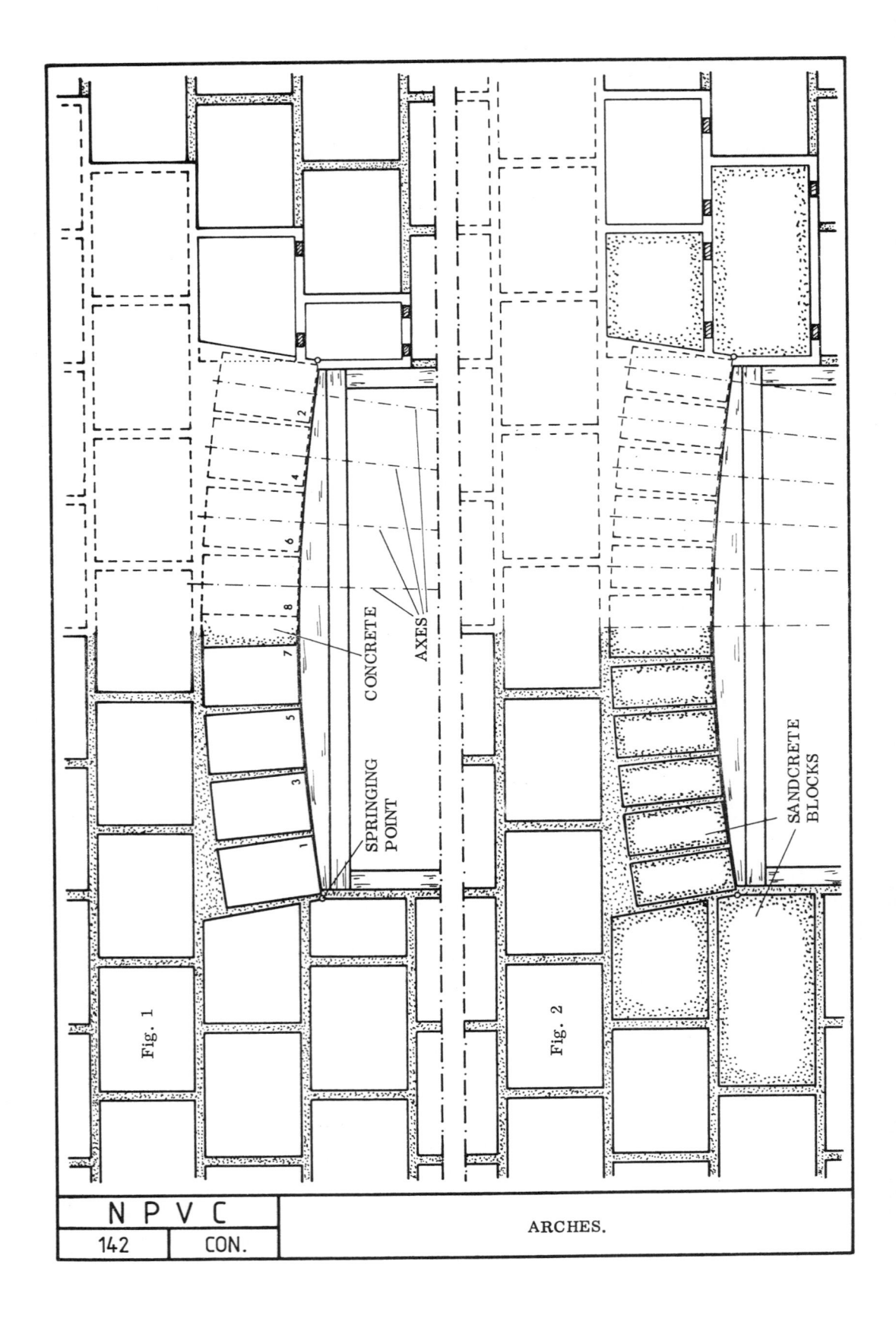

Fig. 1

Fig. 2

SPRINGING POINT

CONCRETE

AXES

SANDCRETE BLOCKS

NPVC		ARCHES.
142	CON.	

LAYING THE ARCHBLOCKS

The four blocks which have the skewbacks are marked and cut before they are laid in mortar. They are set temporarily in place on wooden battens which have the same thickness as the bed joint (Figs. 1 & 2, right sides). Mark the slope of the skewback with the template, and remove the blocks to cut them to shape. Then they can be laid in mortar and the arch can be continued.

Fig. 1 shows a segmental arch built with landcrete blocks. For this purpose, 1/2 blocks have to be cut, or preferably made specially in the Tek block press. This is easily done by putting a 1 cm thick board (for the 1 cm joint) edgewise in the middle of the mould box. The arch is built up by laying the blocks by turns on alternate sides, as indicated in Fig. 1 by the numbers.

The axis of each block should point towards the centre of the arch, with the result that ideally only the middle part of each block should rest on the turning piece, and the lower corners of the blocks do not contact the turning piece; there should be small wedge-shaped gaps. This is because the block faces are straight but the top edge of the turning piece is rounded.

The wedge-shaped joints between the archblocks need special attention. They must be properly filled with a rather wet mortar. If there is a gap left in the centre of the arch it is closed with concrete, which functions as a keystone (Fig. 1).

Segmental arches over a span exceeding approximately 120 cm should be constructed with sandcrete blocks. The skewback blocks which abut the springers should also be sandcrete, so as to better distribute the pressure (Fig. 2).

To meet this requirement specially made blocks are needed. The skewback blocks must be 1 cm smaller in height than normal sandcrete blocks, so that they fit in with the landcrete courses. This is achieved by inserting a 1 cm thick piece of wood into the sandcrete block machine.

The archstones consist of 1/4 blocks which can also be made in the machine by inserting three 2 cm thick boards at a distance of 10 cm apart. The keystone is formed with a normal block of 15 cm thickness, cut to match the thickness of the wall.

If the centring is not combined with the frame, it is slightly lowered on completion of the arch, and removed after four days.

- NOTE: Try to set the centring so that the crown of the arch will be level with the top of the upper abutment course (Figs. 1 & 2) and not somewhere in the middle. The distance between the springing line and crown can be calculated as follows: 0,042 x span, plus height of archblocks.

Fig. 1

SCALE 1 : 10

SIDE BOARDS CAN BE ANY SUITABLE BOARDS - THEY DO NOT HAVE TO BE CUT TO MATCH THE LINTEL HEIGHT.

CUTTING WOULD BE A WASTE OF MATERIAL.

HEAD OF FRAME

INSIDE FACE OF WALL

OUTSIDE FACE OF WALL

TEMPORARY SUPPORT

Fig. 2

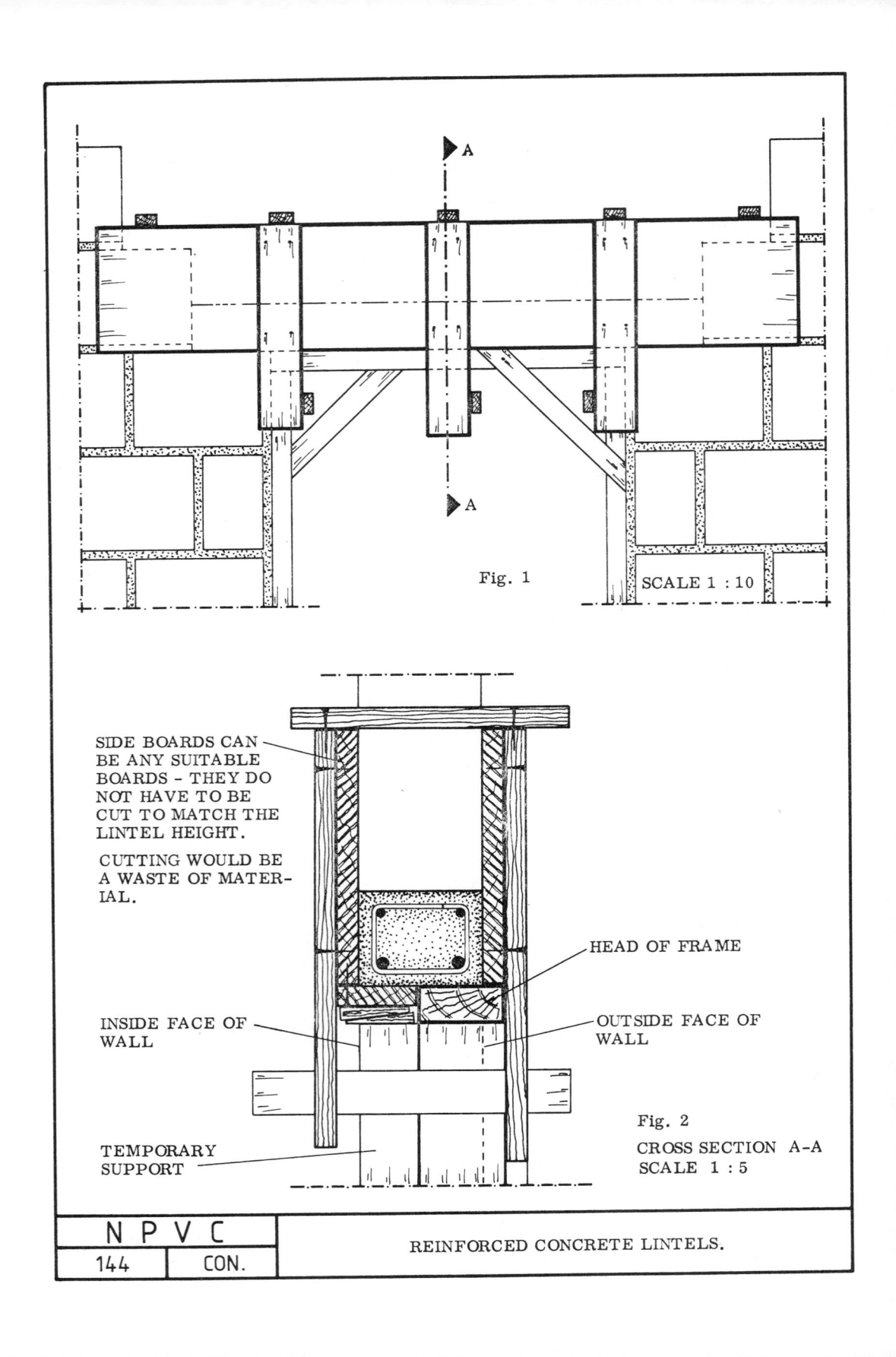

CROSS SECTION A-A
SCALE 1 : 5

NPVC
144 CON.

REINFORCED CONCRETE LINTELS.

REINFORCED CONCRETE LINTELS

FORMWORK

The formwork for cast-in-situ lintels consists of two main parts: the shuttering and the strutting. Precast lintels need only shuttering because they are cast on the ground.

- CONSTRUCTION: The construction of the formwork is shown in Fig. 1, page 167 of the Basic Knowledge book. This illustrates a situation where the frame will be set later or where there is an opening without a frame.

  Since in most cases the frames are set already, the formwork does not require special strutting (Fig. 1). The head of the frame then acts as part of the soffit of the shuttering. The side boards must be fixed so that the vertical sides of the concrete lintel will be flush to the surface of the blockwork (Fig. 2).

- PREPARATION: If the lintel will not be plastered over later, the inside of the shuttering boards which are in contact with the concrete should be planed until they are smooth. This makes the stripping easier and also gives the concrete surface a nicer appearance. The inside of the shuttering may be treated with a special formwork lubricant, if this is available. Old engine oil can serve the same purpose; it should be applied lightly.

  In case the lintel will be plastered later, it is better to leave the contact faces of the shuttering rough-sawn. Instead of oiling, wet the shuttering boards thoroughly with water, so that they don't absorb too much moisture from the freshly cast concrete.

  Before casting the concrete make sure that the inside of the shuttering is clean; all joints, slits and gaps closed; and that the clear cover (the thickness of the concrete covering the reinforcement iron) is the correct thickness everywhere.

NOTES:

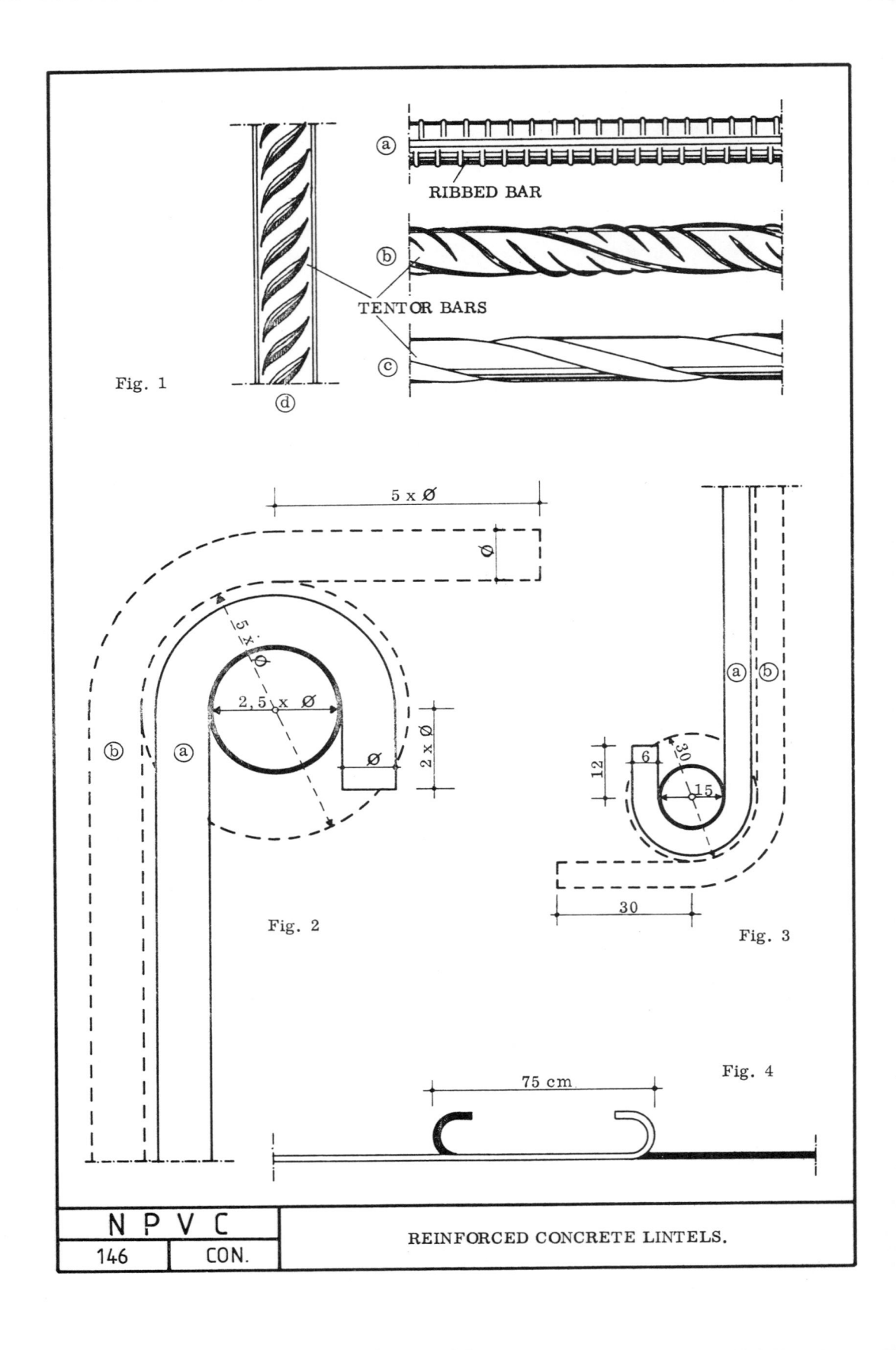

Fig. 1

ⓐ RIBBED BAR

ⓑ
ⓒ
TENTOR BARS

ⓓ

Fig. 2

5 x Ø
2,5 x Ø
5 x Ø
2 x Ø
Ø
ⓐ
ⓑ

Fig. 3

12
6
30
15
30
ⓐ ⓑ

Fig. 4

75 cm

N P V C	REINFORCED CONCRETE LINTELS.
146 CON.	

REINFORCEMENT

The reinforcement for a lintel consists of a combination of different iron rods assembled to form a so-called "reinforcement cage" (see Basic Knowledge book, page 166).

In Rural Building the most commonly used reinforcement iron is round and has a smooth surface. It is available in rods with diameters ranging from 6 mm to 28 mm; but the most common sizes the Rural Builder uses are 6 mm (1/4") rods and 12 mm (1/2") rods.

If ribbed bars or tentor bars are available, these are preferred because the concrete grips better to a rough surface. Figs. 1a, b, c, and d show how the surfaces of some reinforcement bars look. Also see page 171 of the Reference Book.

- BENDING REINFORCEMENT BARS: According to how they will be used, sometimes iron rods must be bent or shaped to a hook form at the ends so that they can be well anchored. This must be done according to the following rules:

  - All circular bars with smooth surfaces must have U-shaped end hooks (see Figs. 2a & 3a). For ribbed bars and tentor bars L-shaped hooks are sufficient (Figs. 2b & 3b).

  - The minimum length for the straight end part of the hooks is: 2 times the rod diameter for circular bars; and 5 times the rod diameter for ribbed bars and tentor bars (Figs. 2 & 3).

  - The smallest permissible bending diameter can be calculated according to the figures in the table below.

DIAMETER OF ROD	CIRCULAR BARS	RIBBED BARS & TENTOR BARS
6 mm up to 20 mm	2,5 x rod diameter	5 x rod diameter
22 mm up to 28 mm	5 x rod diameter	7 x rod diameter
more than 28 mm	----------------	10 x rod diameter

The smallest permissible bending diameter for the 6 mm circular bar in Fig. 3a will be 15 mm; or 2,5 x 6 mm. For the 6 mm ribbed bar in Fig. 3b, the bending diameter will be at least 30 mm, or 5 x 6 mm.

- NOTE: As a rule, main bars should never be extended. If this is necessary, the shortest overlap for bars under tension is 75 cm, and the ends of the bars must have U-shaped hooks (Fig. 4).

REINFORCED CONCRETE LINTELS.

Fig. 1a

Fig. 1b — ERECTION BARS, MAIN BARS

Fig. 2a

Fig. 2b — ERECTION BARS, MAIN BARS

NPVC	REINFORCED CONCRETE LINTELS.
148 CON.	

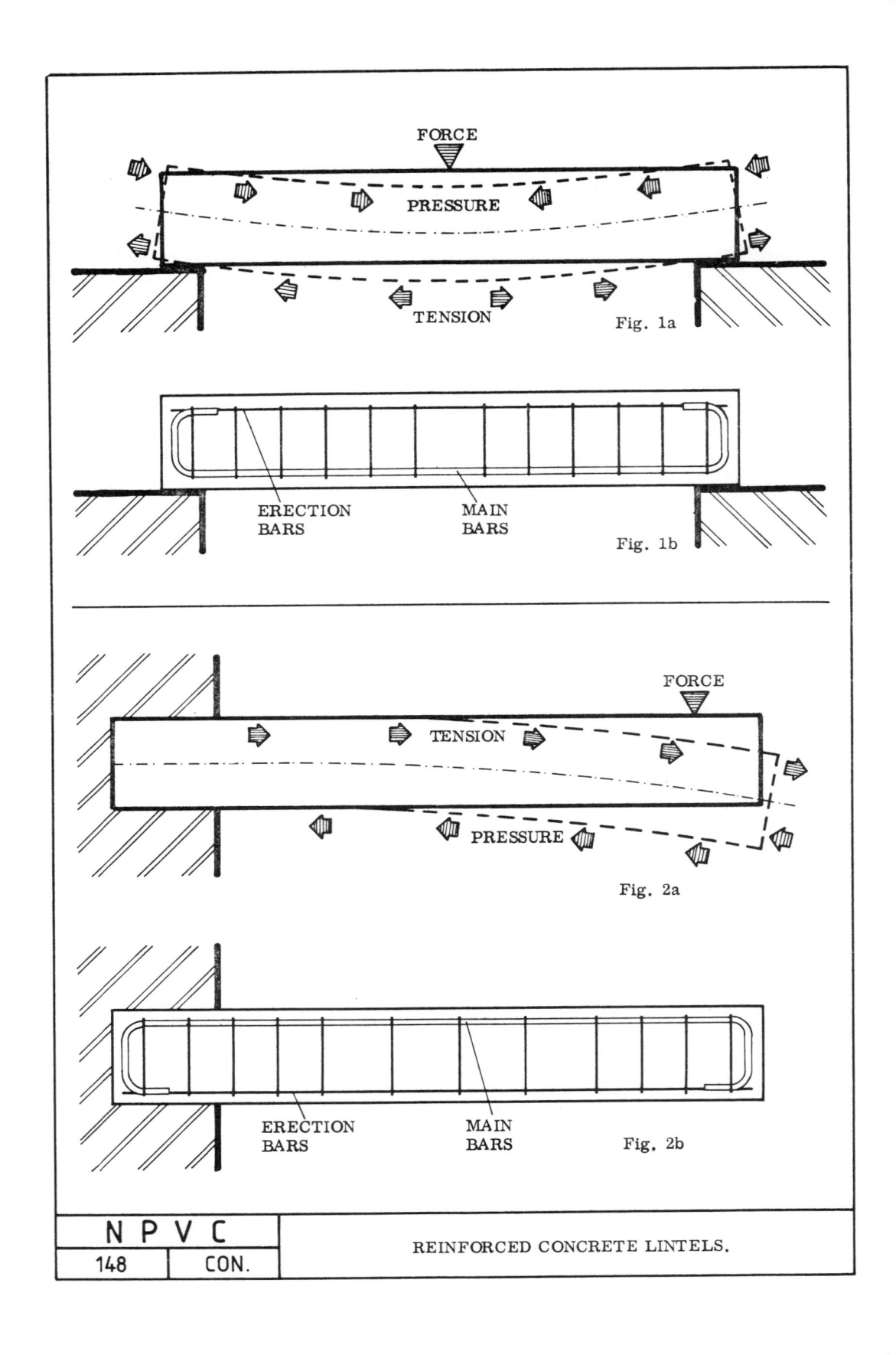

REINFORCEMENT SYSTEMS

The construction of large reinforced concrete members usually requires complicated calculations to determine the diameters, number, shape and arrangement of the rods. This job should be left to a qualified design engineer.

However small and simple members of the structure such as lintels, columns, short span beams, etc. can be constructed by the Rural Builder according to the sketches, particulars and hints given here.

To place the reinforcement rods correctly, it is essential to understand and be able to predict in which part of the concrete member tensile stress (tension) is likely to occur. This is where the member has to be reinforced by the main bars. Fig. 1a shows the bold outline of a lintel; the dotted line shows how it would tend to bend under pressure. The arrows show how the whole upper part is under pressure, while the lower surface is under tension.

Fig. 1b illustrates how the lintel can be reinforced against this tension. Note that the main bars are on the bottom, in the zone which is under tension; and that the ends of the main bars are anchored on top, within the zone that is under pressure.

Fig. 2 shows another situation, where a beam is supported from one side only. This is called a cantilever beam. The unsupported end will tend to bend as shown. Here the main bars are on top to counteract the tension in the top surface (Fig. 2b).

NOTES:

REINFORCED CONCRETE LINTELS.

FORCE

Fig. 1a

"BENT-UP" BARS   MAIN BARS

Fig. 1b

STIRRUP

"BENT-UP" BARS

x = CLEAR COVER

Fig. 2
CROSS SECTION A-A

N P V C	REINFORCED CONCRETE LINTELS.
150	CON.

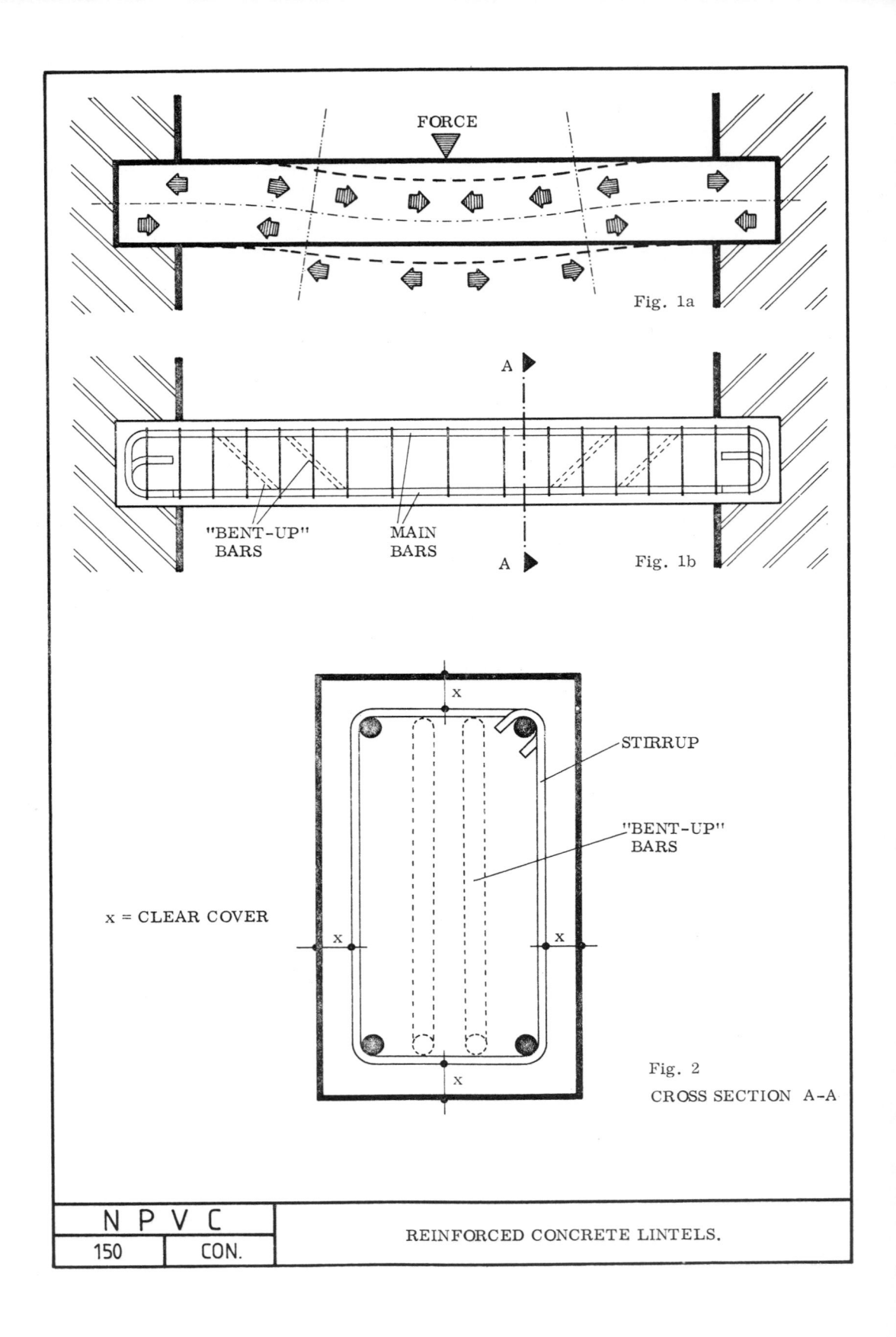

The longer reinforced beam or lintel shown here in Figs. 1a and 1b is reinforced with main bars on the top and bottom.

The cross section in Fig. 2 shows how the clear cover is measured from the outside of the stirrups to the outside face of the concrete. In other words, the clear cover or concrete cover is the distance between the concrete surface and the nearest reinforcement bar, regardless of the function of the bar.

The distance between the stirrups is never more than 20 cm. For safety, the Rural Builder is advised to put the stirrups no more than 15 cm apart. The stirrups closer to the supporting walls (50 cm away and closer) should be no more than 10 cm apart (see the illustrations here and on the next page).

So-called "bent up" bars are sometimes used as part of the reinforcement system (Figs. 1 & 2) but these are not used in Rural Building because of the difficult calculations required to place them correctly; also they are not necessary for the small structural members which are required in Rural Building.

NOTES:

FORCE

FORCE

PILLAR

MAIN BARS

STIRRUPS

"BENT-UP" BARS

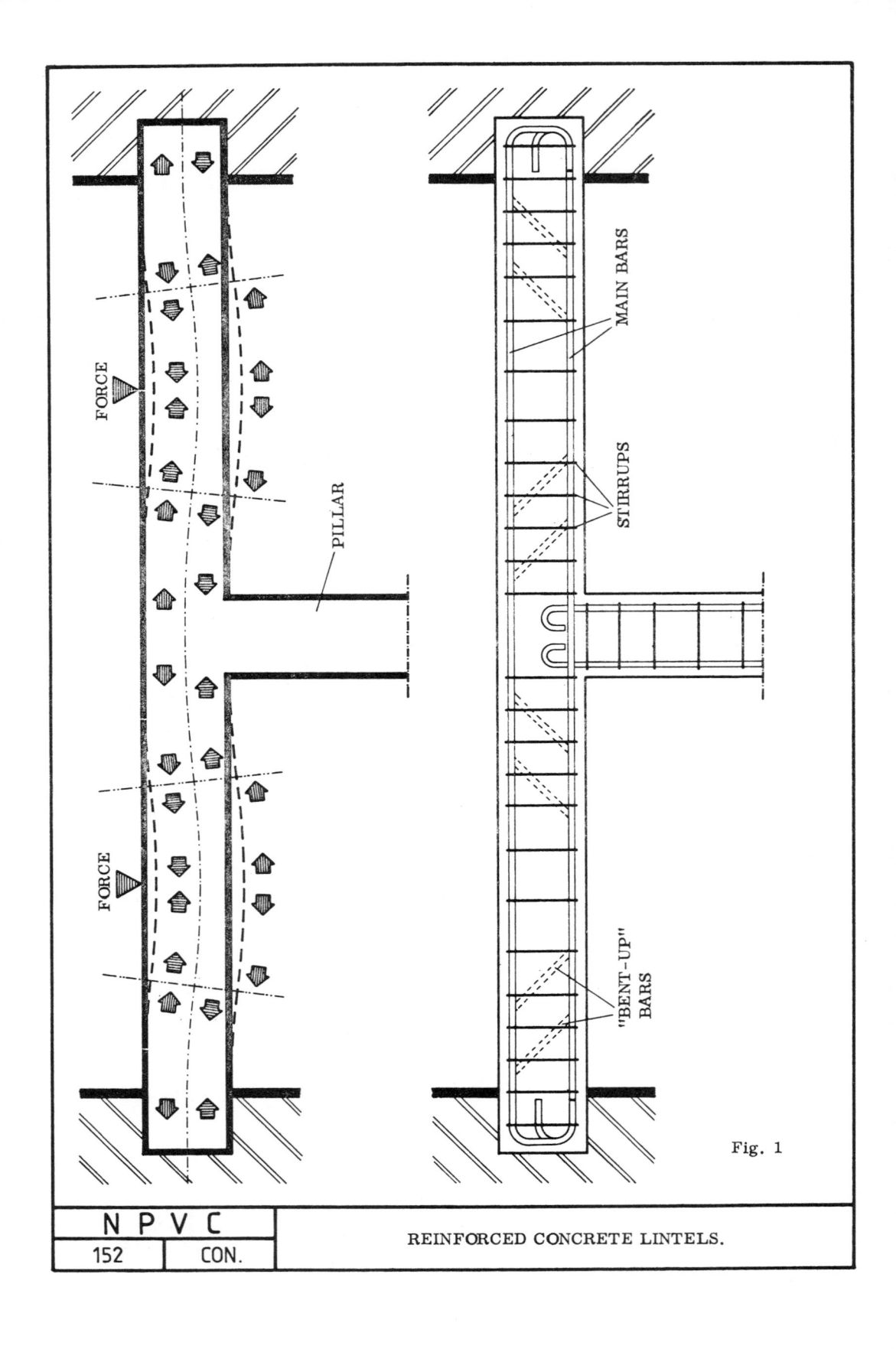

Fig. 1

N P V C	
152	CON.

REINFORCED CONCRETE LINTELS.

On the left is a drawing of a reinforcement system for a continuous beam; this might occur for example in an eave beam supported by a reinforced pillar (Fig. 1).

CORROSION IN REINFORCED CONCRETE

It may seem strange, but the weakest part of a reinforced concrete structure is the hard iron, because it is easily broken down if it is not protected well. The structure remains stable only as long as the iron is protected against corrosion.

The main problem with reinforced concrete is the development of very small cracks in the concrete, which means that the iron is exposed to air and becomes rusty. On the concrete surface below the cracks, reddish brown rust stains appear. Once these are visible it is usually too late to save the member.

Unless immediate protective measures are taken, the rust develops more and pressure builds up between the iron and concrete, causing the concrete cover to break open; this speeds up the rusting process and eventually the reinforcement bars become too weak to carry loads or to withstand other forces. The system then collapses.

- PREVENTION OF CORROSION: Here we must include a strong warning against over-reliance on concrete strength. Not only laymen, but also many construction experts believe "the more cement, the harder and the better the concrete", and they tend to improve the mix proportions by using greater amounts of cement. This is wrong, because very hard concrete is not only expensive but it is also not necessarily the best protection against rust. Therefore, the Rural Builder is advised to follow the correct mix proportions.

  If possible use high quality steel for the reinforcement as this does not rust as easily as common circular bars.

  The surface of the concrete should be protected against the weather by the application of a waterproof cement paint (see Reference Book, page 201). However, such a paint coat is wasted if rust stains are already present. The paint does not seal the cracks and only makes the stains disappear for a short time, if at all.

- NOTE: Never apply paint on concrete surfaces that have not set hard yet. Possible shrinkage of the concrete during the hardening process would cause tiny cracks in the paint coat, making the paint useless as protection against the penetration of water and air.

7,5

5

WELDED

approx. 38 cm

Fig. 1

Fig. 2

Fig. 3

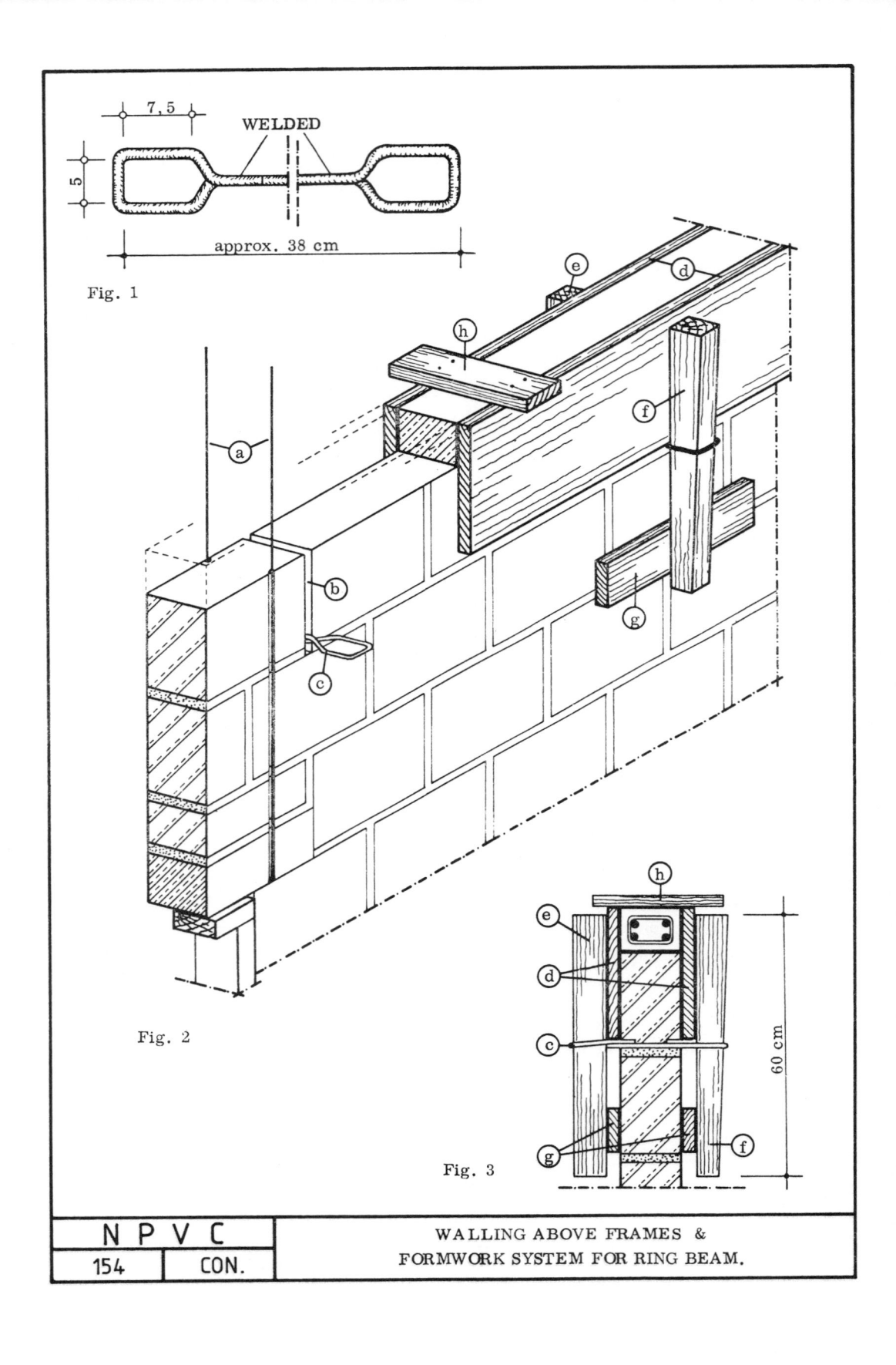

60 cm

NPVC	WALLING ABOVE FRAMES &
154　CON.	FORMWORK SYSTEM FOR RING BEAM.

WALLING ABOVE FRAMES

The last part of the walls above the frames should consist of at least three courses of blocks, so that the roof construction can be well anchored. Experience has shown that anchoring the trusses in the ring beam alone is not sufficient for them to withstand strong wind forces. Therefore, the anchorage irons are placed in the first bed joint just above the frames (see Anchorage, page 197).

If the exact positions of the trusses are not known yet, the irons can be set later, in the finished wall. This is done by passing the iron through a hole chiselled in the bed joint (Fig. 2 a). It is generally better and easier though to plan ahead and place the irons during the wall construction. The irons are set in grooves so that they will be flush to the wall surface and will be covered and protected later by the plaster or render.

FORMWORK SYSTEM FOR RING BEAM

If a reinforced concrete ring beam will be made, every third cross joint in the last course is left open (Fig. 2, b) so that formwork clasps can be inserted.

These special clasps have been developed and manufactured at N P V C in order to save strutting materials and to ease the construction of the formwork for a ring beam. Formerly long standards and braces in great number were needed to construct the formwork. With the clasp however, only a few short boards are necessary for the bracing. The formwork clasps are easy to handle and have passed all tests at N P V C. They are made from a piece of iron rod bent as shown in Fig. 1, and welded to hold the loop ends closed.

Fig. 2 shows the formwork system, and Fig. 3 shows the cross section. When the clasp is inserted in the open cross joint (Fig. 2, c) the remaining open space must be closed with paper (empty cement bags) to make it possible to remove the clasp later. The side boards (d) rest on the clasps and are pressed against the wall by straight boards (e) on one side, and wedge-shaped boards (f) on the other side. Short pieces of board below the side boards (g) act as distance pieces to keep the shuttering boards plumb. Cleats or spreaders (h) may also be used to maintain the correct width of the beam. There are no nails used, except for fixing the cleats, which reduces the damage to the boards.

The Rural Builder is advised to supply himself with at least four clasps and to carry them along with his set of tools.

Fig. 1

Fig. 2

Fig. 3

Fig. 4

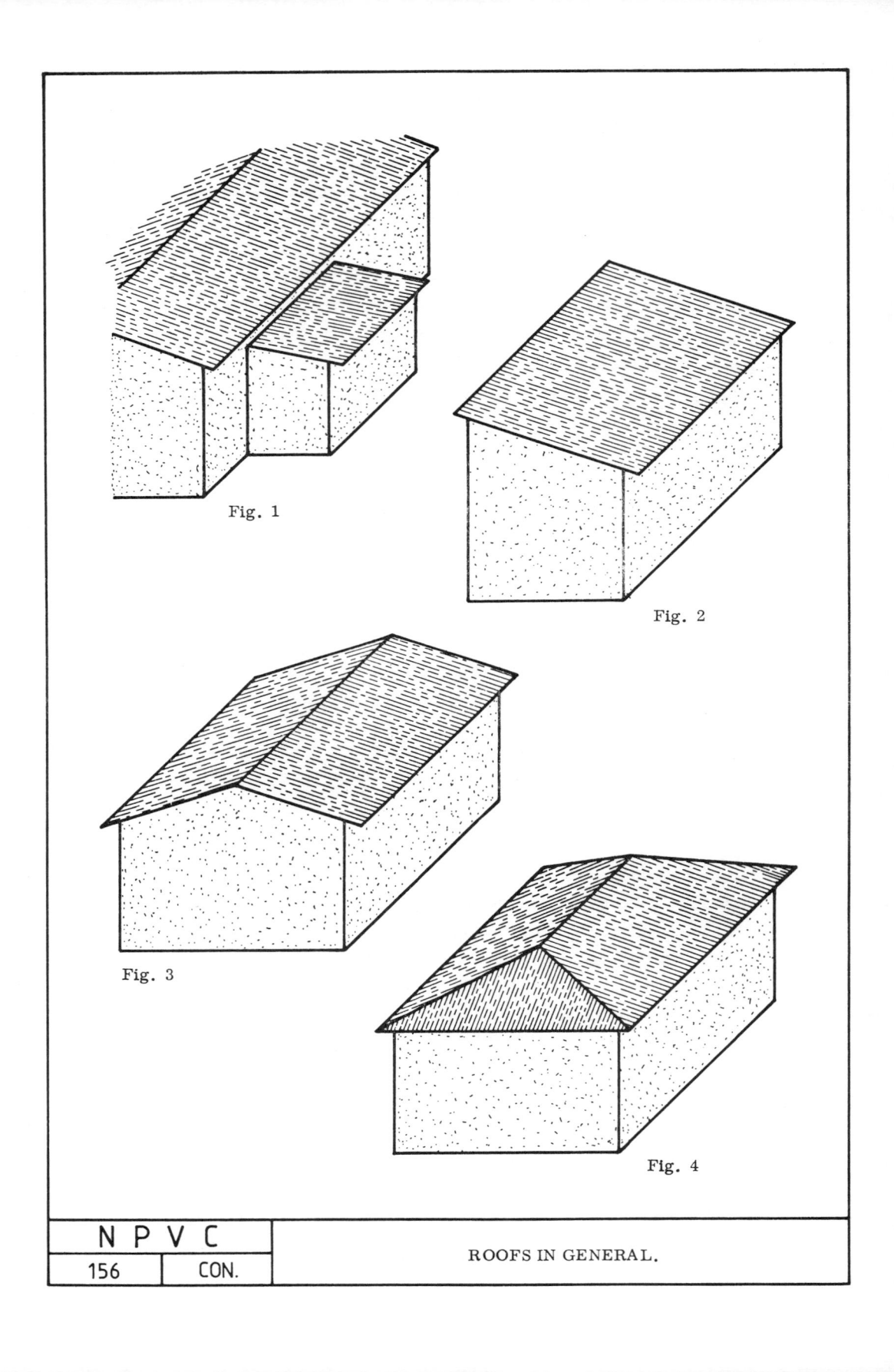

N	P	V	C		
156		CON.		ROOFS IN GENERAL.	

ROOFS IN GENERAL

The roof is a very important part of the building structure. It performs several essential functions:

- It gives shelter to people.
- It provides shade.
- It isolates the building from cold and heat.
- It keeps out dust and dirt.
- It protects the interior of the building.
- It sheds rainwater.

ROOF TYPES

There are many different types of roofs. In Rural Building we deal only with the following four types:

- The lean-to roof (Fig. 1)
- The pent roof (Fig. 2)
- The gable roof (Fig. 3)
- The hipped roof (Fig. 4).

These roofs will be described in detail in the following sections of the book. In the next few sections we will describe some factors which are important in the construction of all roofs. These must be understood well before we can go on to describe the construction details of the particular roof types.

NOTES:

SIZE OF THE ROOF

The cost of roofing sheets will be a significant part of the cost of the whole building. Therefore it is the size of the roofing sheets which will determine the size, and especially the width, of the whole building.

For this reason, we make an outline design of the roof before we determine the other measurements of the building. We cannot design the building first and later fit a roof on it. The outline design tells us the width that our building should have so that we can fit a roof on it without unnecessary and wasteful cutting and trimming of the sheets (Reference Book, pages 239 & 240).

To make the outline design of the roof, we need to know:

- the pitch of the roof
- the effective length of the sheets
- the distance the roof will project past the outside walls of the building.

ROOF PITCH

The angle of the slope of the roof is called the pitch. Most roof types have a standard pitch. If corrugated sheet materials (Reference Book, pages 230 and 231) are used, the pitch angle should be between 15 and 20 degrees (Fig. 1).

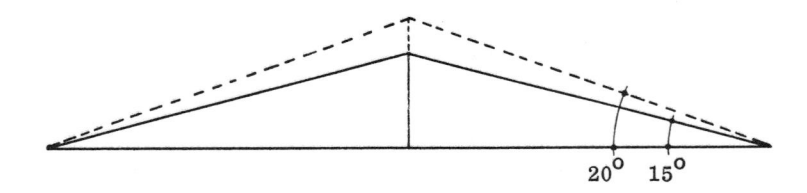

Fig. 1

ROOFS IN GENERAL.

EFFECTIVE LENGTH OF THE SHEETS

The effective length of roofing sheets is the length of the sheet (x) minus the overlap (y) between the sheets (Fig. 2).

The minimum overlap in the length for corrugated sheeting materials is 15 cm. The most common length for roofing sheets in Ghana is 244 cm.

In order to use the sheet materials as economically as possible, we use either 1, $1\frac{1}{2}$, 2, $2\frac{1}{2}$, or 3 (and so on) sheets to cover the distance from the highest point of the roof to the lower edge. Thus the effective length will be:

- for 1 sheet   - 244 cm
- for $1\frac{1}{2}$ sheets   - 244 cm + 122 cm - 15 cm = 351 cm
- for 2 sheets   - 244 cm + 244 cm - 15 cm = 473 cm
- for 3 sheets   - (244 cm x 3) - (15 cm x 2) = 702 cm

and so on.

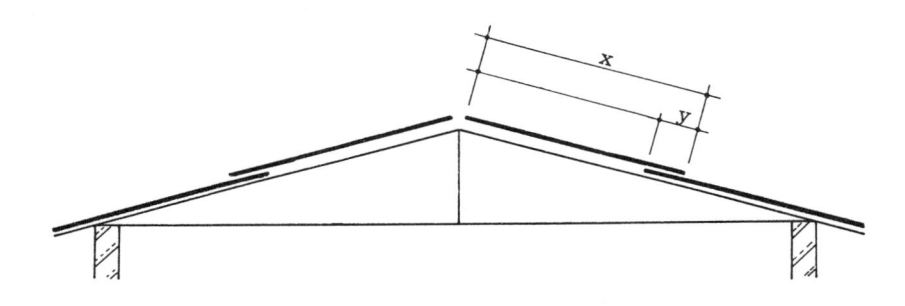

Fig. 2

NOTES:

ROOFS IN GENERAL.

ANGLE OF PITCH

ROOFS IN GENERAL.

NPVC 160 CON.

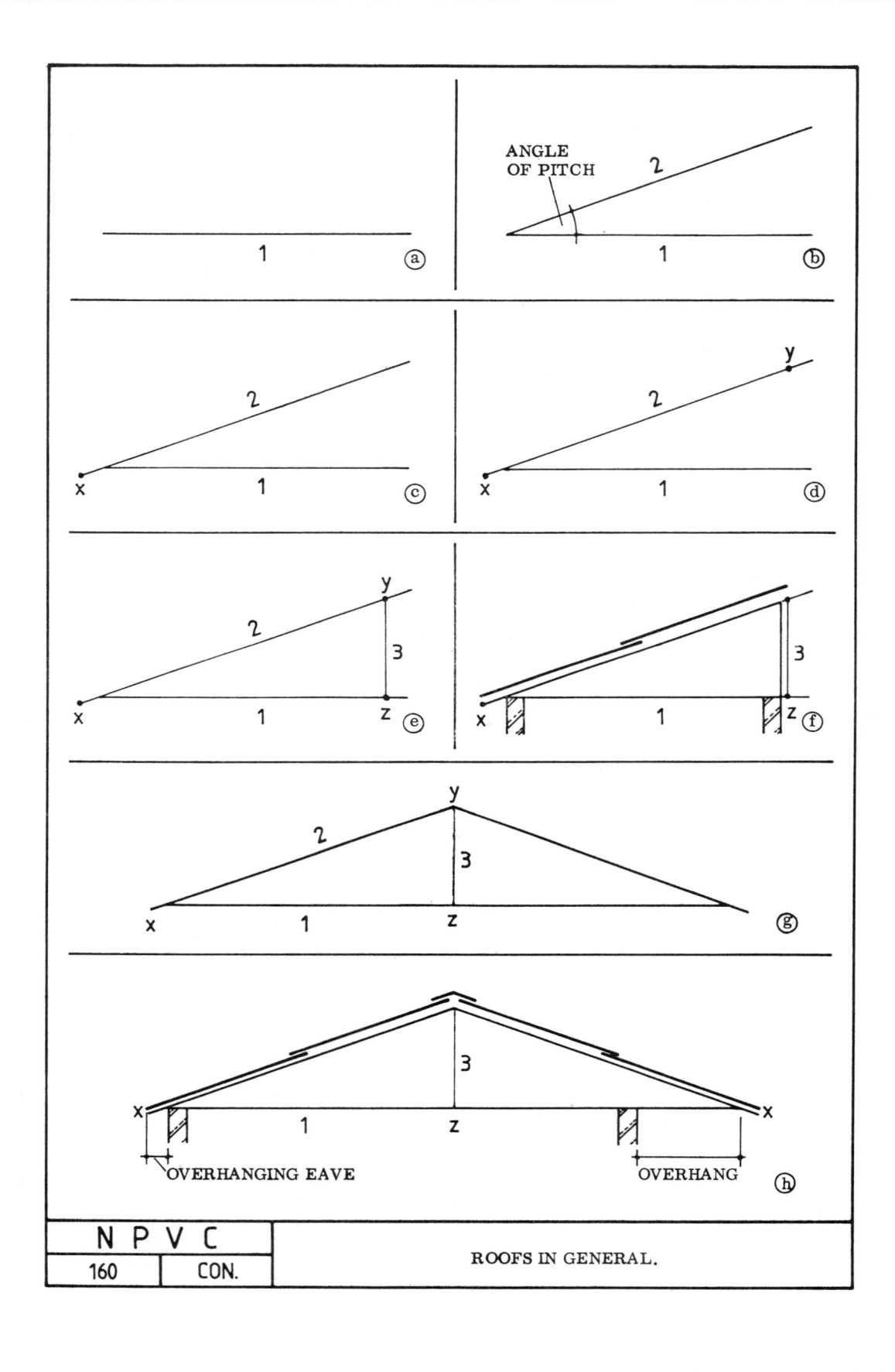

OVERHANGING EAVE

OVERHANG

OUTLINE DESIGN

The outline design should be made in as large a scale as possible, since it is used to find some measurements for the future building design.

a. Draw a horizontal line (line 1).

b. Draw line 2. The angle between line 1 and line 2 should be the angle of the pitch of the roof.

c. Extend line 2 to show the projection of the roof beyond the wall of the building. Mark point x at the end of line 2.

d. Measure the effective length of the sheets from point x, remembering to subtract the overlap. Mark point y on line 2.

e. Make a line perpendicular to line 1 and passing through point y. This is line 3. Mark point z on line 1.

f. If a lean-to or pent roof is planned, the length of line 1 up to point z gives the width of the building. Now indicate the supporting walls and the sheets with the overlap. In a lean-to roof, remember that the end of the sheet at the top is enclosed in the wall; and for the pent roof, the sheets project past the wall on the top side.

g. If a gable roof is planned, draw the other half of the roof in the same manner. The extended line 1 will give the width of the building.

h. Indicate the supporting walls, sheets and the overlap.

- TERMS: The point where the sloping lines meet, point y, is called the "ridge".

Line 3 is the "rise" of the roof.

If the horizontal line (line 1) projects beyond the supporting walls, the projecting part is called an "overhang".

If the sloping part projects beyond the wall, the projecting part is called an "overhanging eave".

The lowest part of a sloping roof is called the "eave" (point x).

NOTES:

ROOFS IN GENERAL.

ROOF COVERING
ROOF CONSTRUCTION
CEILING
CEILING FANS

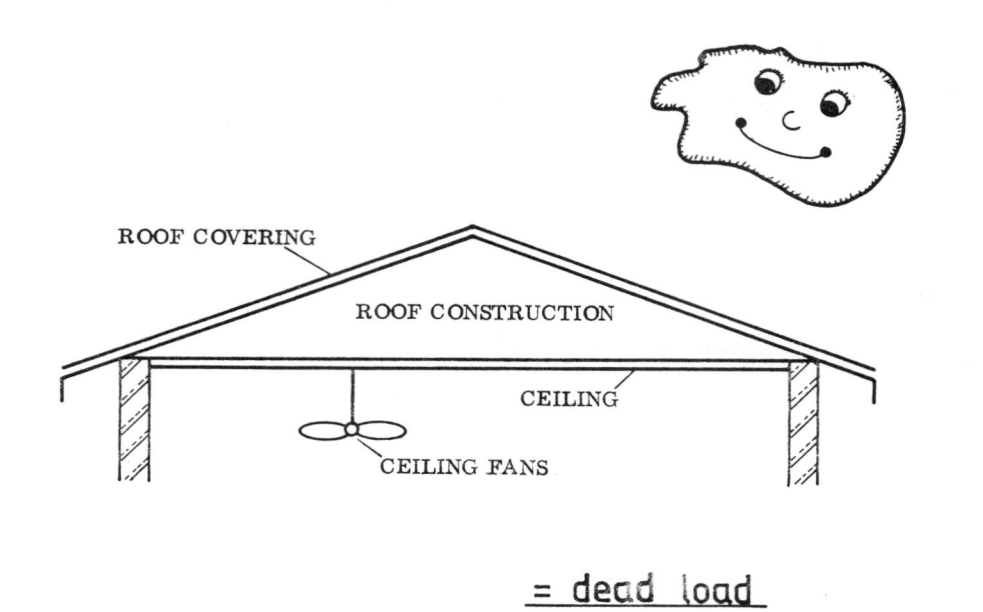

= dead load

Fig. 1

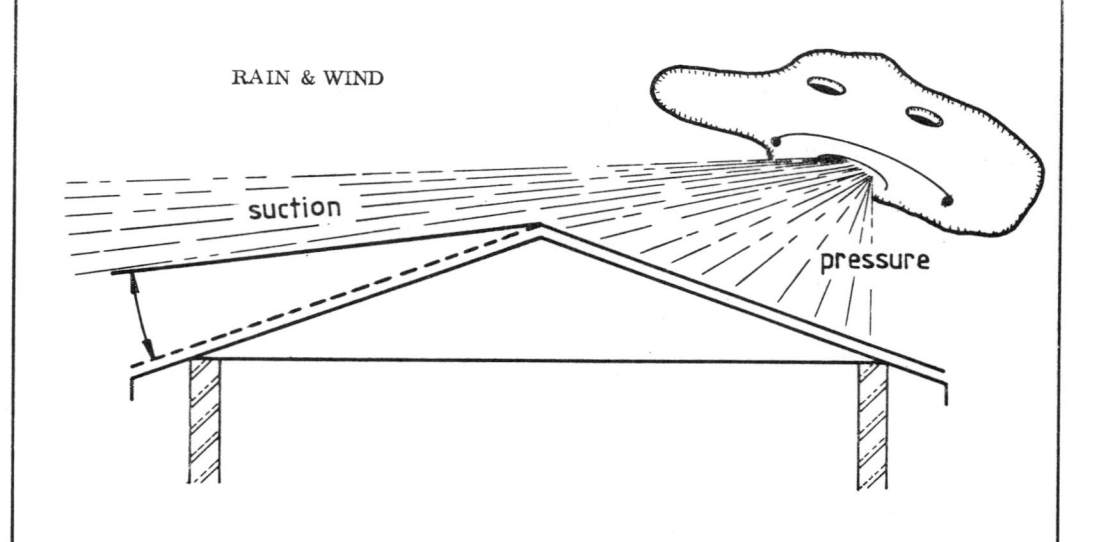

RAIN & WIND
suction
pressure

= external load

Fig. 2

N P V C
162 CON.
ROOFS IN GENERAL.

LOADS

In order to build a good strong roof, it is necessary to take into account certain forces that will affect it. We call these forces "loads".

Some of the loads come from the weight of the roof itself and the ceiling. These are called "dead loads". They consist of: (see Fig. 1)

- the weight of the roof covering
- the weight of the roof construction
- the weight of the ceiling
- the weight of anything attached to the ceiling such as ceiling fans or light fixtures.

Other loads are caused by external (outside) forces such as rain or wind. These are called "external loads". External loads are: (see Fig. 2)

- windloads
- rainloads

All loads, external and dead loads, must be taken into account in the planning and construction of a roof.

The most dangerous load in Ghana is the windload. Strong winds or storms can cause a lot of damage to a roof if it is not well made and securely fixed to the building.

One particular danger is the suction on one side of the roof caused by the wind, which creates pressure on one side and suction on the other side of the roof (Fig. 2). This suction on roofs can be demonstrated by an experiment. Fold a sheet of paper in half and blow over the top edge. The other half of the paper will be lifted up by suction.

- SHAPE OF THE ROOF WITH RESPECT TO WINDLOAD: Roof constructions can be severely damaged by storms. Apart from the precaution of anchoring the roof well, which will be discussed later, there are some ways to shape a roof so that it is less vulnerable to the force of the wind:

    - Consider carefully the pitch of the roof. Roofs with a pitch of less than 10 degrees are much more prone to high suction forces.
    - 15 to 20 degrees is a better roof pitch, because it causes less suction.
    - Consider making a hip roof instead of a gable roof. This will decrease the suction even more.
    - See the Drawing Book, pages 105 and 106, for more suggestions.

ROOFS IN GENERAL.

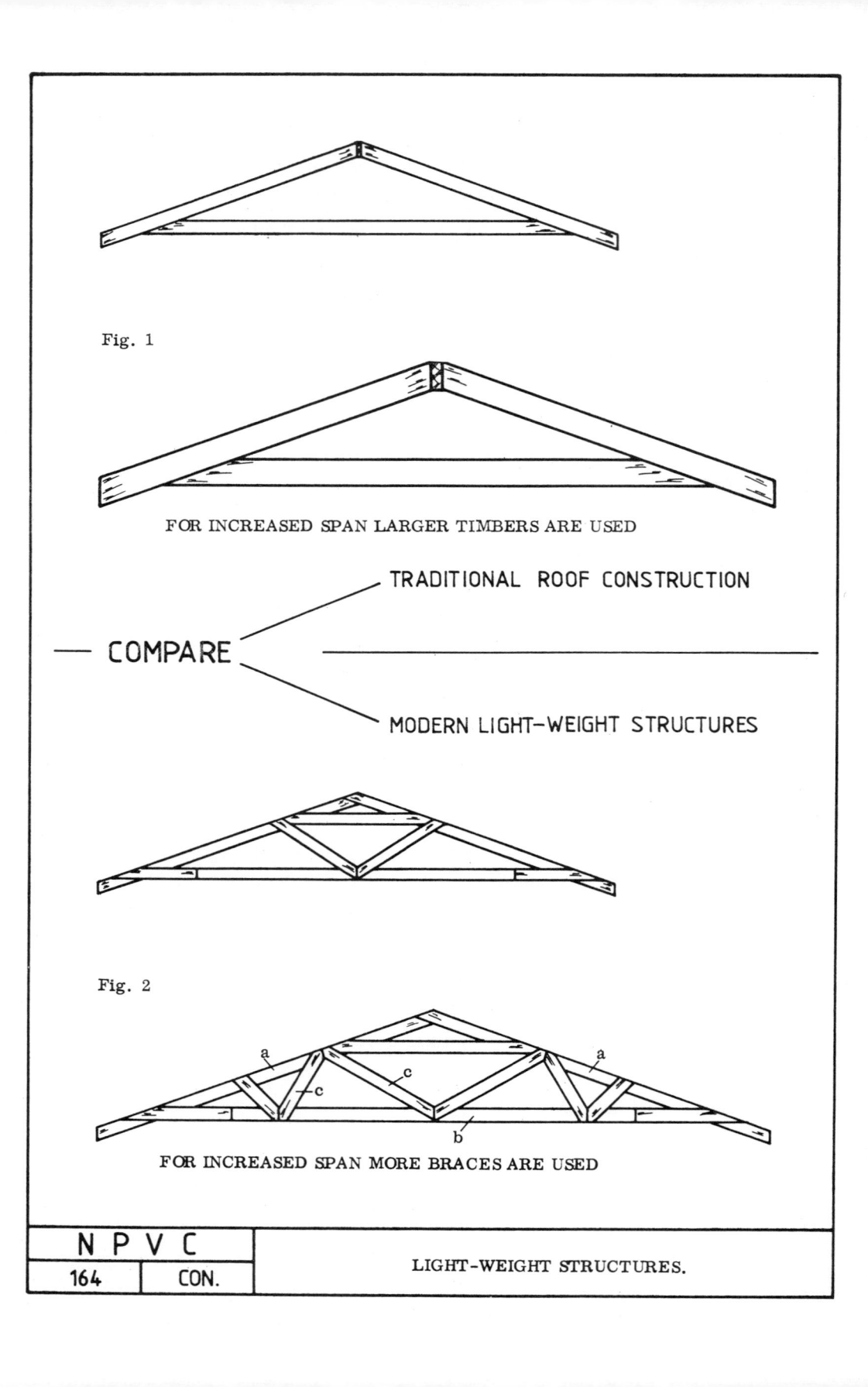

Fig. 1

FOR INCREASED SPAN LARGER TIMBERS ARE USED

COMPARE — TRADITIONAL ROOF CONSTRUCTION
— MODERN LIGHT-WEIGHT STRUCTURES

Fig. 2

FOR INCREASED SPAN MORE BRACES ARE USED

N P V C
164 CON.

LIGHT-WEIGHT STRUCTURES.

LIGHT-WEIGHT STRUCTURES

As the span (width) of the building increases, the strength of the roof construction must also be increased. Formerly this was done by using thicker and wider timbers (Fig. 1).

This method resulted in very heavy and expensive roof constructions. Large timbers are normally much more expensive than smaller timbers (compare the cost of 5 x 7,5 cm Odum and 5 x 15 cm Odum).

This kind of roof construction is now uneconomical, as well as difficult because of the complicated joints needed for large timbers.

In modern light-weight structures, the strength of the roof is increased for larger spans by building up brace structures in the shape of triangles (Fig. 2), instead of using larger timbers. This construction uses small timbers, which are cheaper and lighter in weight; therefore we get a light-weight structure.

Light-weight structures are less costly, and result in a smaller dead load on the roof. Another advantage is that we can use simple joints instead of complicated roofing joints.

Since the construction is built up from small members, it is sometimes called a "built-up structure". This is the only type of roof construction we will learn in Rural Building.

- TERMS: The sloping members are called "rafters" (Fig. 2, a).

    The horizontal member is called the "tie beam" (b).

    The members strengthening the construction are called "braces" (c).

NOTES:

FORCE

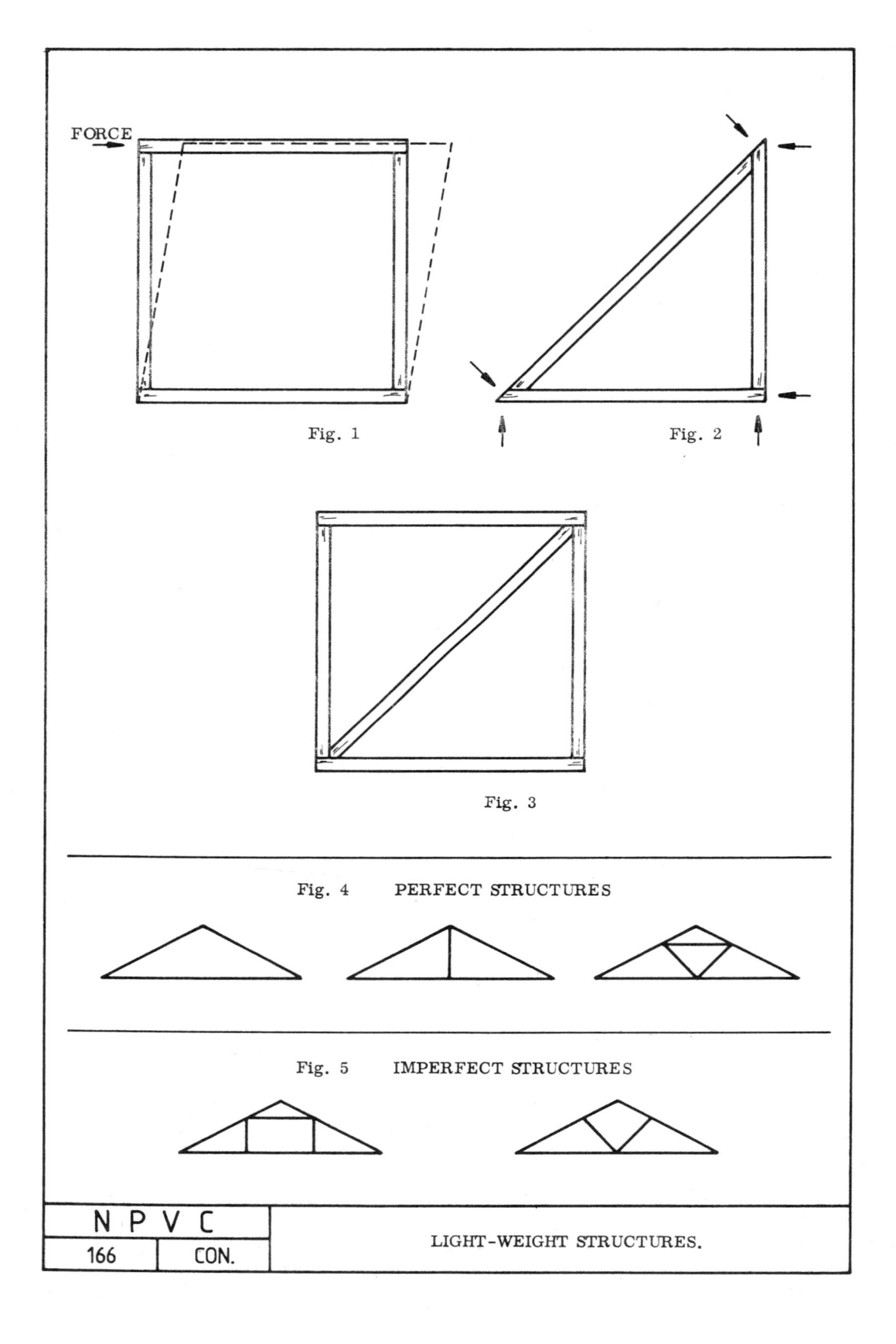

Fig. 1

Fig. 2

Fig. 3

Fig. 4   PERFECT STRUCTURES

Fig. 5   IMPERFECT STRUCTURES

N P V C		
166	CON.	LIGHT-WEIGHT STRUCTURES.

BASIC SHAPE OF THE ROOF CONSTRUCTION

It is essential that the roof construction remains rigid. To ensure this, we need to find a shape for it which cannot be distorted.

- NOTE: The only construction shape that cannot be distorted is a triangle.

The strength and efficiency of the triangular construction can be proven by experiment:

For example, a square shaped construction (Fig. 1) can easily be distorted. The only resistance to the distortion is in the corner joints. This construction is called a "non-perfect" or imperfect structure.

A triangle shaped structure cannot be distorted (Fig. 2). When force is applied to any part of it, the frame remains rigid. This construction is called a "perfect" structure.

As soon as a diagonal brace is introduced to a square shaped frame (Fig. 3), it becomes rigid. The brace changes the square shape into two triangles, making it a perfect structure.

If the brace were to be installed vertically or horizontally instead of across the diagonal, the structure could still be distorted because it would still be non-triangular, and thus an imperfect structure.

There are many examples of triangular constructions in our surroundings. Look at a bicycle frame, scaffold braces or braces for a door frame, for example.

In Fig. 4 are shown some perfect structures.

In Fig. 5 are shown some imperfect structures.

In roof construction, it is essential to make only perfect structures.

TECHNICAL TERMS

- TIE BEAM: This is the horizontal member of the roof structure which ties together the feet of the rafters (Fig. 1, next page).

- RAFTERS: These are the sloping members which give support to the purlins (Fig. 1, next page).

- BRACES: These are the members which strengthen the construction.

- ROOF TRUSS: This is the structure made up of the rafters, tie beam and braces, which forms the main load carrying unit in some kinds of roofs.

- PURLINS: These members lie across the rafters and support the roofing sheets.

- RIDGE: This is the highest point of a roof construction.

- RISE OF THE TRUSS: This is the vertical height of the truss (Fig. 1, a), measured between the highest point of the truss and the soffit of the tie beam.

- SPAN OF THE TRUSS: This is the clear horizontal distance between the internal faces of the rafters (b) at the point where they meet the soffit of the tie beam.

- SPAN OF THE BUILDING: The building span is the clear horizontal distance between the inside faces of the walls which support the roof (c).

- OVERHANG: When the tie beam projects beyond the supporting wall, the projecting part is called the overhang (d). The overhang is measured square to the wall.

- OVERHANGING EAVE: When the rafters project beyond the supporting wall, the projecting part is called an overhanging eave (e). The width of the overhanging eave is measured square to the wall.

- EAVE: This is the lowest part of the overhang or overhanging eave (point x).

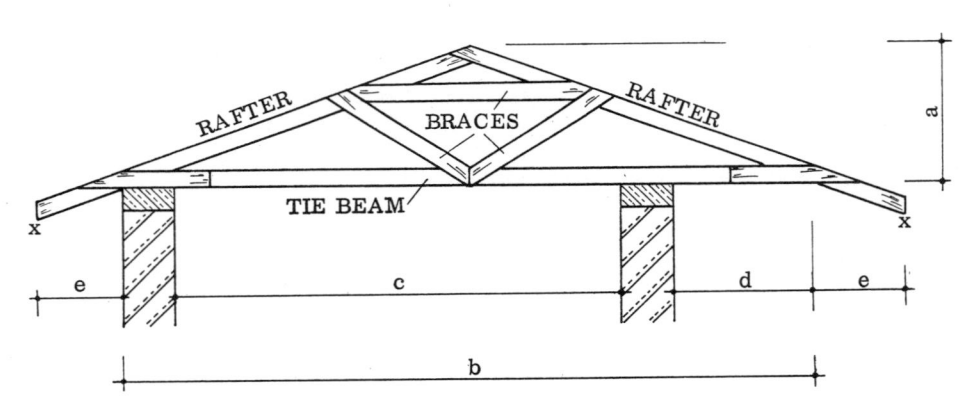

Fig. 1

NOTES:

N P V C	LIGHT-WEIGHT STRUCTURES.
168 CON.	

LENGTHENING JOINTS FOR LIGHT-WEIGHT STRUCTURES

Always select the timbers before starting work on the roof construction. Keep the long straight pieces for the rafters, tie beams, and purlins; and the short pieces for braces.

Sometimes pieces of the right length may not be available. Then it may be necessary to join the members in order to make them longer.

All joints used for lengthening the members of the roof construction will be fish joints with one or two fish plates. These will be discussed in detail in the next section.

Keep in mind that each lengthening joint in the roof construction will weaken the roof. Therefore, try to select the timbers so that you don't need to use many lengthening joints.

WHERE TO PLACE A LENGTHENING JOINT

It is often difficult to judge where to place a lengthening joint. In general, one can say that it should be as close as possible to a support such as a wall or pillar, but we must also keep in mind that the joint is weak, so it cannot be in a place where it is under strain.

For example: A 6 m long tie beam is needed. The only long piece of wood you have is 5,70 m long. If you join that to a 30 cm piece to make 6 m, then the joint will indeed be near a load-carrying wall -- but in relation to the entire structure the joint will be in the wrong place, because most of the weight of the truss is concentrated in that area, where we now have a weak joint.

In this case, a better solution is to join two shorter pieces to get the required length.

Lengthening joints should also not be placed where the braces meet the member.

NOTES:

ONE NAILED FISH PLATE

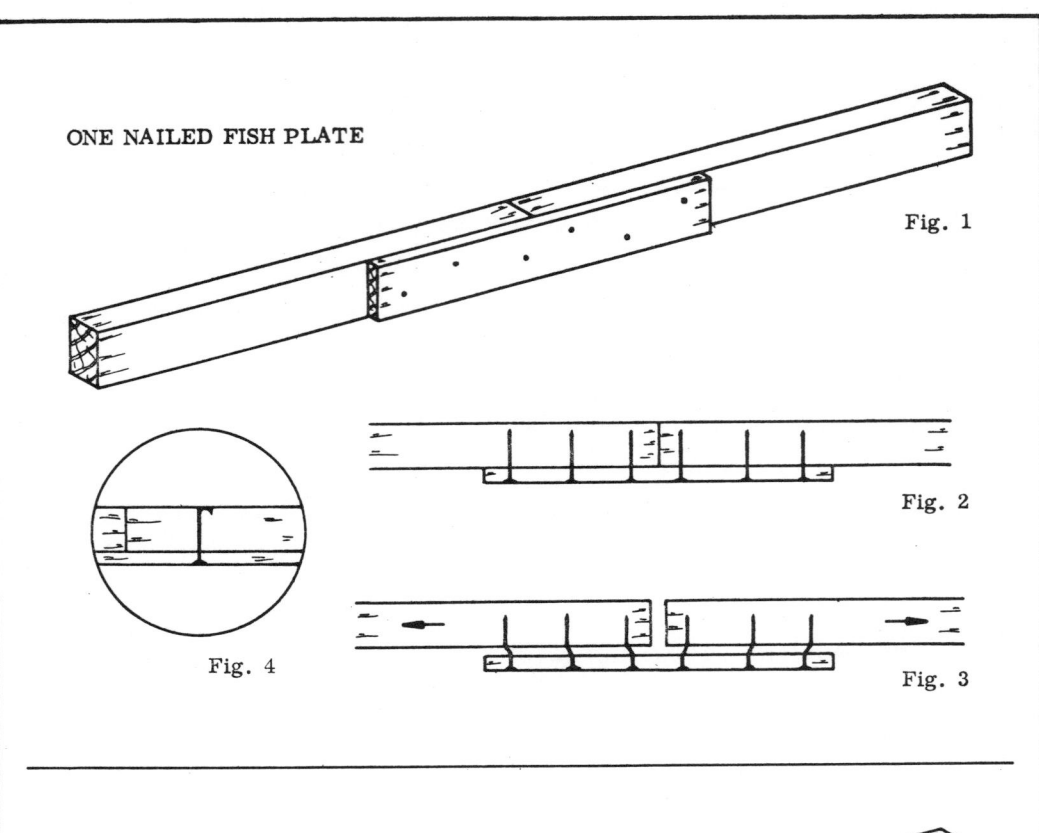

Fig. 1

Fig. 2

Fig. 3

Fig. 4

TWO NAILED FISH PLATES

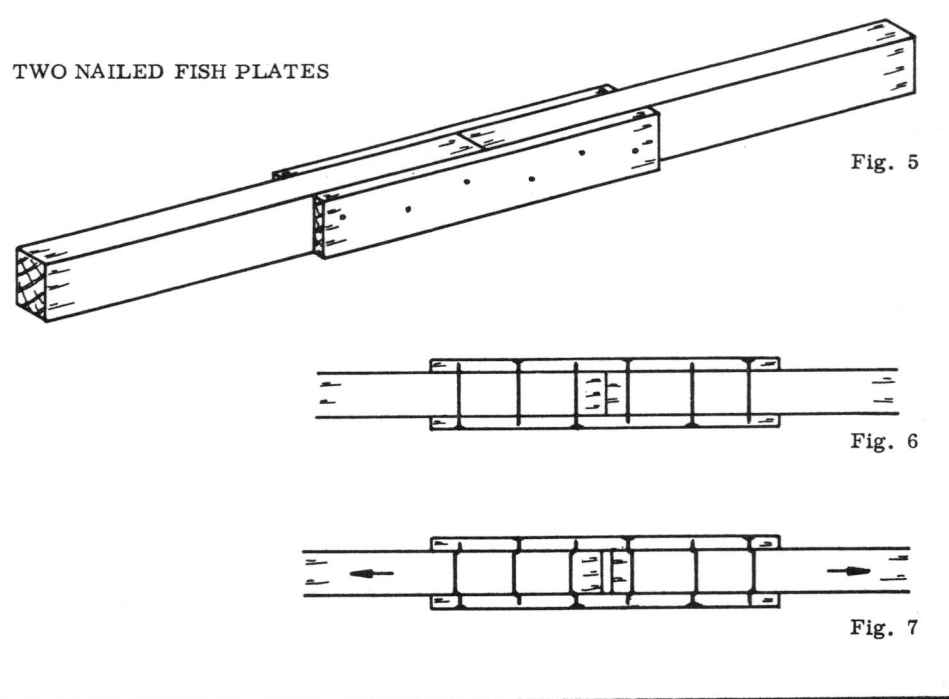

Fig. 5

Fig. 6

Fig. 7

N P V C	
170	CON.

LENGTHENING JOINTS FOR LIGHT-WEIGHT STRUCTURES.

FISH JOINTS

We have said that a fish joint will be sufficiently strong for most parts of the roof construction. However, when we assemble a fish joint, we have to follow certain guidelines to end up with a strong joint.

A fish joint will usually be assembled with nails. The strength of the nailed fish joint depends first on the number of nails used; and second on whether one or two fish plates are used. The length of the fish plates should be at least 5 times the width of the joined members.

- SHEAR STRESS IN A JOINT WITH ONE FISH PLATE: If one nailed fish plate (Fig. 1) is used, the nails tend to shear off (break) at the joint line between the members and the fish plate because of the force exerted as the members try to move apart (Fig. 3). The joint is said to be in single shear (Fig. 2).

  In order to prevent the members from moving apart and shearing the nails, it is necessary to clinch the nails, so that they are anchored firmly (Fig. 4).

- SHEAR STRESS IN A JOINT WITH TWO FISH PLATES: When two nailed fish plates are used (Fig. 5) to join the members together, the nails tend to shear at two points. The joint is said to be in double shear (Fig. 6).

  In this joint there is less tendency for the members to pull apart, since there are two fish plates (Fig. 7). The nails must go through both fish plates.

- NOTE: The strength of a joint in double shear (with two fish plates) is approximately twice that of a similiar joint in single shear (with one fish plate).

- NAILING THE FISH JOINT: We have seen that the strength of a nailed fish joint depends on the number of nails used in it and on the number of fish plates. It is important that the nails are fixed correctly, or the joint will be weakened.

  - The nails should be evenly distributed over the entire fish plate.

  - Nails should always be staggered.

  - Nails should always be blunted.

  - If thick nails are used, they have a tendency to split the wood. Often staggering and blunting cannot prevent this splitting. In this case it is necessary to drill holes for the nails. The holes should have a slightly smaller diameter than the nails.

  There are certain rules for spacing the nails on the fish plate. These are:

  - In the direction of the grain, the distance between the nails should be 10 times the thickness of the nail (Fig. 1, next page).

- The distance to the margin of the fish plate should be: in the direction of the grain, 10 times the thickness of the nail; and across the grain it should be 5 times the thickness of the nail (Fig. 1).
- See the Reference Book, page 207, and Basic Knowledge, pages 92 to 94, for more details about nails and nailing.

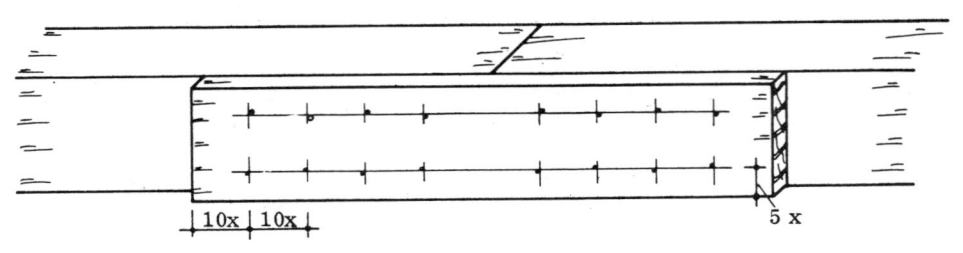

Fig. 1

OTHER LENGTHENING JOINTS

Other joints besides the butt joint with fish plates are sometimes used to lengthen the members of a roof construction. These joints require more work to make and they are not necessarily better than the ordinary fish joint. They are used only if extra strength is required.

These joints are:

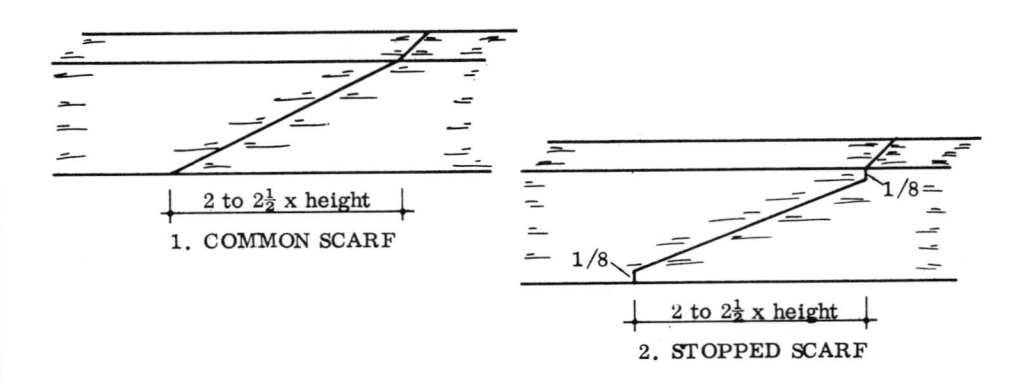

1. COMMON SCARF — 2 to 2½ x height

2. STOPPED SCARF — 2 to 2½ x height, 1/8

LENGTHENING JOINTS FOR LIGHT-WEIGHT STRUCTURES.

3. STOPPED HOOK SCARF — 2 to 2½ x height

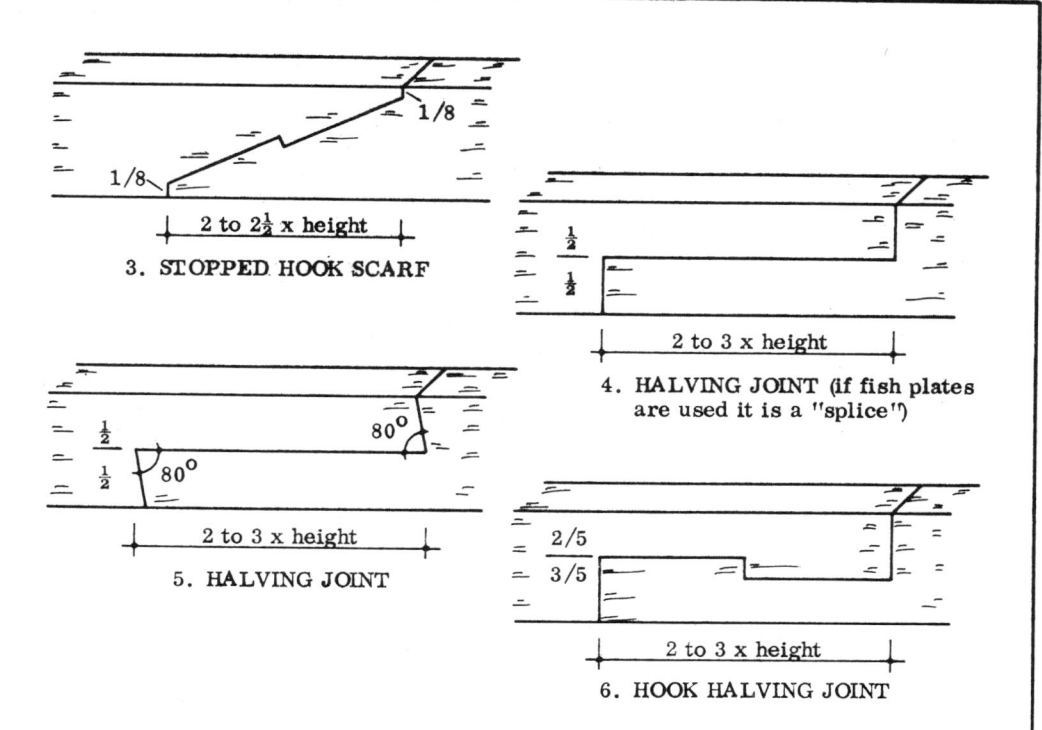

4. HALVING JOINT (if fish plates are used it is a "splice") — 2 to 3 x height

5. HALVING JOINT — 2 to 3 x height

6. HOOK HALVING JOINT — 2 to 3 x height

Joints 1 - 6 are normally used for joining rafters, tie beams or purlins in the length. They can be used with one or two fish plates.

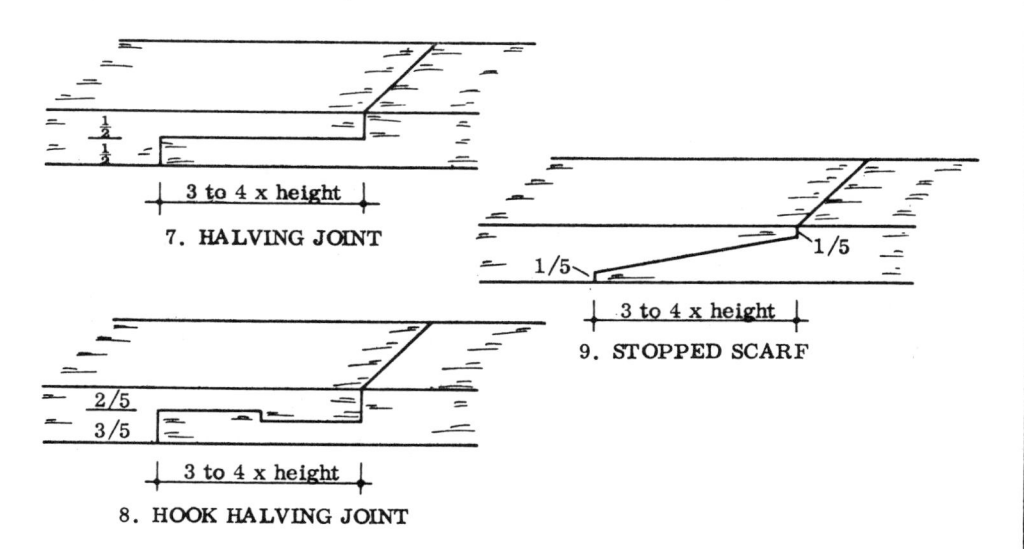

7. HALVING JOINT — 3 to 4 x height

9. STOPPED SCARF — 3 to 4 x height

8. HOOK HALVING JOINT — 3 to 4 x height

Joints 7 - 9 are used mostly for joining wall plates in the length. The corner joint for a wall plate can also be a halving joint.

LENGTHENING JOINTS FOR LIGHT-WEIGHT STRUCTURES. NPVC CON. 173

Fig. 1

Fig. 2

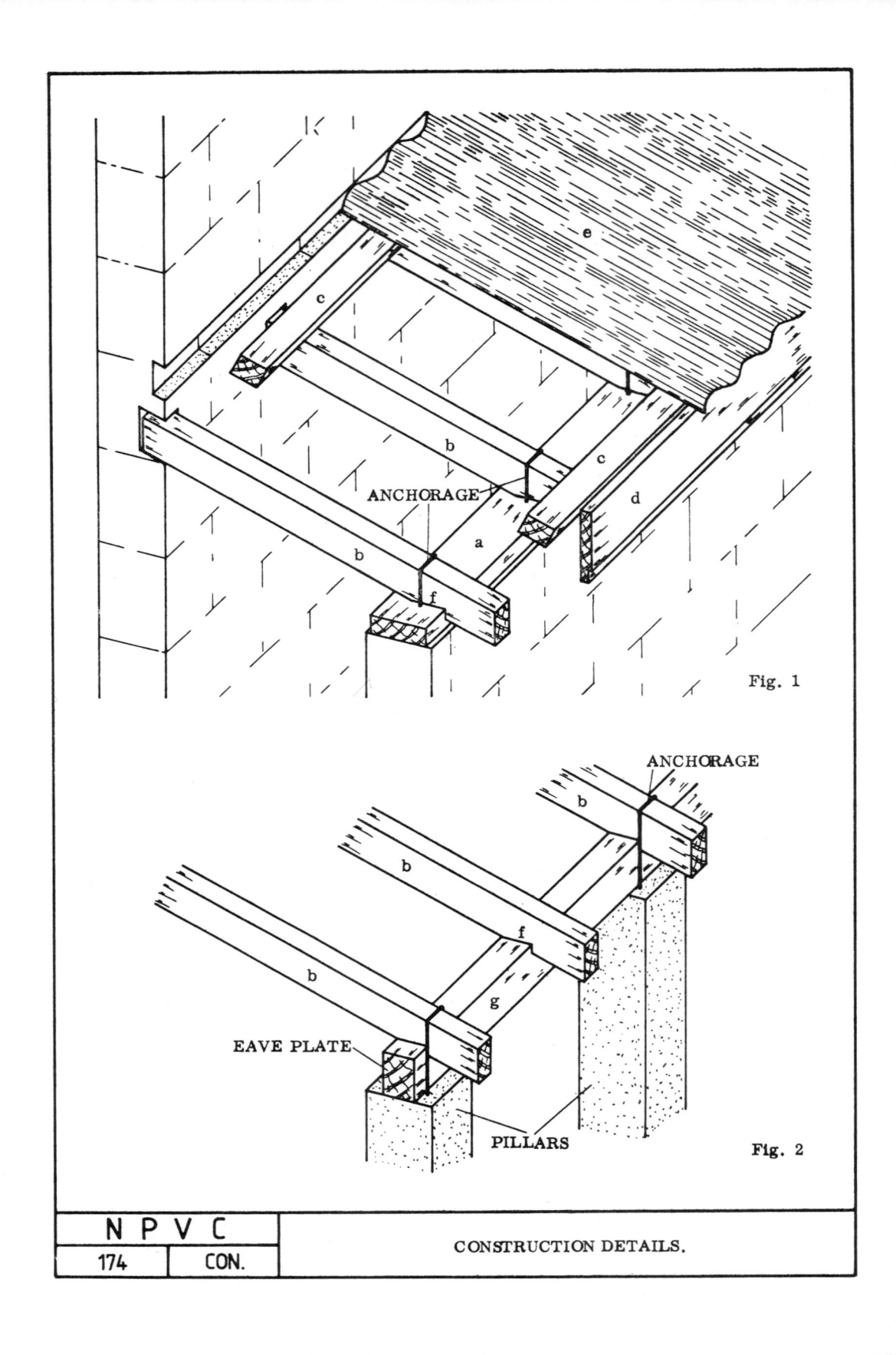

ANCHORAGE

EAVE PLATE

PILLARS

N P V C		CONSTRUCTION DETAILS.
174	CON.	

CONSTRUCTION DETAILS

LEAN-TO ROOF

A lean-to roof is a sloping roof attached to the wall of another building. It is "leaning" against the building. It is usually used for a small store or similiar building which is attached to an existing building. The main members are: (Fig. 1)

- the wall plate (a)   5 x 10 to 15 cm
- the rafters (b)   5 x 7,5 to 10 cm
- the purlins (c)   5 x 7,5 cm
- the fascia board (d)   2,5 x 20 to 30 cm
- the sheet material (e)

The above measurements can be used as a guide in selecting timbers for this kind of roof.

Below are some rules for the construction.

- A good method for fixing rafters onto the main wall is to chisel holes in the wall into which the rafters can be fitted. The holes should be deep enough to ensure a safe rest for the rafters (Fig. 1).

- The lower part of the rafter should rest on a wall plate (a) or a concrete belt. The rafter and wall plate should be firmly anchored through at least 3 courses of blocks.

- The rafters should each be fitted with a "bird's mouth" near the end (f). The depth of the bird's mouth should be 1/5th of the width of the rafter. This is so that the rafters rest securely on the horizontal support.

- The purlins (c) should be well secured to the rafters.

- To prevent water from entering between the wall and the sheeting material, the part of the sheet which attaches to the main wall should fit into a recess in this wall, so that the ends of the sheets can be plastered over later (Fig. 1).

- In an open fronted structure, the roof is supported by pillars on the open side, instead of by a wall. The plate which rests on the pillars and supports the rafters is called the "eave plate" (Fig. 2, g). The size of the eave plate should be about 5 x 7,5 to 10 cm.

NOTES:

LEAN-TO ROOF, continued

- Remember that in an open construction the pillars must not only support the dead load of the roof construction, but also must be strongly anchored in the ground so they cannot be pulled out by a strong wind catching the roof from underneath.
- Note that in an open-fronted construction side walls are to be avoided. If they are required, there must be openings in them to permit wind to escape. A building with an open front should never face in the direction of the prevailing wind (the direction from which the wind usually comes).

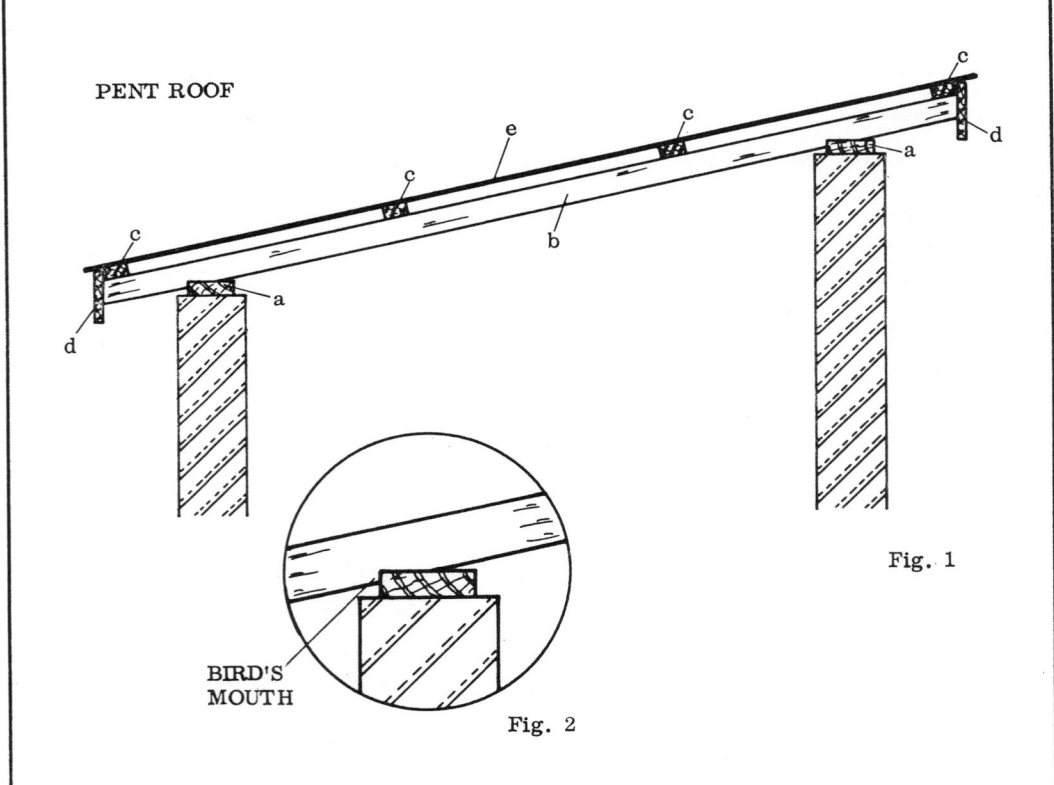

PENT ROOF

Fig. 1

BIRD'S MOUTH

Fig. 2

N P V C	CONSTRUCTION DETAILS.
176	CON.	

PENT ROOFS

A pent roof is a roof which slopes to one side. It differs from a lean-to roof in that it is not attached to the wall of another building, but is supported by its own walls.

In Rural Building, we deal with two types of pent roof: the ordinary pent roof and the enclosed or parapetted pent roof.

- ORDINARY PENT ROOF: In this roof, the rafters and purlins project beyond the outside walls and a fascia board is fixed all around the building. The pitch of this roof will usually be about 15 degrees. The main members are: (Fig. 1)

    - the wall plate (a)   5 x 10 to 15 cm
    - the rafters (b)   5 x 15 cm
    - the purlins (c)   5 x 7,5 cm
    - the fascia (d)   2,5 x 20 to 30 cm
    - the sheet materials (e)

The above timber measurements can be used as a guide in selecting the timbers for an ordinary pent roof; that is, a pent roof with a span of less than 3,5 m. The distance between rafters should be 1 to 2 m, and the distance between the purlins depends on the size of the sheet material.

There are certain rules for constructing this type of roof:

- The rafters should always be fitted with a bird's mouth so that they rest securely on the wall plate (Fig. 2).
- Wall plates, rafters, and purlins should be well anchored (see the section on anchorage which comes later in this book).

The pent roof is often used because it is cheaper to construct than other roofs, since only rafters and purlins are used, and no tie beam, braces, etc. are needed.

NOTES:

CONSTRUCTION DETAILS.   NPVC   CON. 177

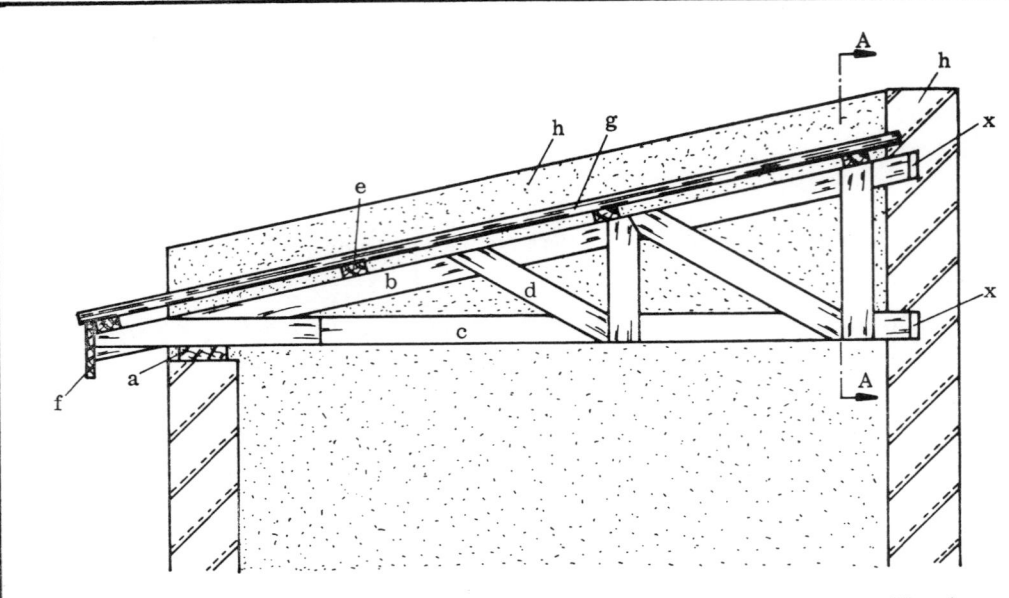

Fig. 1

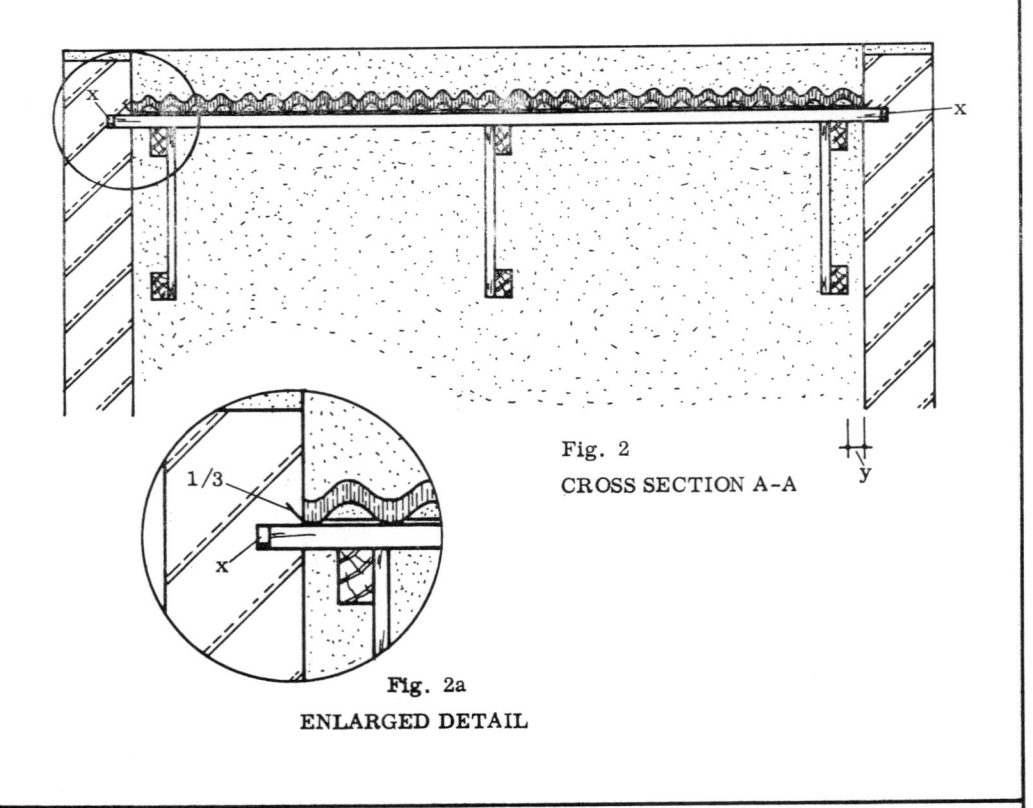

Fig. 2
CROSS SECTION A-A

Fig. 2a
ENLARGED DETAIL

N P V C	CONSTRUCTION DETAILS.
178 CON.	

- ENCLOSED OR PARAPETTED PENT ROOF: In this roof, the higher wall and the two sloping walls enclose and protect three sides of the roof. The parts of the walls which project above roof level are called parapets. Parapets help to reduce suction on the roof and to keep the sheets in place. The pitch of this roof will be about 15 degrees (Fig. 1). The parts of the roof are:

- the wall plate
  or concrete belt (a)
- the rafters (b)   5 x 7,5 cm
- the tie beam (c)   5 x 7,5 cm
- the braces (d)   2,5 x 7,5 cm
- the purlins (e)   5 x 7,5 cm
- the fascia (f)   2,5 x 20 to 30 cm
- the sheet material (g)
- the parapet (h)

The measurements above can be used as a guide for constructing this kind of roof. The distance between the rafters should be 1 to 2 m, and the distance between the purlins will depend on the size of the sheet material.

Keep in mind the following construction pointers:

- A built-up structure is used for this roof (a half truss). We will learn about how to make the truss in a later lesson.

- Before the half truss is erected, all the walls should be built to the level of the tie beam. The truss is then erected and the walling continued.

- There should be an expansion gap at the end of each rafter and tie beam (Fig. 1, x) where they fit into the wall; and at the ends of the purlins (Fig. 2, x). The gap prevents the wall from being cracked when the wooden member expands.

- The sheet material should be fixed so that about 1/3rd of a corrugation on each side of the roof is enclosed in the wall (Fig. 2). To keep water out, this edge should be pointing upwards (Fig. 2a).

- Leave about 10 cm space between the last truss and the wall so that there is enough space for plastering (Fig. 2, y).

NOTES:

CONSTRUCTION DETAILS.

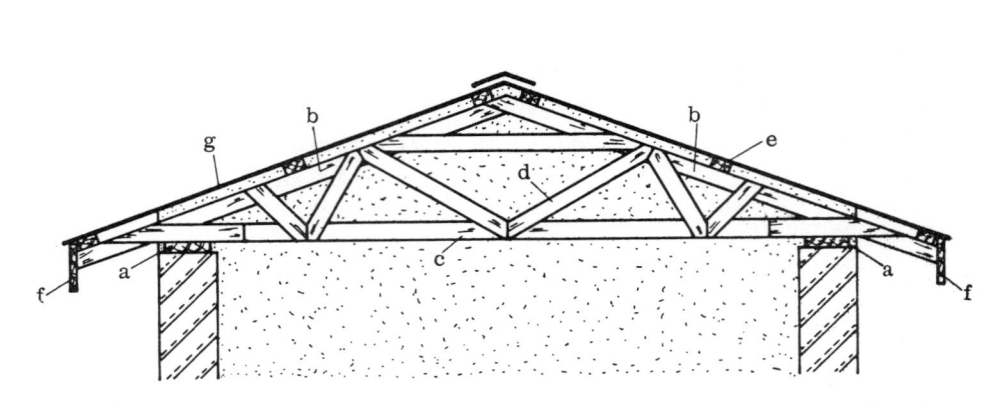

Fig. 1

Fig. 2

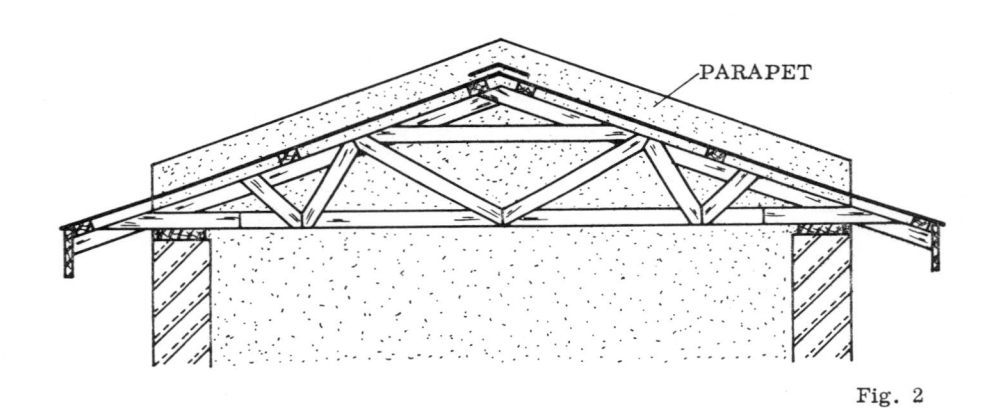

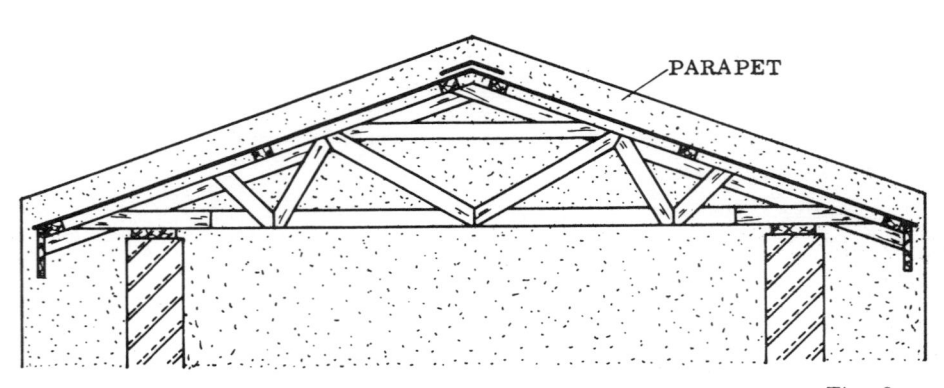

PARAPET

Fig. 3

PARAPET

N P V C	CONSTRUCTION DETAILS.
180 CON.	

GABLE ROOF

This is a roof which slopes down on the two sides of the ridge and has a gable on one or two end walls. The gable is the triangular shaped part of the end wall where it comes up to the sloping edges of the roof.

The advantage of the gable roof over the pent roof is that it can be constructed to permit cross ventilation. It can be used for large or small spans.

In Rural Building, the kind of gable roof we make is constructed with built-up trusses. The main parts and members of the gable roof are (Fig. 1):

- the wall plate or concrete belt (a)
- the rafters (b)   5 x 7,5 cm
- the tie beam (c)   5 x 7,5 cm
- the braces (d)   2,5 x 7,5 cm
- the purlins (e)   5 x 7,5 cm
- the fascia (f)   2,5 x 20 to 30 cm
- the sheet materials (g)

The above measurements should be used as a guide when you make this type of roof. The distance between the trusses will be about 200 cm. The last truss should be close to the wall, except if parapets are used. In that case leave about 10 cm space for mortaring.

The gable ends can be constructed so that the purlins project beyond the gable and the fascia boards are nailed onto them. The end sheets should be nailed so that the last corrugation folds down over the top of the fascia.

Another way of constructing the gable is to put a parapet on top of the gable in the same way as for a parapetted pent roof. Again, the sheets should be enclosed in the parapet wall by about 1/3rd of a corrugation, with the edge pointed up.

The parapet can be on top of the gable wall only (Fig. 2), or the wall can project with the parapet beyond the ends of the sheets (Fig. 3).

NOTES:

CONSTRUCTION DETAILS.

NPVC
CON. 181

DESIGNING BUILT-UP TRUSSES

Built-up trusses are used in gable roofs to support the purlins and the roof covering.

In constructing these trusses, we must remember the principle of "perfect structures" (see Light-Weight Structures), otherwise the roof construction may get distorted.

The following points will show why roof trusses are built up in the way they are.

1. Figure 1 shows that a roof construction of only rafters tends to spread the walls. This is because the roof doesn't have a triangular shape, so it is not a perfect structure. This construction could cause damage to the walls, the roof, and the whole building.

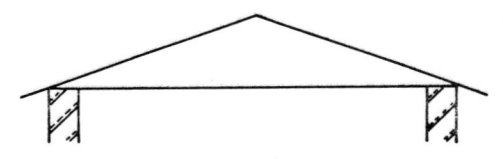

2. As we see in Fig. 2, as soon as a tie beam is introduced we get a triangular shaped, perfect roof structure.

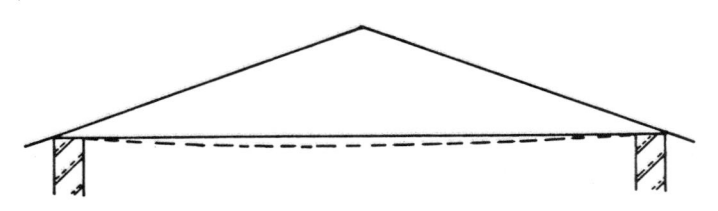

3. As the span of the roof increases the tie beam tends to sag.

CONSTRUCTION DETAILS.

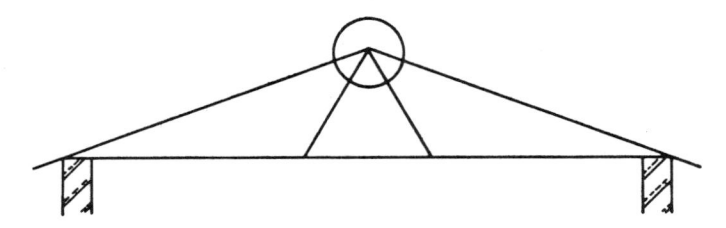

4. To prevent the sagging of the tie beam, braces are introduced. Because there are 4 members (2 rafters and 2 braces) that meet at the ridge, there is a danger of over-nailing the joint, which would weaken the truss. Therefore we need to find another way of fixing the braces while keeping a perfect structure.

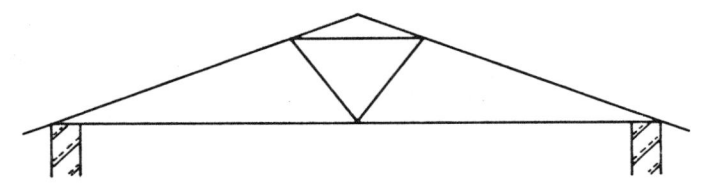

5. This is a possible solution to prevent overnailing.

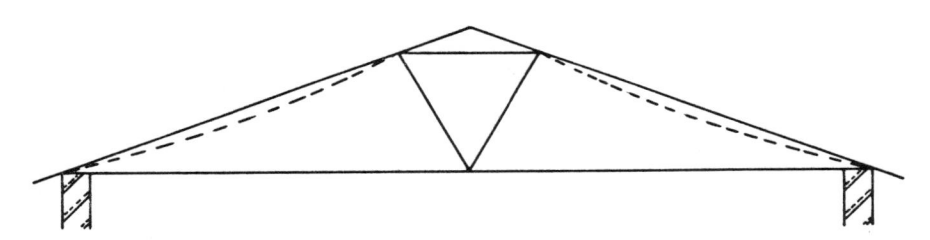

6. As the span increases still further the rafters will also be longer and they will tend to bend.

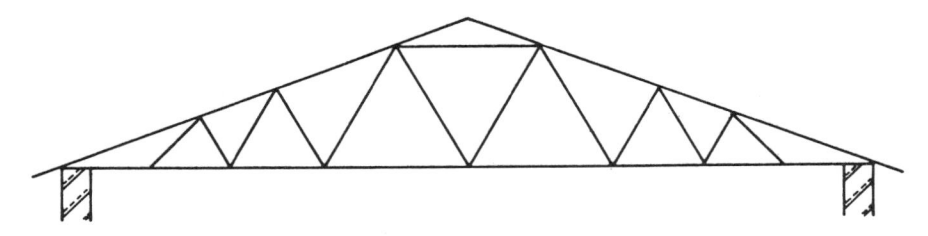

7. To prevent the rafters from bending, more braces are introduced.

CONSTRUCTION DETAILS.

NPVC
CON. 183

Fig. 1

Fig. 2

Fig. 3

Fig. 4

NPVC	CONSTRUCTION DETAILS.
184 CON.	

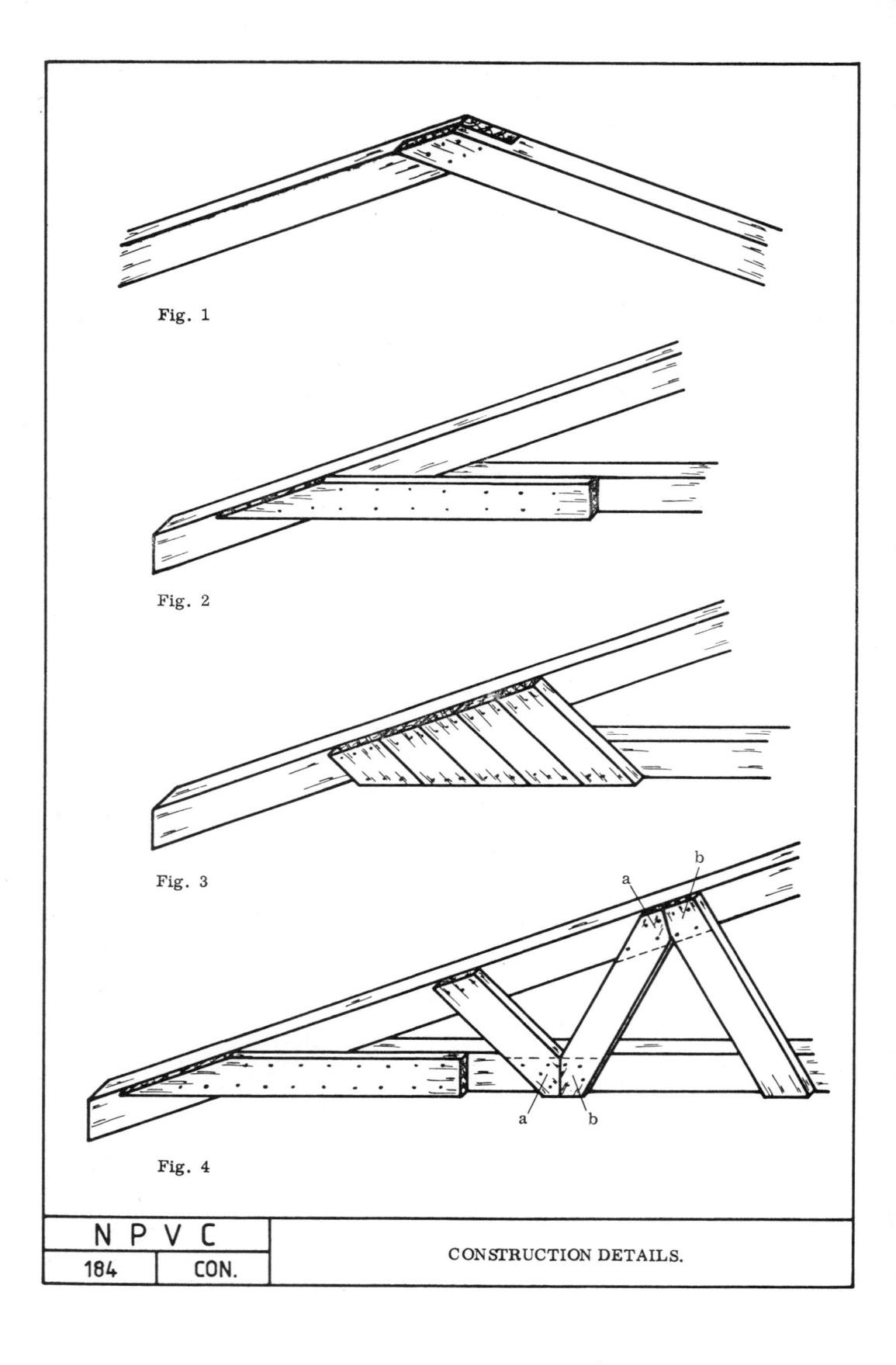

JOINTS FOR THE TRUSS

At the ridge, two rafters must be joined together. This is done with a halving joint (Fig. 1).

The halving joint is assembled with a few long nails. They should be long enough that they can be clenched on the opposite side. Take care not to overnail the joint.

The rafter - tie beam joint can be a fish joint with two nailed fish plates (Fig. 2). After the fish plates are fixed, they must be cut so that they don't project past the edge of the rafter.

If large trusses are made, short braces can be used to strengthen the joint instead of fish plates (Fig. 3).

The braces are added according to the required pattern of the truss. At the place where the two braces meet, they should be cut so that the areas where they are nailed to the truss (areas a and b) are as large as possible and of equal size on both braces (Fig. 4).

This is done by placing them on top of each other, marking and cutting during the assembly of the truss.

All of these reinforcements can be added on either one or both sides of the truss, depending on the strength that is needed for the construction.

NOTES:

CONSTRUCTION DETAILS.

N P V C	CONSTRUCTION DETAILS.
186	CON.

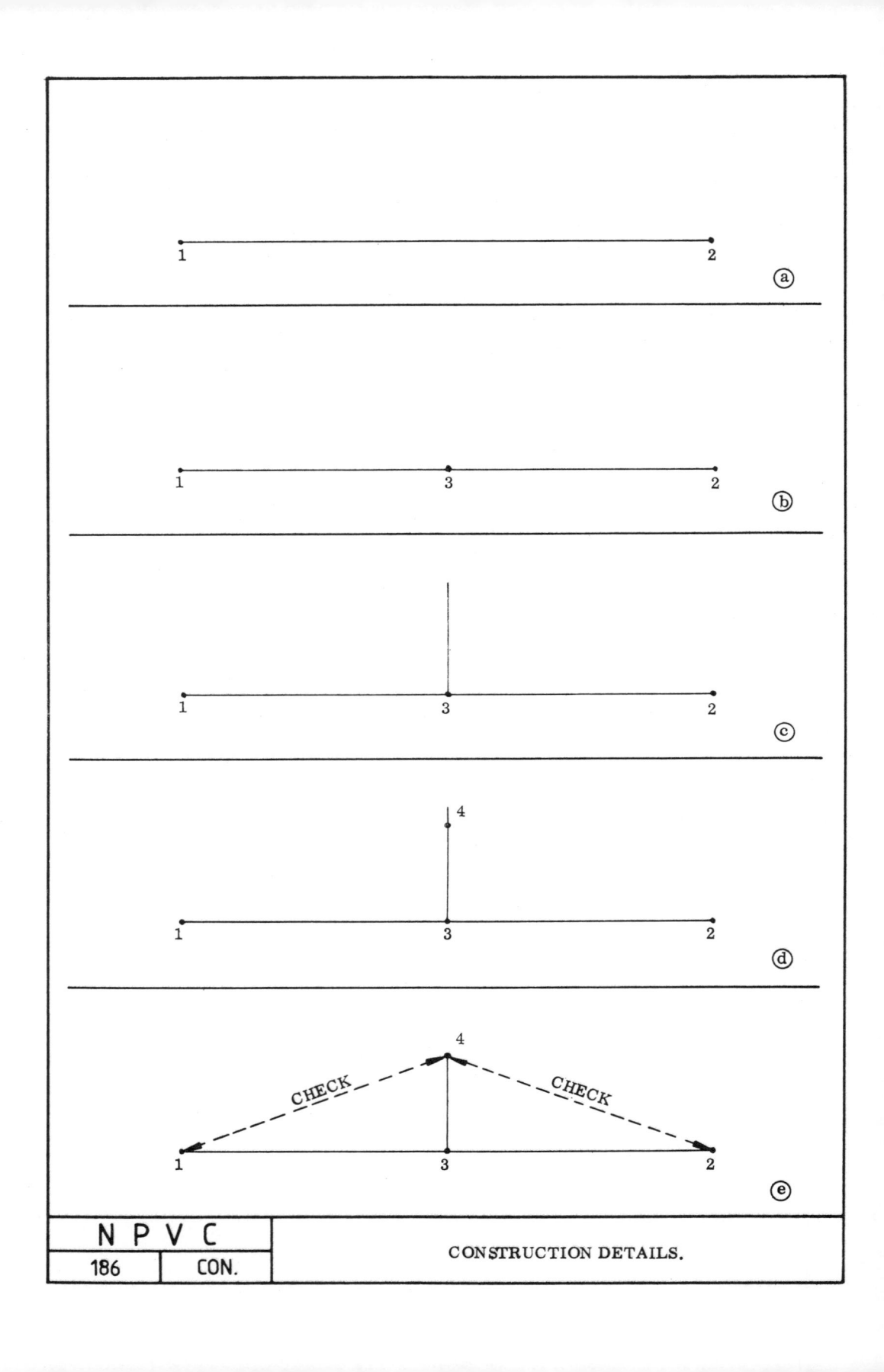

MARKING OUT BUILT-UP TRUSSES

Before you start to mark out the trusses, measure the span of the building at several points and find out the span of the truss. Make sure that the truss is long enough to cross even the widest span of the building.

Do not forget the difference between the span of the building and the span of the truss. You must take into account any verandahs or overhangs in the span of the truss. The outline design will help you to find out the different spans for the building and the truss.

The marking out is done in a cleared, flat area of the building site. After the area is cleared, the shape of the trusses is laid out with iron pins knocked in the ground and with the mason line.

- SEQUENCE OF OPERATIONS:

   a. Place pegs 1 and 2 at a distance equal to the span of the truss, and stretch a line between them.

   b. Divide the distance from point 1 to point 2 in half, and mark point 3.

   c. From point 3, erect a line perpendicular to line 1-2 (use a large square or the 3-4-5 method).

   d. Place peg 4 along the line from point 3, at the distance of the required rise of the truss.

   e. Check that the distances 1 - 4 and 2 - 4 are equal.

   f. Check that the roofing sheets will fit as planned, including a sufficient overlap. Check this by measuring from peg 1 to peg 4, then add the overhang and the projection of the fascia board.

NOTES:

CONSTRUCTION DETAILS.

N P V C
CON. 187

CONSTRUCTION DETAILS.

NPVC 188 CON.

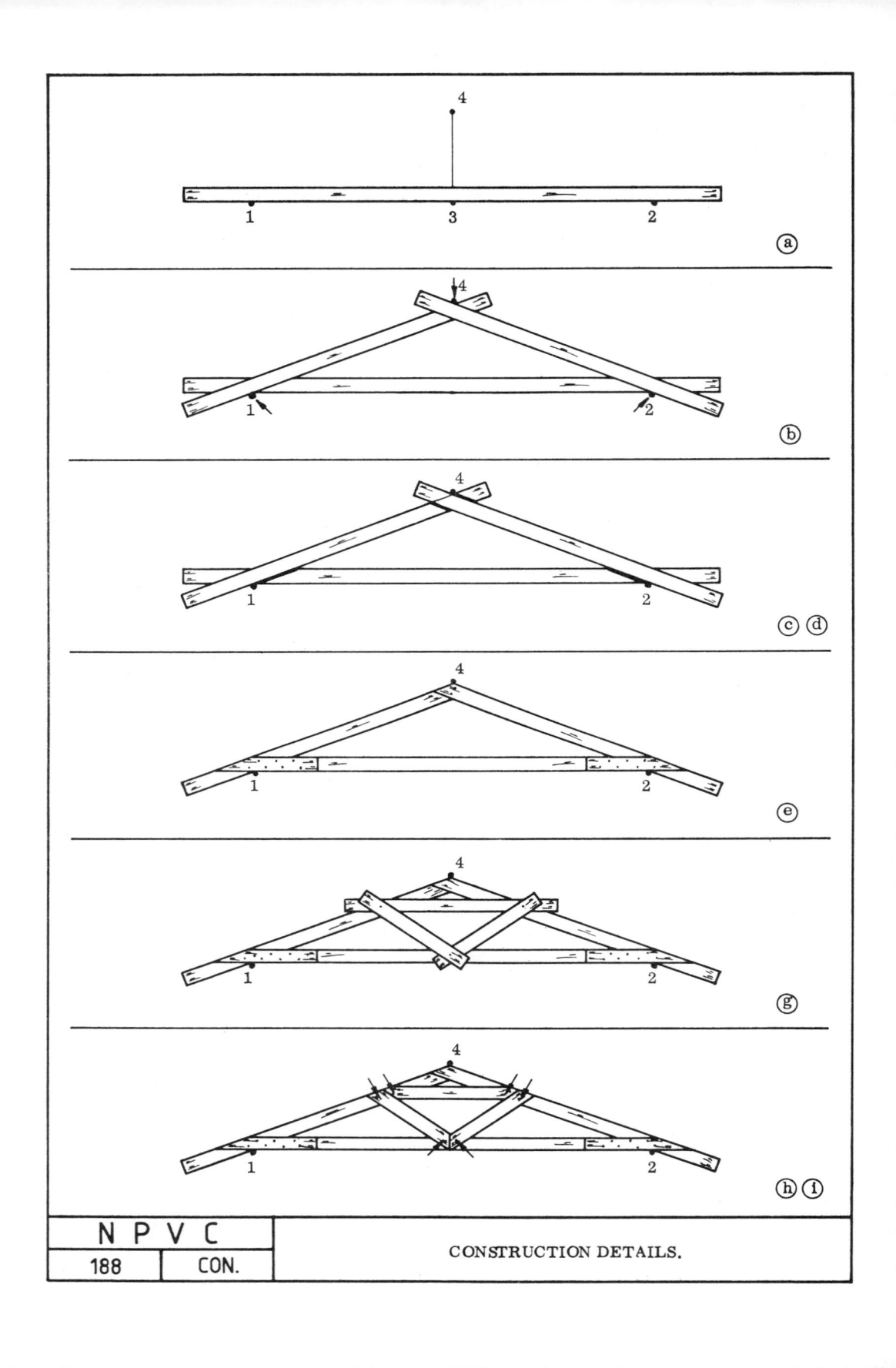

ASSEMBLY OF BUILT-UP TRUSSES

Before you assemble the truss select the timbers, measure them against the setting out, and if necessary join pieces to get members with the required lengths (see Lengthening Joints, pages 169 to 173).

- SEQUENCE OF OPERATIONS:

   a. Lay the tie beam along the line from peg 1 to peg 2.

   b. Place the two rafters between pegs 1 and 4; and between pegs 2 and 4. Note the positions of pegs 1, 2 and 4 with respect to the boards. This is very important. Rafters should be slightly longer than needed. This allowance is cut off after the trusses are erected.

   c. Mark the halving joint on the rafters at the ridge and cut it.

   d. Mark the cutting lines on the tie beam and cut the ends.

   e. Assemble the ridge joint with a few nails; and the rafter to tie beam joint with fish plates.

   f. Mark the positions of the braces according to the plan, on the rafters and tie beam.

   g. Lay the braces one over another on top of the tie beam and rafters. Use a single nail to fix them temporarily.

   h. Cut the braces to the correct shape, keeping in mind that the size of the area where each is nailed to the truss must be the same for all of them (arrows).

   i. Fix the braces with nails.

   j. After the braces are fixed, if necessary a second fish plate can be fixed on the other side of the rafter-tie beam joint. Turn the truss over and fix this plate. Sometimes it is also necessary to fix braces on both sides.

   k. Mark the positions of the purlins on the rafters. This will make work easier later on.

   l. The other trusses can be assembled on top of the finished first truss.

- NOTE: Always use dovetail nailing, throughout the truss.

NOTES:

CONSTRUCTION DETAILS.	N P V C	
	CON.	189

EAVE BEAM

Fig. 1

EAVE BEAM

LINE

Fig. 2

N P V C		CONSTRUCTION DETAILS.
190	CON.	

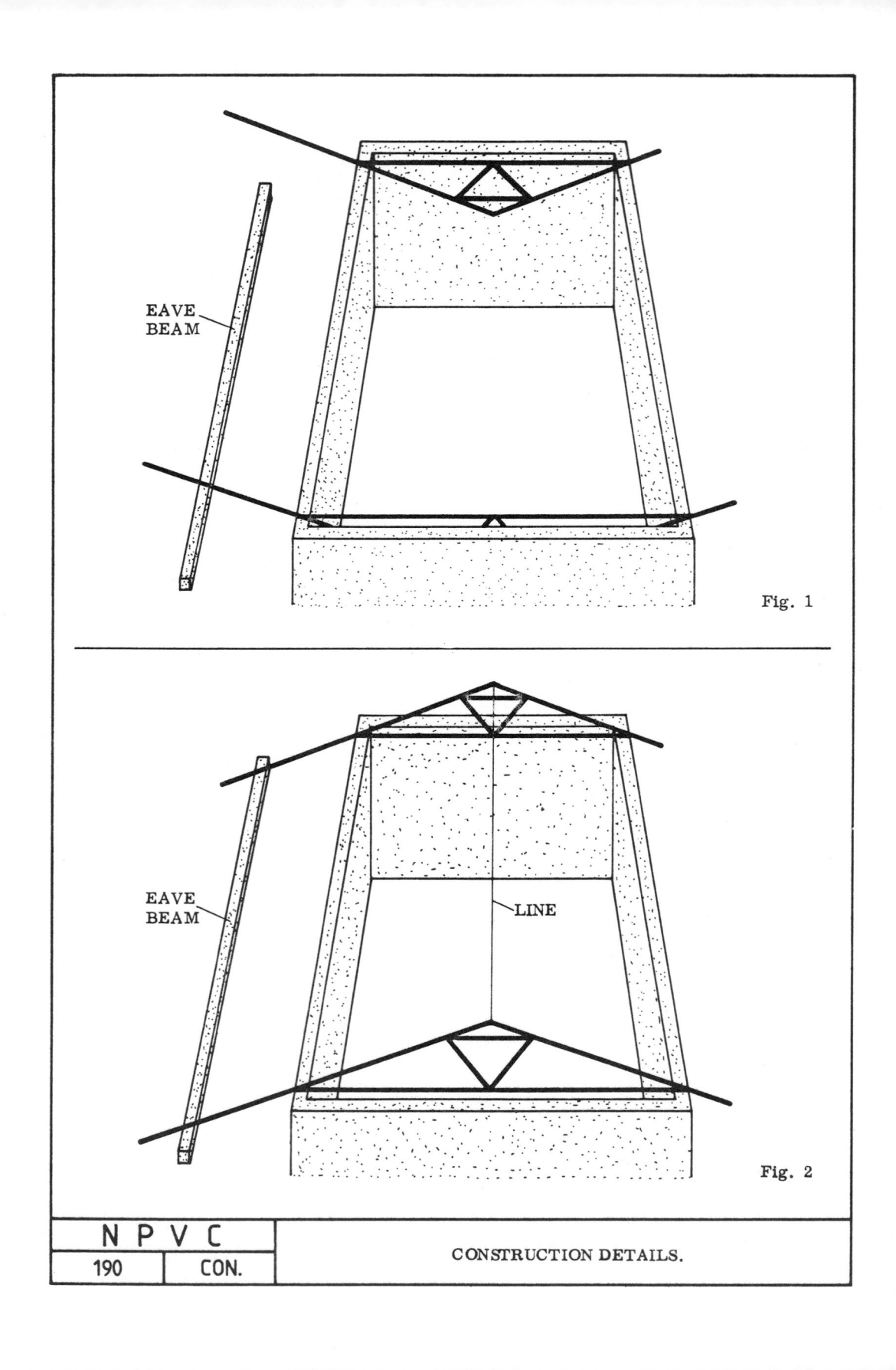

ERECTING A ROOF STRUCTURE FOR A GABLE ROOF

As soon as the trusses are finished, the roof can be erected. If wood preservatives are used they should be applied before the truss is erected. Creosote is often used as a preservative for trusses. If the roof will not be erected immediately, keep the trusses dry and in a shady place.

- SEQUENCE OF OPERATIONS:

   a. Hang a truss upside-down across the span of the building, at each gable end (Fig. 1).

   b. Turn the trusses over and plumb them (Fig. 2).

   c. Secure the trusses at the bottom with anchor rods and brace them temporarily from the rafter to the wall plate.

   d. Fix a line from one ridge to the other (Fig. 2).

   e. Hang the other trusses, one after the other, upside-down between the walls. Turn each one over and plumb it according to the line before hanging the next one. Brace them temporarily.

   f. Nail the purlins according to the measurements of the roof covering (the sheets) and secure them (a section on Anchorage is included later in this book). If the rafters overhang to cover a verandah, make sure that the eave plate is also well secured.

   g. Fix a line at the bottom end of the rafters, on the gable ends of the roof, at the level of the required overhang. Cut off the ends of the rafters according to the line.

   h. Cut off the purlins to the correct length.

   i. Build up the gables.

   j. Nail the fascia boards on the purlins and rafter ends.

Once the roof structure is erected, cover it as soon as possible to keep out the sun and rain; otherwise the wood may start to twist and the joints will loosen. If the trusses rest on a concrete belt, put roofing felt between the concrete belt and the tie beam to help prevent dampness.

NOTES:

CONSTRUCTION DETAILS.	N P V C	
	CON.	191

EAVE PURLIN

REBATE

Fig. 1

REBATE

Fig. 2

FISH PLATE

Fig. 3

Fig. 4

N P V C	
192	CON.

CONSTRUCTION DETAILS.

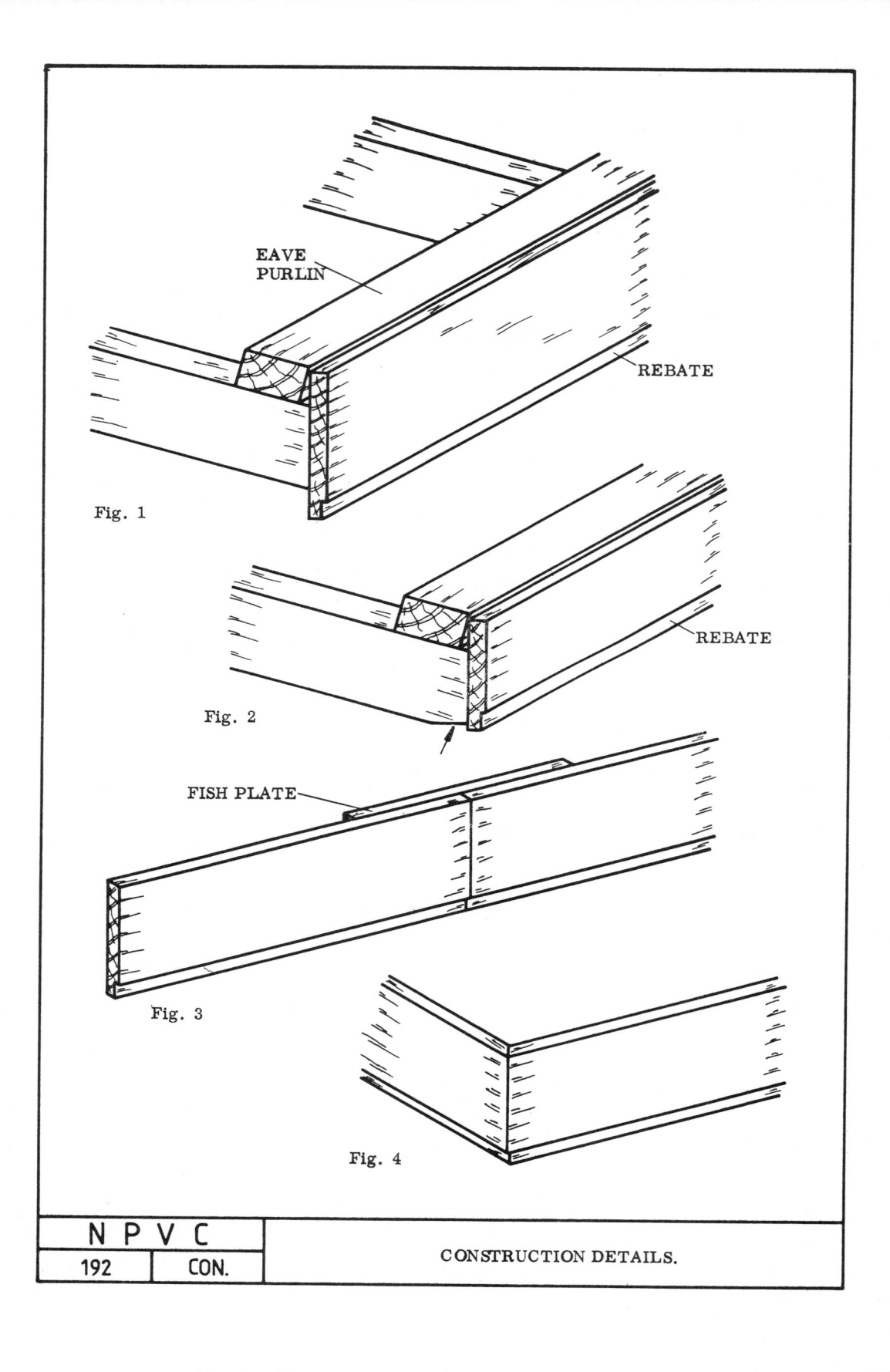

PURLINS

The function of the purlins is to support the roofing sheets and to stiffen the whole roof structure. They are supported by the trusses and sometimes the gable ends. The lengthening joints for purlins can be fish joints.

The size of the purlins should be 5 by 7,5 cm.

The lowest purlin of the construction is called the eave purlin (Fig. 1). The highest one is called the ridge purlin.

To make a wider area for nailing the sheets, lay the purlins flat on the rafters. If larger distances must be bridged, lay the purlins on edge. This reduces the nailing area, but it prevents the purlins from sagging.

The distance between the purlins depends on the size and type of sheets which are used for the roof covering. If thick sheets with the normal 244 cm length are used, then one purlin at each end of the sheet plus one in the middle will be sufficient. For very thin sheets one extra purlin should be added for each sheet.

If the building will require more than 2 sheets to cover the length of the rafter, be careful that the purlins are set at the correct distance apart.

It makes work much easier if you mark the positions of the purlins on the rafters as you assemble the trusses.

FASCIA BOARDS

Fascia boards are often added to the roof construction. They are wide boards which are fixed on the rafter ends and the purlin ends at the gables of the roof construction.

Fascia boards are about 20 to 30 cm wide and 2,5 cm thick. A small rebate can be made into the lower edge of the board to give it a better appearance (Fig. 1).

If the rafter projects below the fascia it can be cut off as shown in Fig. 2.

The eave purlin should be at the correct height so that it does not interfere with the fascia. The fascia boards are fixed before the sheets are laid.

A butt joint is sufficiently strong to use for lengthening the fascia boards, if a fish plate is nailed behind it for strength (Fig. 3). The corner joints should all be mitre joints so that the end grain of the boards is not exposed (Fig. 4).

Before the boards are fixed they should be painted, since it will be difficult to reach some parts later, especially the top edge. Pay special attention to the places where they are joined together. The end grain should be soaked with paint to prevent water from entering there.

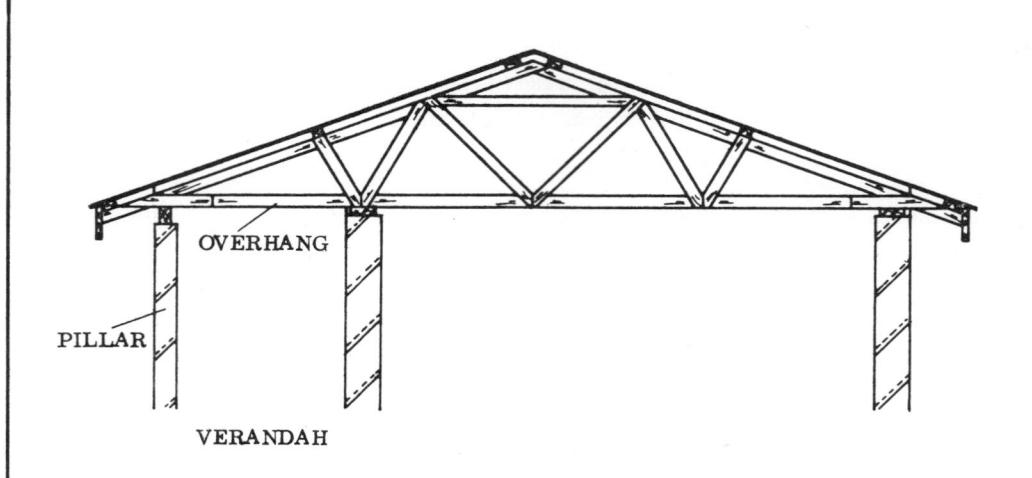

Fig. 1
EXTENDING THE TIE BEAMS

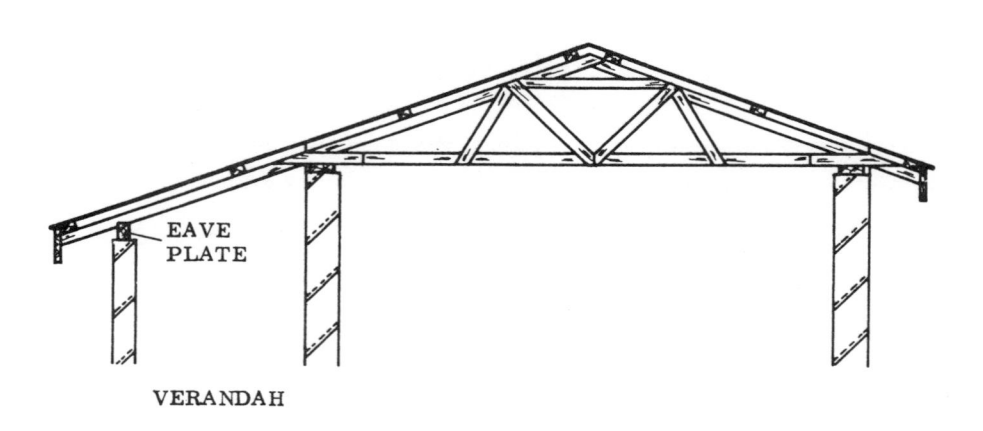

(labels: OVERHANG, PILLAR, VERANDAH)

Fig. 2
EXTENDING THE RAFTERS

(labels: EAVE PLATE, VERANDAH)

N P V C		CONSTRUCTION DETAILS.
194	CON.	

VERANDAHS

When a building with a gable or pent roof has a verandah, there are two possible methods for constructing a roof to cover it:

- By extending the tie beams to cover the verandah (Fig. 1)

- By extending the rafters to cover the verandah (Fig. 2).

- REMEMBER: Open verandahs should not face the direction of the strongest winds.

- EXTENDING THE TIE BEAMS: The part of the tie beam that covers the verandah is called the overhang (Fig. 1). For a small overhang, supporting pillars are not necessary, but it should be braced very well. If the overhang is wide, then pillars are needed to support it (Fig. 1).

If the building has a gable roof, an ordinary truss construction can be used, with the span of the truss including the width of the verandah.

Since the parapetted pent roof has a half truss construction with a tie beam, this construction may also be used with it.

- EXTENDING THE RAFTERS: The overhanging rafters cover the verandah. Note that the building must be high enough so that the lower edge of the verandah is not inconveniently low (Fig. 2). This construction can be used with either an ordinary pent roof or a parapetted pent roof as well as with a gable roof.

The feet of the rafters are supported by an eave plate or a concrete eave beam which rests on pillars. The joint between the rafters and the eave plate should be a bird's mouth to ensure proper support for the rafters. The depth of the bird's mouth should be about 1/5th of the width of the rafter.

The eave plate is normally set edgewise on the pillars. In most cases 5 x 10 Odum is large enough for the eave plate. Take care that it is well anchored on the pillars.

There are certain lengthening joints which are often used for joining eave plates. These are described in the chapter on "Lengthening Joints".

The pillars themselves have to be well anchored in the ground, since they not only have to carry the weight of the structure but also have to hold the roof down in heavy storms (see Lean-to Roofs).

At times the builder may choose to construct the verandah roof using a combination of projecting rafters and projecting tie beams.

CONSTRUCTION DETAILS.

ANCHORAGE ROD

CONCRETE BELT OR WALL PLATE

GROOVE

Fig. 1

IF THERE IS A CONCRETE LINTEL HERE, ANCHOR THE ROD IN THE LINTEL

RAFTER

ANCHORAGE ROD

WALL PLATE OR CONCRETE BELT

Fig. 2

RAFTER

ANCHORAGE ROD

WALL PLATE OR CONCRETE BELT

Fig. 3

HOLE

N P V C
196 CON.

CONSTRUCTION DETAILS.

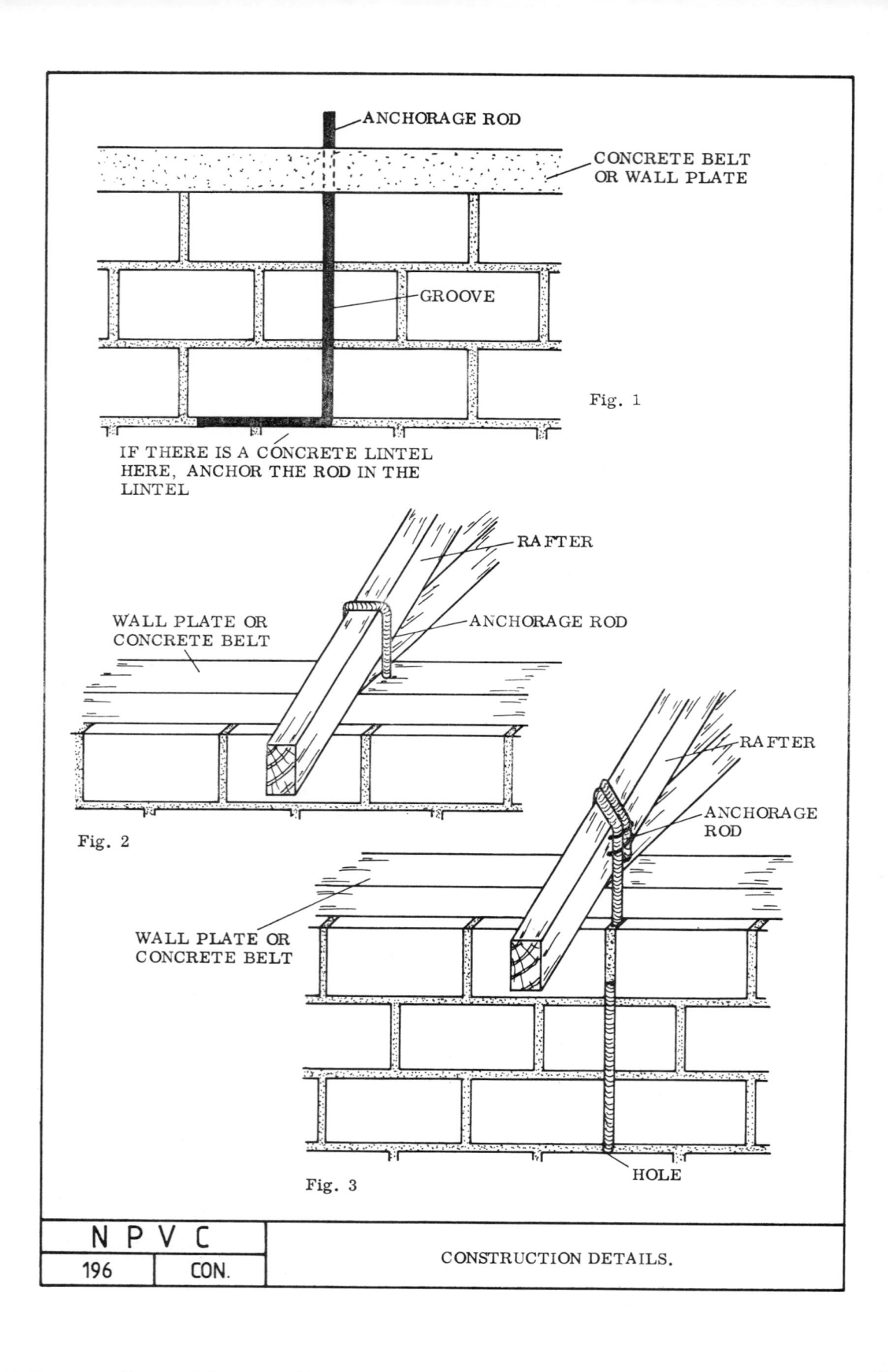

ANCHORAGE

Whatever the form of the roof construction, it must be tied securely to its supporting walls or pillars. If this is neglected, the roof can easily be damaged or even torn off in strong winds.

The following are some points which are important in anchoring a roof.

- ANCHORAGE RODS: The roof is secured to the walls by these iron rods. They can be fixed in the wall during the wall construction if the positions of the trusses have been decided, or else they can be fixed in the finished wall.

  - In the first method, the rod is inserted in the bed joint as the third course below the top of the wall is laid. The rod is bent at a right angle so it can fit into the cross joint (Fig. 1); and it is positioned just inside the face of the wall.

    When the next course is laid, the block above is chiselled out to make a shallow groove (approximately 1 cm deep) for the rod (Fig. 1).

    The rod fits into the next cross joint and is bent gradually inwards toward the top centre of the wall, coming out where it can pass later through the wallplate or concrete belt (Figs. 1 & 2).

    When the truss is erected, the rod is bent around the rafter and tie beam both, and secured with nails (Fig. 2).

  - With the second method, the rod is inserted through a hole chiselled in the bed joint of the third course from the top, after the wall is complete. The ends of the rod are bent up and over the tie beam and rafter, overlap each other, and are secured on both sides with nails (Fig. 3).

  The second method is chiefly used when the roof of an existing building has to be secured. If you are constructing a whole new building, it is best to plan ahead a bit and use the first method, which is both easier and stronger.

- ANCHOR BEAM: Once the wall is completed, the top of the wall may be covered with a strip of reinforced concrete all around the outside walls of the building. This is the anchor beam, also known as the ring beam or concrete belt. The roof anchorage passes through this belt. The anchor beam performs three functions:

  - It keeps the outside walls of the house together, thus strengthening them.

  - It provides a firm base for the trusses and distributes their weight over the softer landcrete blocks below.

  - It anchors the trusses and adds weight to help keep the roof construction from being lifted up in heavy winds.

CONSTRUCTION DETAILS.	N P V C	
	CON.	197

Fig. 1

Fig. 2

Fig. 3

RAFTER
PURLIN
RAFTER
PURLIN
RAFTER
PURLIN
RAFTER
WRONG !
WRONG !

N P V C	
198	CON.

CONSTRUCTION DETAILS.

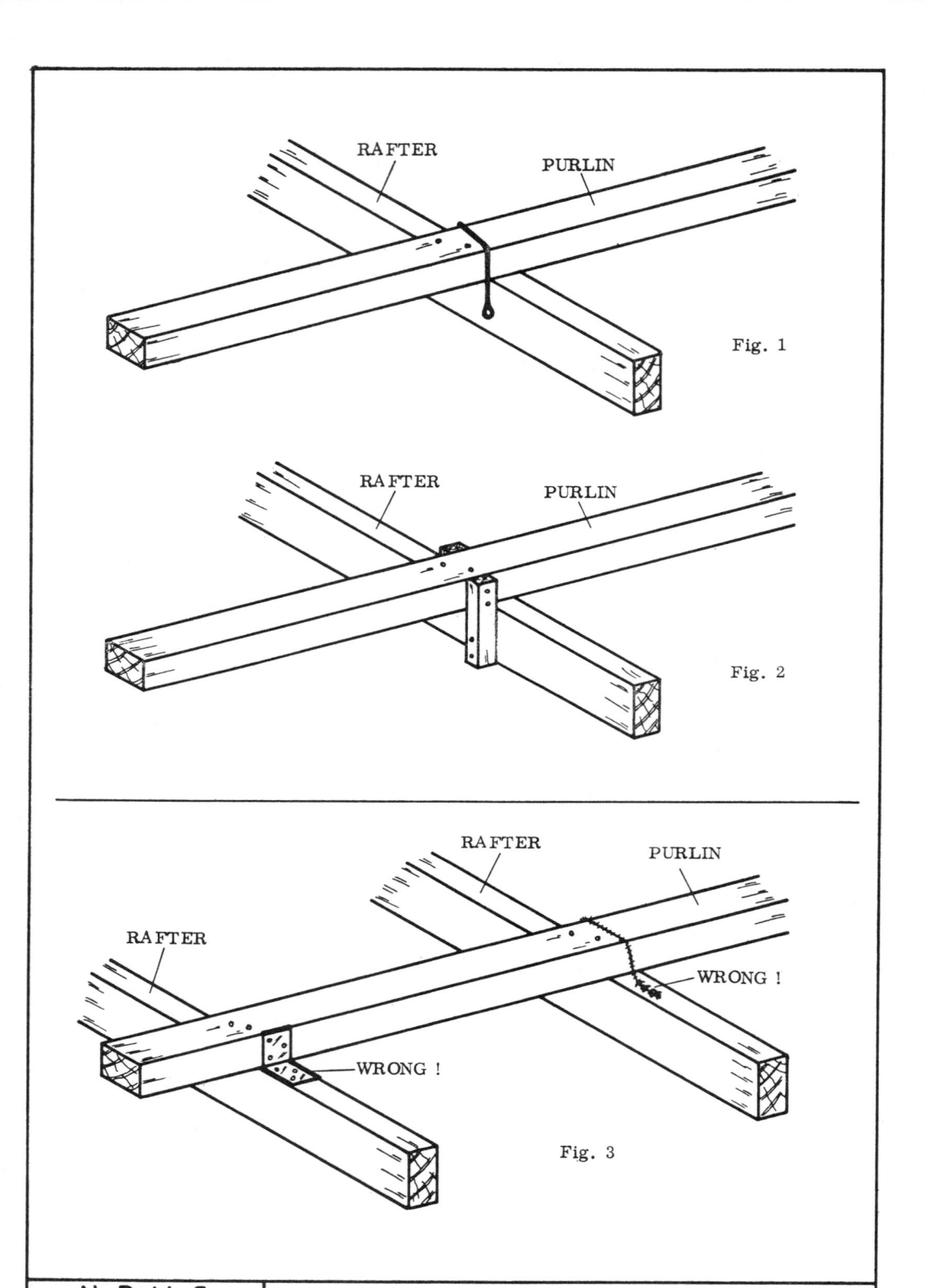

- WALL PLATE: For smaller spans the Rural Builder can install a wooden wall plate instead of an anchor beam. The roof anchorage passes through the wall plate and through at least three courses of blocks.

    There are several joints that may be used to connect pieces lengthwise to make the wall plate (see Lengthening Joints, page 173). The wall plate should be about 5 cm thick by 15 cm wide.

- TRUSSES: The trusses should be evenly spaced across the roof, with one at each gable end and the rest about 2 m apart from each other. Every truss should be anchored to both the walls that it rests upon. The anchorage rod is bent around both the tie beam and the rafter, and secured with nails (Fig. 2, previous page).

- PURLINS: Tie all the purlins to every rafter, preferably with iron rods. These are shaped as shown in Fig. 1, by hammering the rod flat on both ends and drilling holes in the ends, then bending the rod to a U-shape.

    If these fixings can't be made, it is possible to make wooden fixings (Fig. 2). Two wooden pieces are used, one on either side of the purlin.

    Barbed wire can also be used as an alternative to the iron rods. Wrap it firmly around and nail it securely.

    Whatever fixing you use, take care to fix the nails so they are not pulled out by the force of the wind on the roof. Fig. 3 shows two badly nailed fixings.

NOTES:

WRONG !
WATER CAN ENTER

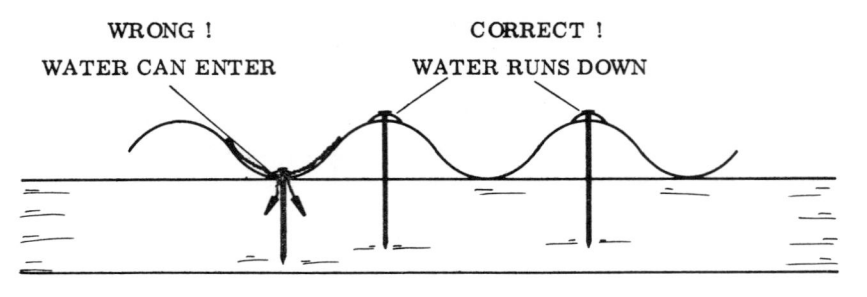

CORRECT !
WATER RUNS DOWN

Fig. 1

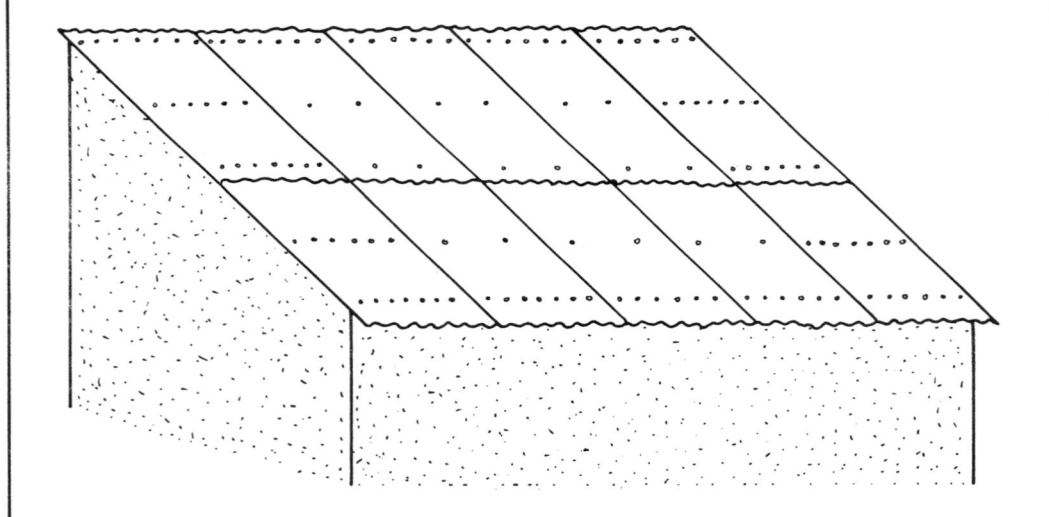

Fig. 2

NOTES:

| N P V C | ROOF COVERING. |
| 200 CON. | |

ROOF COVERING

See the Reference Book, pages 207, 230 and 231 for descriptions of the sheet materials and roofing accessories; also see pages 239 and 240, Tables of Figures.

ALIGNMENT OF THE SHEETS

If possible, always start laying the sheets from one end of the roof so that the free ends of the sheets face away from the direction of the wind. This reduces the danger of the sheets being blown away as they are being installed.

Start laying from one end of the building to the other. As each new sheet is laid, lift the edge of the previous one so that it overlaps the new sheet by 2 corrugations. Each sheet is thus held in position by the one previously fixed, so they are more easily aligned in the correct position.

Exact sidelaps (2 corrugations) and endlaps (15 cm) are essential to make the roof waterproof.

- NAILING: When you nail corrugated roofing sheets to purlins, always nail through the top of the corrugation. This is so that rain will tend to run away from the nail (Fig. 1).

    Each nail should be driven in until the roofing felt just touches the sheet. If it is driven further, the nail will flatten the corrugation and distort the sheet. This can cause the roof to leak.

    The sheets should be nailed to all the purlins. Nail every second corrugation in the sheets along the eave purlin and along the ridge purlin; and also on the end sheets at the gables. Over the rest of the roof, nail at every third corrugation over the purlins (Fig. 2).

- HOW TO LAY THE SHEETS: Before the sheets can be laid, the gables must be built up to the correct height. Make sure that all the nail heads of the wooden construction are punched well below the surface.

    Fix a line on the eave to indicate the desired projection of the sheets. Note that the sheets should project about 2 cm over the fascia board in order to provide a drip overhang.

    It helps to station a helper at eave level to observe if the general line of the sheets is straight and that they project uniformly. Another helper can check that the sheets are straight along the corrugations from the eave to the ridge.

Fig. 1

Fig. 2

RIDGE

EAST GABLE END

EAVE

NPVC	ROOF COVERING.
202 CON.	

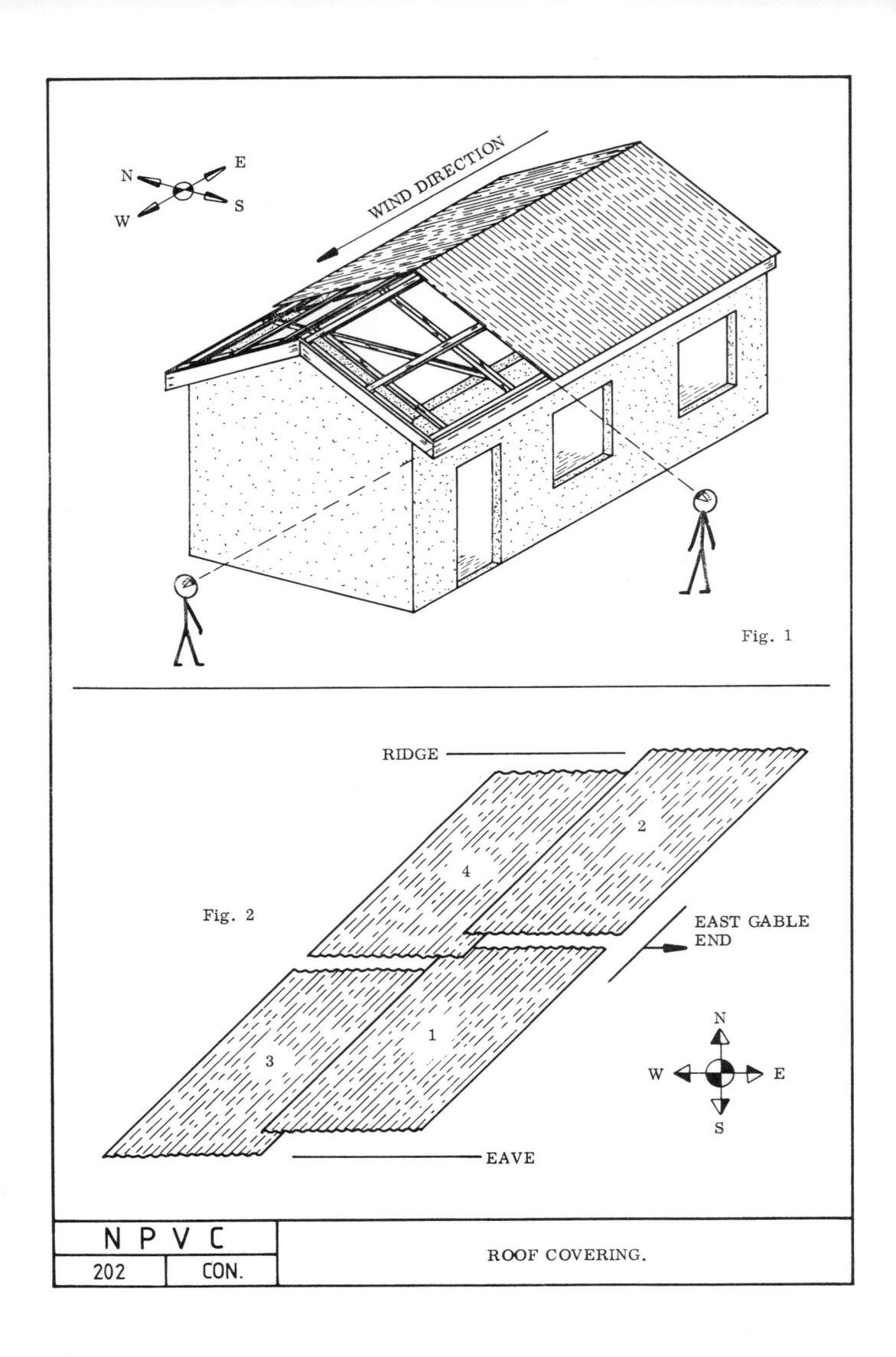

The helpers can also check from both sides of the roof to see if the corrugations match at the ridge line (Fig. 1).

For nailing the sheets it is best to have one man at each purlin.

- SEQUENCE OF OPERATIONS: In order to join the overlaps correctly, follow this sequence, starting from the end of the roof which is in the direction where the strongest winds are coming from (here it is the east) (Fig. 1).

   a. Place sheets 1 and 2 loosely along the gable end, with sheet 2 overlapping sheet 1 by 15 cm in the length (Fig. 2).

   b. Align them with the help of the two observers on the ground.

   c. Secure the sheets with one nail in the middle. One man should do the securing, or the sheets might shift because of the shocks of the hammer blows.

   d. Have the observers check that the sheets are still straight in both directions.

   e. Nail the sheets home with all of the required nails except the nails at the edges where the next sheet will be fixed under.

   f. Place sheet 3, lifting up the edge of sheet 1 until it overlaps sheet 3 by two corrugations (Fig. 2).

   g. Place sheet 4, lifting sheet 2 until it overlaps sheet 4 by two corrugations. Sheet 4 should overlap sheet 3 in the length by 15 cm (Fig. 2).

   h. Align sheets 3 and 4 with the help of the two observers on the ground.

   i. Secure the sheets with one nail each at the centre.

   j. Recheck the straightness in two directions.

   k. Nail the sheets home with all the required nails except the ones at the edge where the next sheet fits.

   l. Continue with this sequence until the whole roof is covered.

   m. Proceed to build the parapets if they are required.

If no parapets are made, the last corrugation at the gable ends can be lapped over the fascia board, turned down and nailed onto the face of the fascia. Place the nails at intervals of about 10 cm.

- NOTE: In northern Ghana, the strongest winds are from the east, so you should start laying the sheets from the east end of the building. In areas where the wind comes from another direction lay the sheets accordingly.

ROOF COVERING.

NPVC
CON. 203

PURLIN

Fig. 1

STRAIGHT-EDGED BOARDS

WORK BENCH

SHEET

Fig. 2

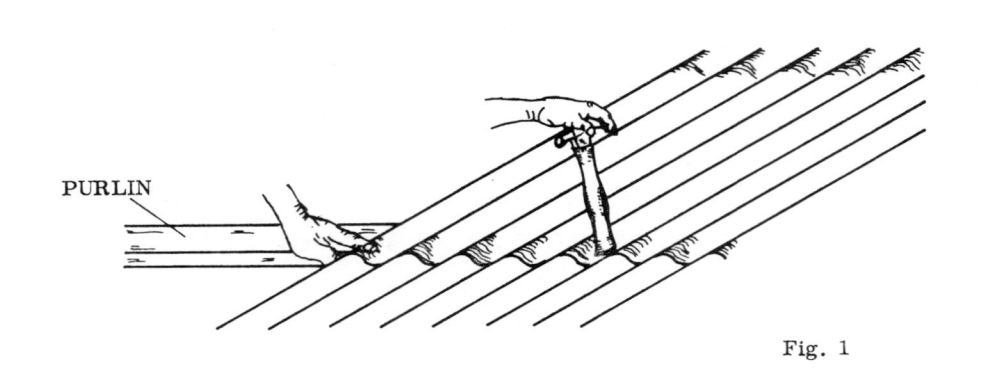

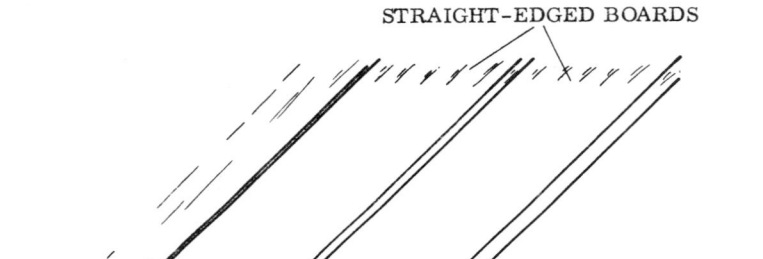

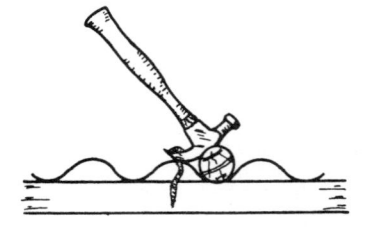

FILED TO A
CUTTING EDGE   Fig. 3

Fig. 4

NPVC
204 CON.

ROOF COVERING.

SIDELAPS

Since corrugations often get distorted or flattened, it can sometimes be difficult to get an exact and watertight sidelap between the sheets.

To straighten the sheet and get a good sidelap, fix all the required nails in the middle corrugation after first securing the sheet with one nail and checking the straightness. Then press the sheet to the correct position and nail the rest of the sheet (Fig. 1).

Do not try to get a sheet in the correct position by nailing the corners where the 4 sheets meet. This will distort the sheet even more.

RIDGE CAPS

If no ready-made ridge caps are available, they can be made on the site with aluminium or galvanized iron sheets. With the help of two straight-edged boards, the sheets are bent to the required angle on the work bench (Fig. 2).

It is better to make the ridge cap so that the bend goes across the corrugations, rather than along them. This ensures a tighter fit, but the corrugations on both sides of the roof have to match exactly (see Reference Book, pages 231 & 239).

HELPFUL HINTS FOR ROOF COVERING

- Put all nails into the washers with the roofing felt before you start laying sheets, to prevent delays.

- The roofing felt can be cut with an iron pipe that has been filed to a cutting edge at one end (Fig. 3).

- Use soap to make driving nails easier in hard wood.

- There should be people on the ground ready to hand up the sheets.

- Before you put a sheet in place, make a mark on the previous sheet to show where the purlin is for nailing.

- Always use a line at the eave, with supports if necessary to keep it straight.

- Have a long, slim and pointed punch ready. It can be made locally from an iron rod. This punch is used to pierce holes through in places where the nail must pass through 4 sheets. Without piercing the hole for the nail first, there is a danger of flattening the corrugation and distorting the overlap while nailing it.

- If a nail begins to bend and should be pulled out, use a round piece of timber or metal as shown in Fig. 4.

ROOF COVERING. NPVC CON. 205

- Be careful in handling the sheets. They have sharp edges which can cut you deeply.
- If a corrugation has been badly damaged through pulling out a nail, try to restore the area around the nail hole to as near as possible the original shape. A round piece of metal can be put under the corrugation to help in this.
- If the sheets have to be cut, do not leave the cut-off pieces lying around. They can still be used for other work.
- Do not leave sheets lying where people might step on them and flatten the corrugations.
- Be sure, especially in seasons where there is a danger of storms, that no sheets are left loosely nailed to the roof, or left on the ground without being weighted down or otherwise secured. A sheet blown around by the wind can easily kill someone.

NOTES:

HEAT PENETRATION - INSULATION

One of the main goals of building in the tropics is to reduce heat penetration into the building. There are a number of building constructions which help to make the building cooler.

- Build in an east-west orientation, so the sun hits directly on only the end walls of the building.

- Construct a verandah to shade the wall.

- Construct overhanging eaves or overhangs to shade the walls.

- Use an open roof construction (no ceilings) on the verandahs, overhangs and overhanging eaves; to permit cross ventilation between the roof and ceilings in the rooms.

- Make ceilings in the rooms so that the sun cannot heat the room directly through the roofing sheets and so that there is cross ventilation between the sheets and the ceiling.

- When possible, use a combination of any or all of the above constructions.

LIGHTENING

To prevent lightening from striking a building, you can install a lightening conductor. This is a pointed copper rod, which is fixed above the highest point of the roof and connected to a copper rod driven into the ground. Long buildings need more than one lightening rod. Follow the manufacturer's instructions to install the conductor.

NOTES:

HEAT PENETRATION - INSULATION & LIGHTENING.   N P V C   CON. 207

HIP RAFTER

FULL TRUSS

HIP RAFTER

HALF TRUSS

Fig. 1

RIDGE PURLIN

PURLINS

EAVE PURLIN

HIP TRUSS

HALF TRUSS

HIP TRUSS

Fig. 2

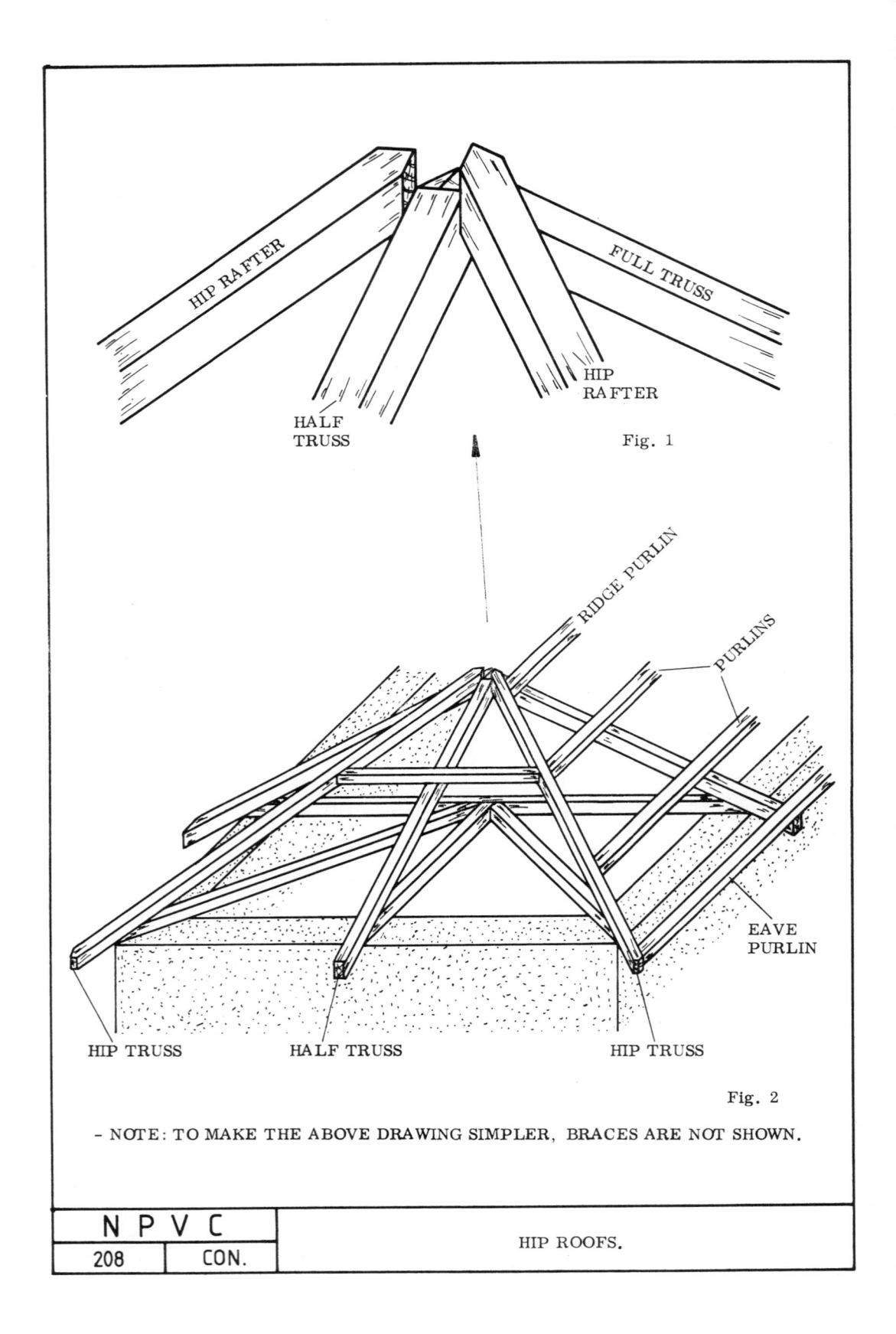

- NOTE: TO MAKE THE ABOVE DRAWING SIMPLER, BRACES ARE NOT SHOWN.

N P V C	
208	CON.

HIP ROOFS.

HIP ROOFS

A hip roof is often chosen as an alternative to a gable roof. It has a nicer appearance, and it is less vulnerable to suction from heavy winds. However it takes more work to construct it, and requires more timber than a gable roof. Also there is a certain amount of waste involved in cutting the sheets at the corners. The hip roof is more apt to leak unless it is constructed exactly right.

The main part of the construction is the same as for the gable roof. In addition, a half truss and two hip trusses are needed for each hip (Fig. 2).

The two hip trusses run diagonally from the ridge of the last full truss to the corners of the building. The hip rafter has to sit higher than the other rafters, so that the purlins butt into it at the side instead of lying on top of it. This is so that the sheets will be able to fit smoothly around the hip (Fig. 1).

The half truss is set out in the same way as the full truss, with only half the span. If the tie beams of the other trusses will project beyond the walls, the half truss tie beam has to project by the same amount (Fig. 3).

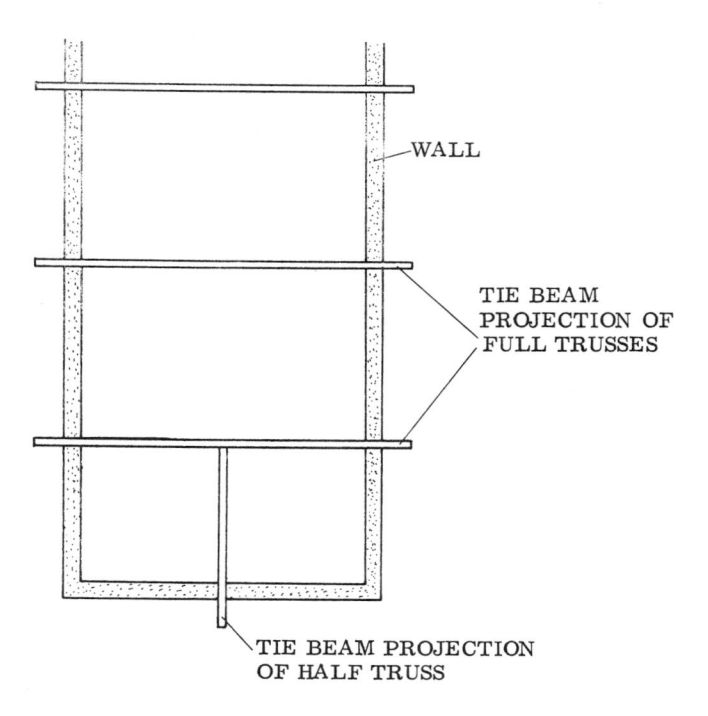

Fig. 3

HIP ROOFS.

N P V C
CON. 209

Fig. 1

- NOTE: TO MAKE THE DRAWINGS SIMPLER, BRACES ARE NOT SHOWN.

PURLINS

RISE

SPAN

Fig. 3

HIP RAFTER

LINE

LINE

Fig. 2

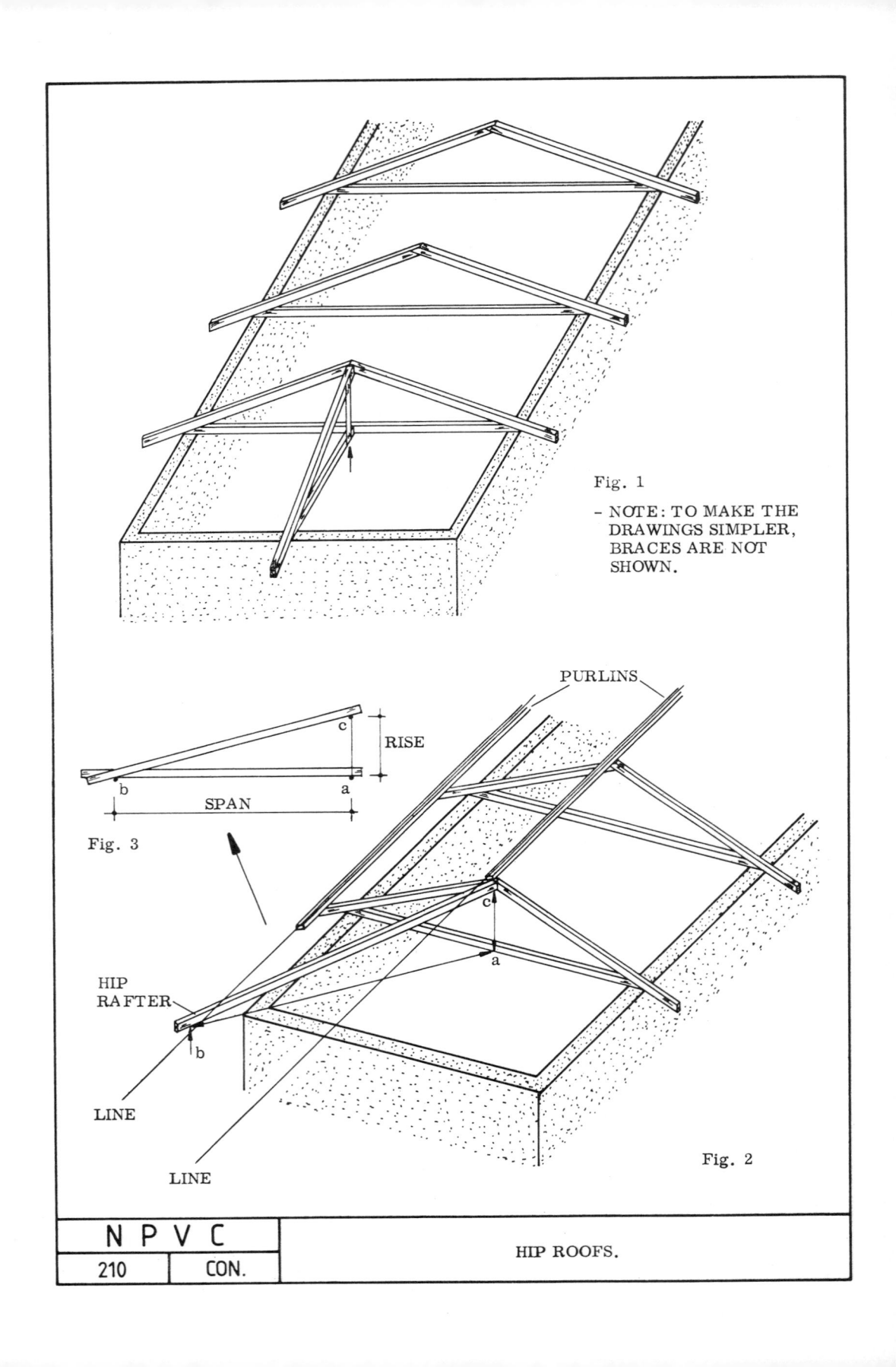

N P V C		
210	CON.	HIP ROOFS.

SETTING OUT AND CONSTRUCTING THE HALF TRUSS

Before the half truss is fixed, all the other trusses have to be in place.

To assemble the half truss, use the same setting out which was made for the full trusses. Simply construct half of a full truss. The last brace near the ridge should be set back about 10 cm from the end, to leave space for fixing the trusses together (Fig. 1, arrow).

When the half truss is ready, fix it temporarily into place against the full truss. Remember that if the other tie beams project past the walls, the half truss tie beam must also project, by the same distance.

POSITION AND MEASUREMENTS OF THE HIP TRUSS

Follow the sequence below to find the measurements for the hip truss, as well as its correct position.

a. Lay two purlins flatwise near the eave and ridge, across the last 3 full trusses of the roof. Make sure that they are aligned parallel to the wall, and nail them temporarily (Fig. 2).

b. Fix a line along the top surface of each purlin. The lines should extend past the ends of the purlins to the end wall of the building, where they can be held taut by a temporarily fixed pole. The lines must be taut and parallel to the side wall. Each line should touch the whole length of the top surface of the purlin (Fig. 2)

c. Have two men put the hip rafter in position, holding it so that one end is at the ridge and the other end is above the corner of the building. The top surface of the hip rafter should just touch both lines. Thus, the top of the hip rafter is on the same level as the top of the purlins.

d. Now another man can measure the length of the hip truss tie beam, which will be the horizontal distance from the centre of the full truss tie beam (Fig. 2, point a) to the soffit of the hip rafter outside the wall (point b). This distance is the span of the tie beam for the hip truss (see Fig. 3).

e. To obtain the rise of the hip truss, measure the distance from the soffit of the full truss tie beam to the soffit of the hip truss rafter (Figs. 2 & 3, point a to point c).

Now that you have the measurements for the hip truss, you can go ahead to construct it. Measure the other hip truss for the opposite side of the roof in the same manner.

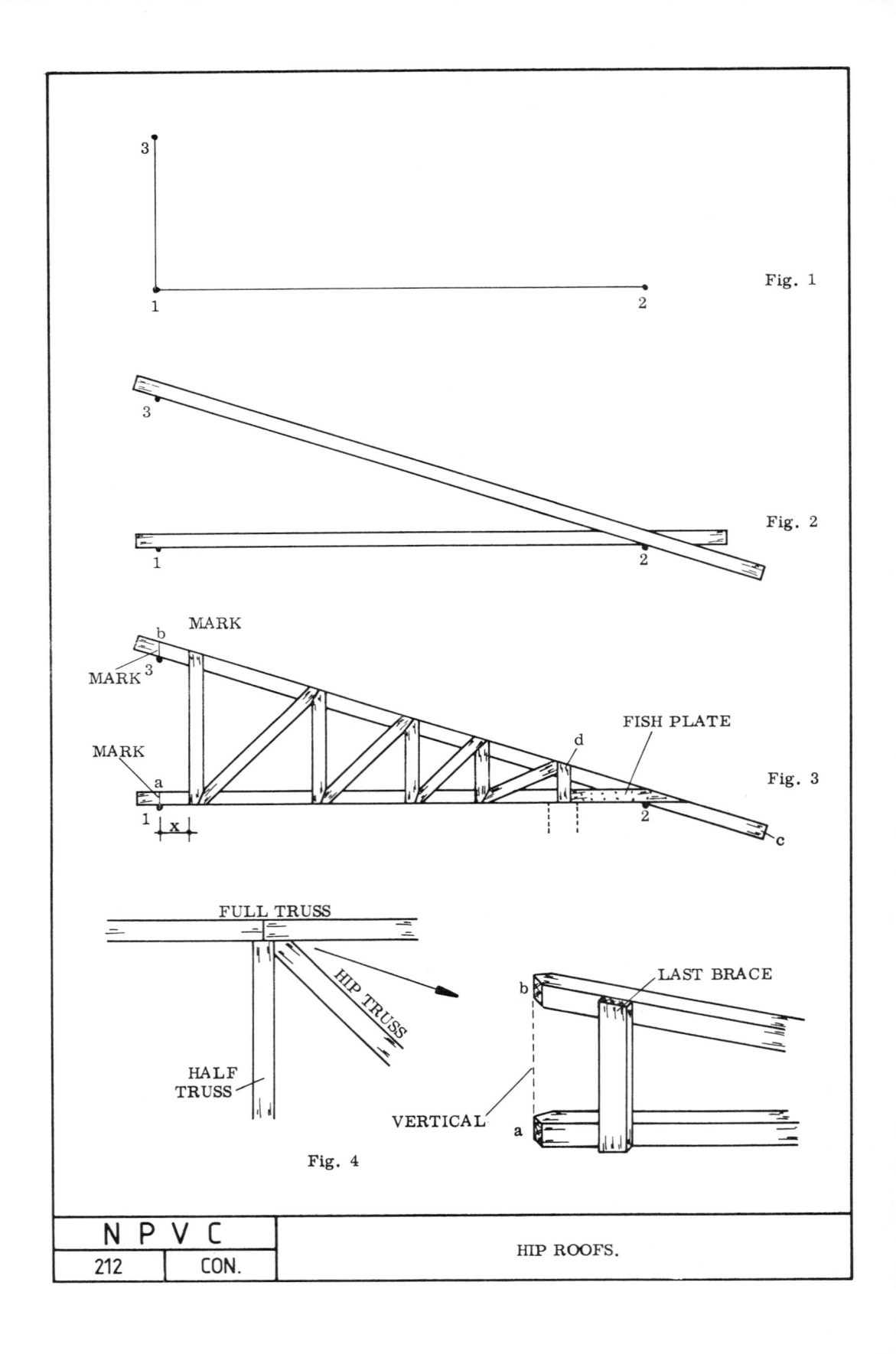

Fig. 1

Fig. 2

MARK b
MARK 3
MARK a
FISH PLATE
Fig. 3

FULL TRUSS
HIP TRUSS
HALF TRUSS
LAST BRACE
VERTICAL
Fig. 4

NPVC 212 CON. HIP ROOFS.

CONSTRUCTING THE HIP TRUSS

Look at the hip roof design in the Drawing Book, page 108.

- SETTING OUT:

    a. Drive pegs 1 and 2 in the ground at a distance equal to the span of the hip truss (the span is measured as described on the previous page). Fix a line between the pegs (Fig. 1).

    b. Fix a line perpendicular to line 1-2, starting from peg 1 (Fig. 1).

    c. Drive peg 3 at a distance equal to the rise of the hip truss (Fig. 1).

- ASSEMBLING THE HIP TRUSS:

    d. Lay the hip tie beam against pegs 1 and 2 (Fig. 2)

    e. Lay the hip rafter against pegs 2 and 3 (Fig. 2). Make sure that the tie beam and rafter are correctly positioned with respect to the pegs.

    f. Cut the rafter-to-tie beam joint. Mark the inside end of the tie beam (a) and the inside end of the rafter (b). Make sure that the rafter has a long projection on the eave end (Fig. 3). This projection is trimmed only just before the fascia board is fixed (c).

    g. Nail the braces and the fish plates. The last vertical brace on the ridge end is set back (Fig. 3, x) so that there is space to fit the trusses together later. Nail a short vertical brace (Fig. 3, d) where the tie beam meets the supporting wall.

    h. Cut the ends of the rafter and tie beam (ends a and b) at a 45 degree angle, as shown in Fig. 4.

    i. Assemble the second hip truss on top of the first one.

NOTES:

HIP ROOFS.

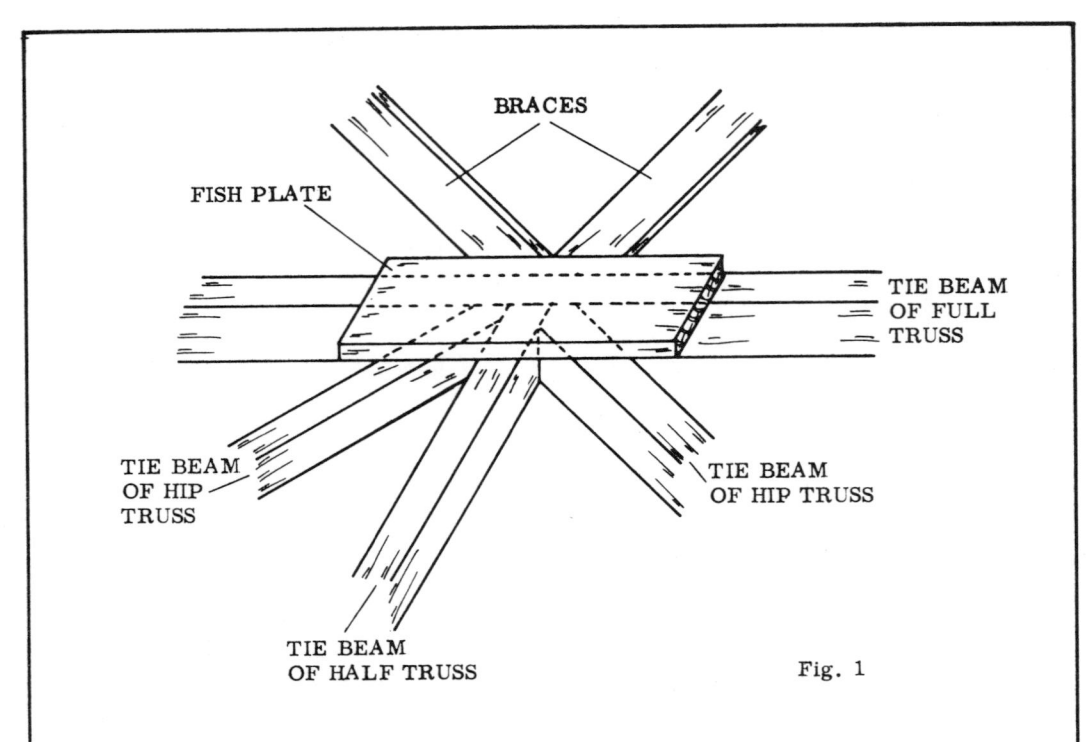

BRACES

FISH PLATE

TIE BEAM OF FULL TRUSS

TIE BEAM OF HIP TRUSS

TIE BEAM OF HIP TRUSS

TIE BEAM OF HALF TRUSS

Fig. 1

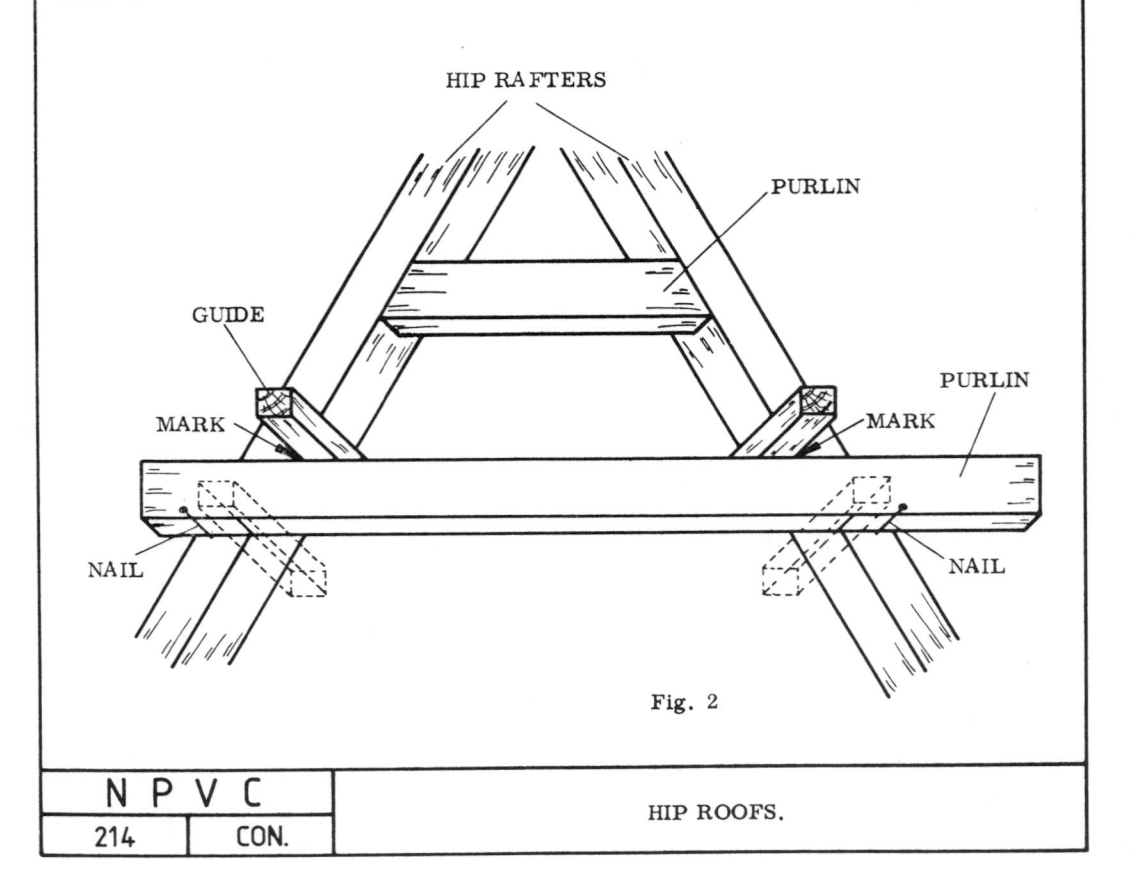

HIP RAFTERS

PURLIN

GUIDE

MARK

MARK

PURLIN

NAIL

NAIL

Fig. 2

NPVC		HIP ROOFS.
214	CON.	

ERECTING THE HIP OF THE ROOF

When the half truss and hip trusses are ready, the hip structure of the roof can be erected. The connection between the tie beams of the half truss, hip trusses, and full truss can be made as shown in Fig. 1. The rafter connection is made as shown in Fig. 1 on page 208.

- FIXING THE PURLINS: The purlins which butt against the hip rafter must be cut at exactly the correct angle so that they will fit snugly against the rafter.

  - Mark the positions of the purlins on top of the hip rafters, and insert a nail at the mark to support the purlin during marking (Fig. 2).
  - Place the purlin against the nails and mark the cutting lines. Use a short piece of wood as a guide (Fig. 2). Make the marks on both sides of the purlin at once, then mark both sides at the other end.
  - Cut the butt joints and fix the purlin.

  Fix the rest of the purlins in the same manner. The layout for the purlins on a roof with a single sheet length on each side is shown below (Fig. 3).

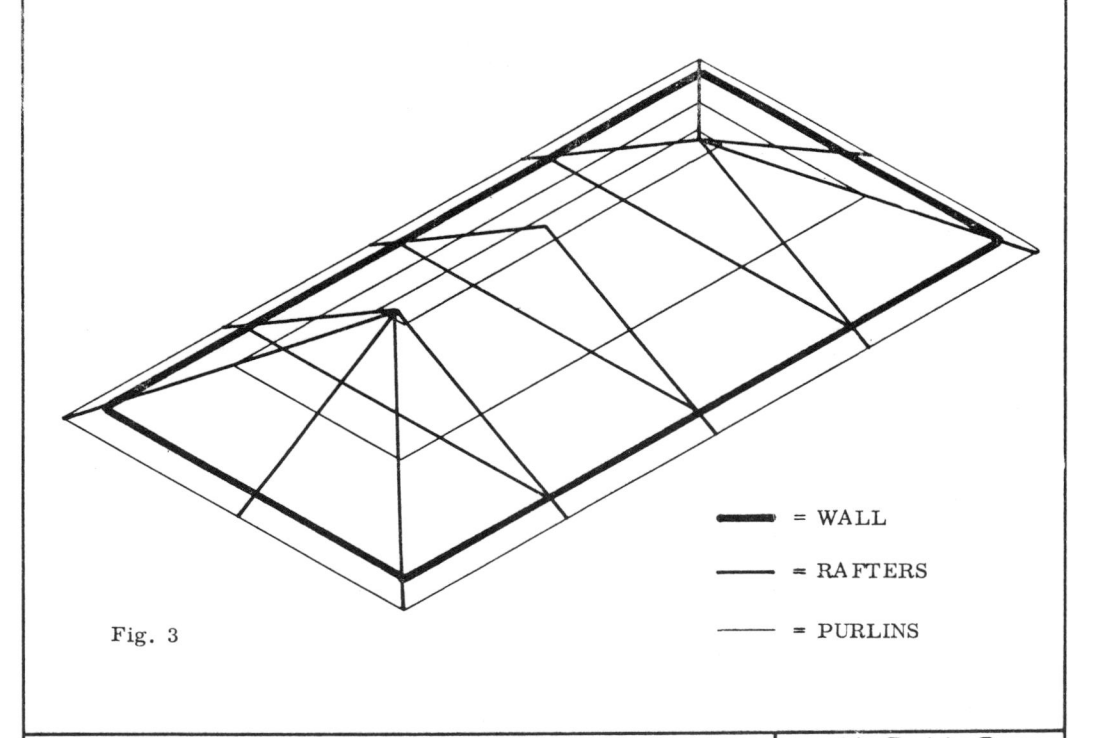

Fig. 3

▬▬ = WALL
──── = RAFTERS
──── = PURLINS

HIP ROOFS.

NPVC
CON. 215

- NOTE: TO MAKE THE DRAWING SIMPLER, BRACES ARE NOT SHOWN.

ADDITIONAL PURLINS

Fig. 1

HIP RAFTER

HIP CAP

ADDTIONAL PURLINS

Fig. 2

NPVC
216 CON.

HIP ROOFS.

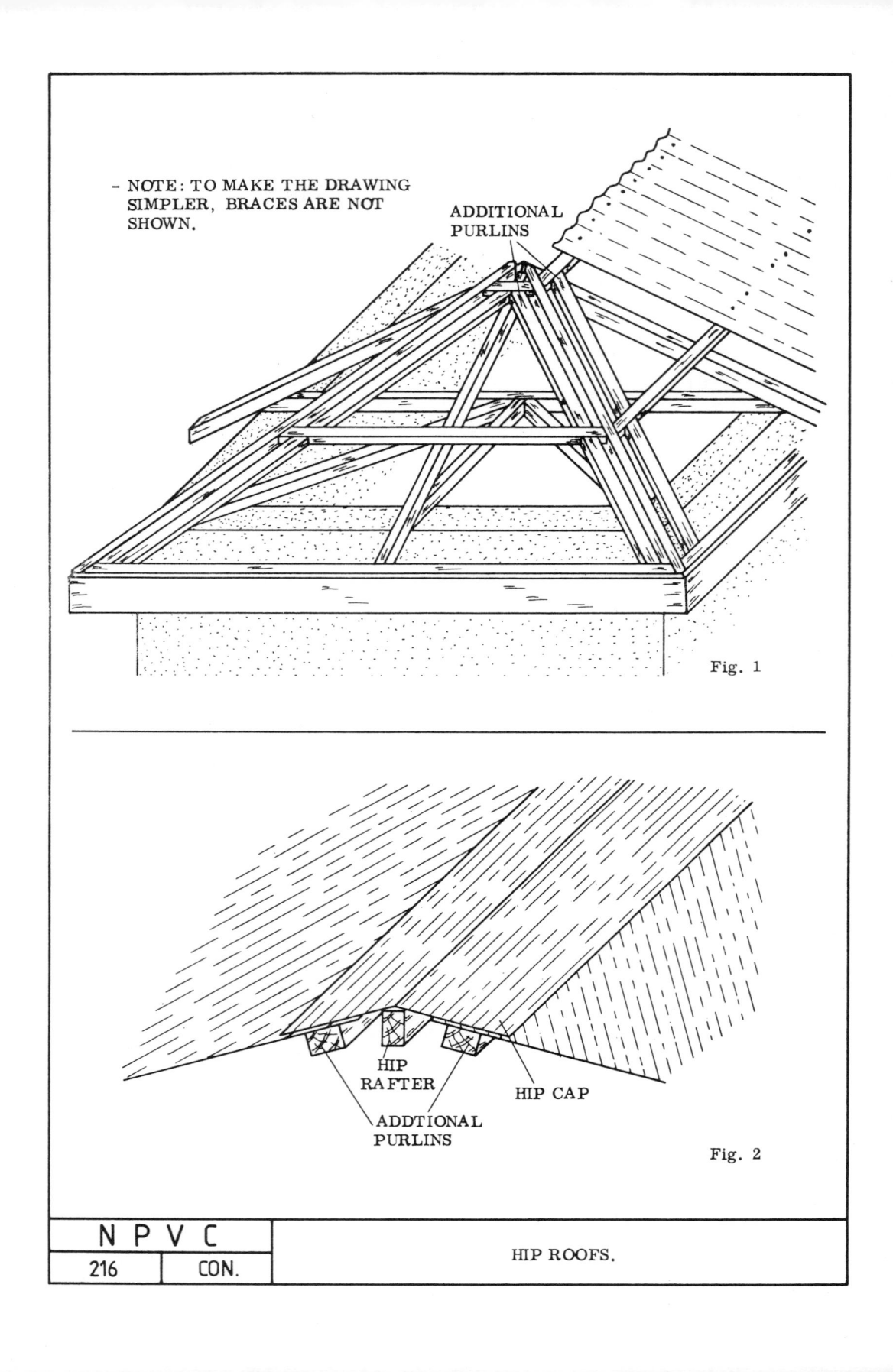

COVERING A HIP ROOF

Nail two purlins parallel to the hip rafters, as shown in Figs. 1 and 2. The extra purlins provide more area for nailing the hip ridge cap.

Lay the first sheet at the hip rafter in position and mark with a straight edge where it should be cut. Cut the sheet and nail it in position before you mark and cut the next sheet. Continue in this way until the hip is covered.

A hip cap, similar to the ridge cap, must be used to cover the hip line, where the sheets meet. If no ready-made caps are available, one can be made as explained on page 205.

NOTES:

HIP SIDE

SPACERS

VALLEY SIDE

Fig. 1

HIP AND VALLEY TRUSS

HALF TRUSS

HALF TRUSS

FULL TRUSSES

Fig. 2

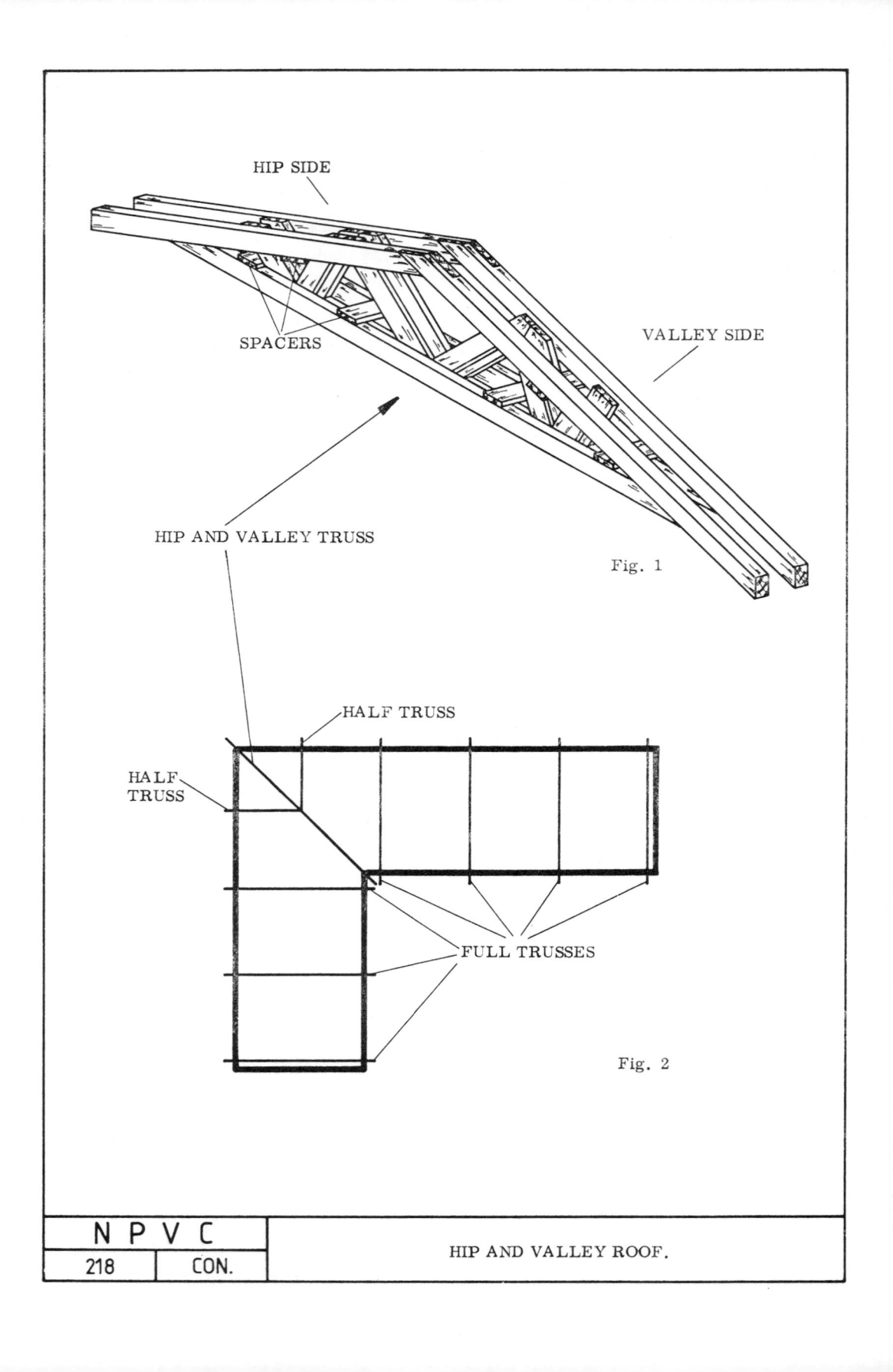

N P V C	
218	CON.

HIP AND VALLEY ROOF.

HIP AND VALLEY ROOF

When a roof is constructed for an L- or U-shaped building, a hip and valley roof is normally used.

The main problem of this construction is to make a truss which provides sufficient nailing area for the valley and also for the hip. This is solved by assembling two trusses, with spacers in between to connect them (Fig. 1). The spacers provide the required distance between the trusses. This distance is important, because the valley should be wide and deep enough so that rain can run down it easily and so it does not get blocked by leaves (Fig. 1).

As in the hip roof, the hip rafters and the valley rafters must be set higher so that the purlins meet them at the sides, and don't lie on top as they do with the other trusses.

If a large span must be covered with a hip and valley truss, it may be necessary to add some half trusses onto the hip and valley truss to provide enough support for the purlins (Fig. 2).

NOTES:

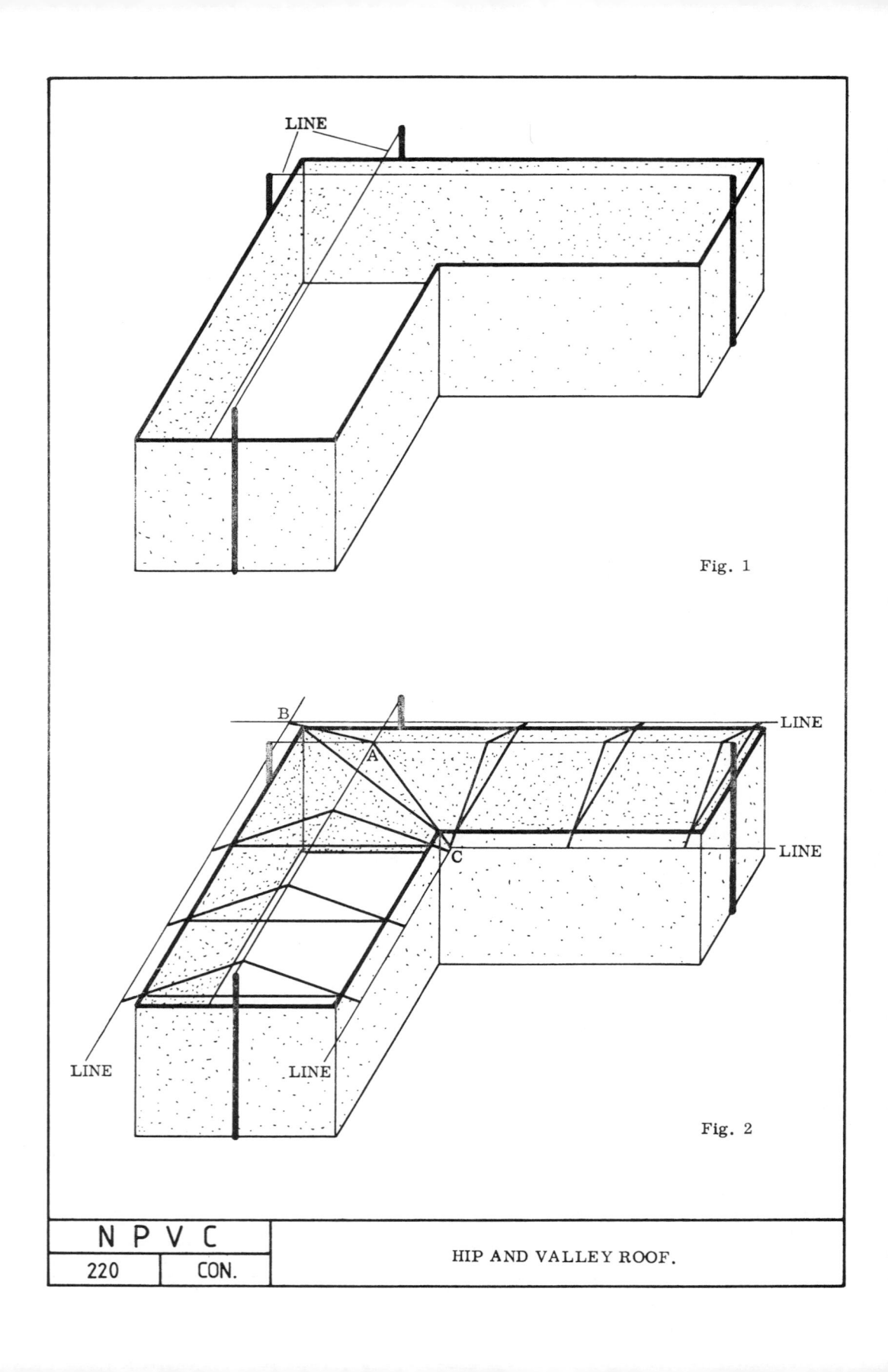

Fig. 1

Fig. 2

LINE

N P V C		HIP AND VALLEY ROOF.
220	CON.	

POSITION AND MEASUREMENTS OF THE HIP AND VALLEY ROOF

- Start by erecting poles in the positions shown in Fig. 1, and fix lines at the ridge.
- Erect the ordinary trusses according to the lines in their correct places (Fig. 2). Refer to the section on erecting a gable roof, page 191.
- Lay purlins temporarily on top of the rafters, as was done with the hip roof (page 211).
- Fix lines parallel to the walls along the tops of these purlins, at eave height on both sides of the roof (Fig. 2).
- Set the ridge lines higher, by the amount of the purlin thickness (Fig. 2).
- Now you will have three points where the lines cross. These are: point A at the ridge; point B at eave height, and point C on the other side at eave height (Fig. 2). These points mark the outer surfaces of the hip and valley truss.
- Next construct a lightweight, temporary model truss. Hold it in position on top of the roof (Fig. 3). Adjust it until points A, B, and C touch the outer surface of the truss.
- Now the model truss will have the correct dimensions and can be used as a model for constructing the two trusses of the hip and valley truss.

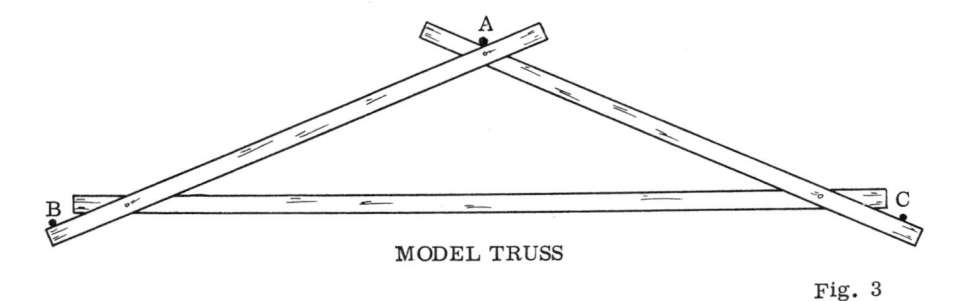

MODEL TRUSS

Fig. 3

NOTES:

HIP AND VALLEY ROOF.

Fig. 1

Fig. 2

BRACES SET BACK (ON VALLEY SIDE ONLY)

VALLEY RAFTER

Fig. 3

N P V C		
222	CON.	HIP AND VALLEY ROOF.

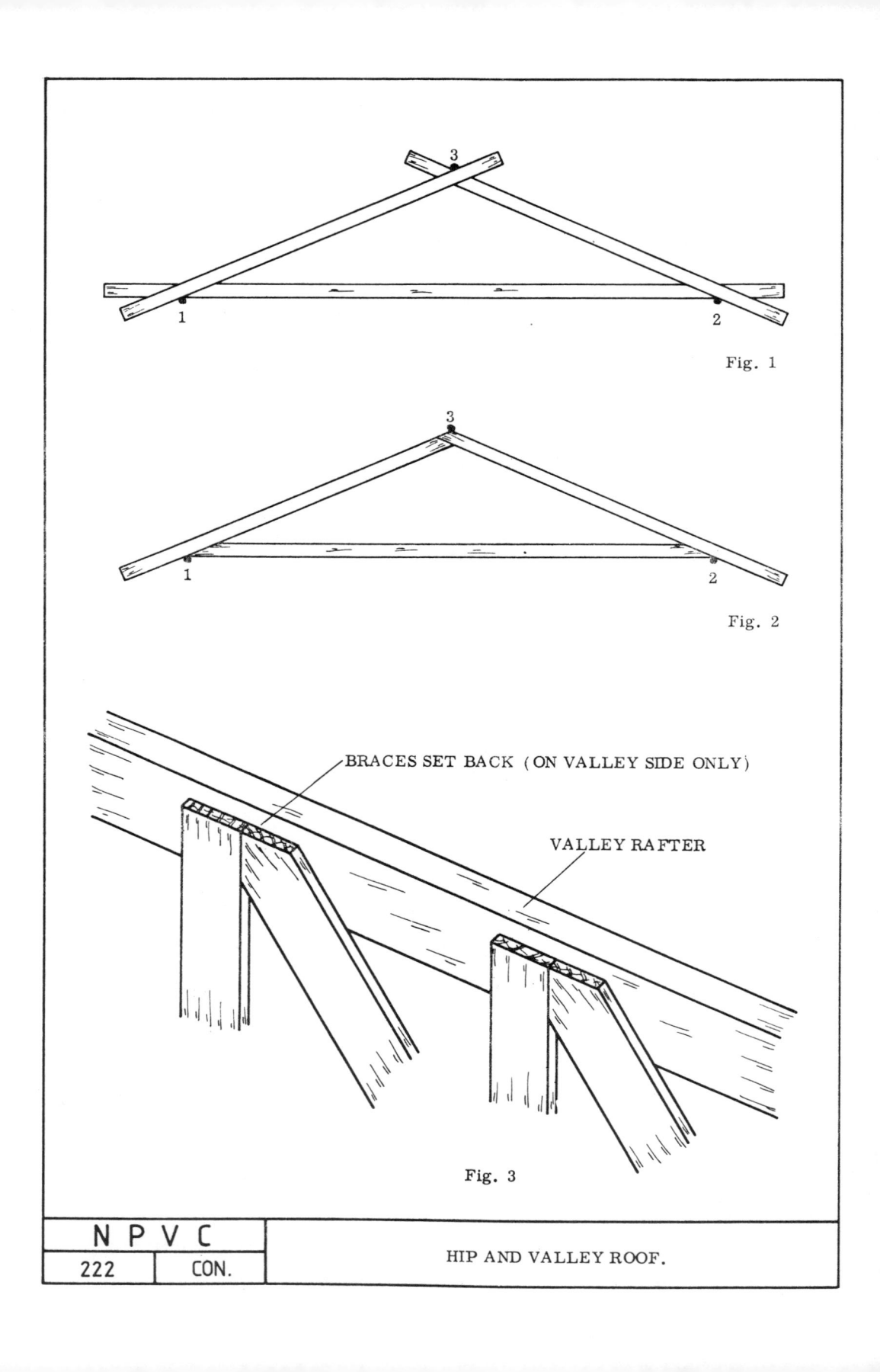

CONSTRUCTION OF A HIP AND VALLEY TRUSS

Take the model truss down and set it on the ground in a cleared and level spot.

- Place an iron peg at each side of the model where the rafters meet the tie beam (Fig. 1, pegs 1 and 2) and one peg at the ridge (peg 3). Remove the model truss.

- Now lay the actual tie beam on the ground against the pegs.

- Lay the two rafters on the ground in the correct positions. Be sure that the members are in the correct positions with respect to the pegs, as in Fig. 1.

- Cut the rafter-to-tie beam joint, and the ridge joint (Fig. 2).

- Nail the braces. Take care that the braces are set back from the top of the rafters on the valley side, so there is room to construct the valley later (Fig. 3).

- Assemble the second truss on top of the first one. Be careful to nail the braces so that later when the two trusses are combined, the braces will all be on the inside (Fig. 4).

- When both trusses are ready, they can be set in place on top of the wall. Then the spacers can be nailed to keep the trusses at the correct distance apart. Remember that the distance between the trusses should be wide enough (approximately 15 cm) to provide a wide valley.

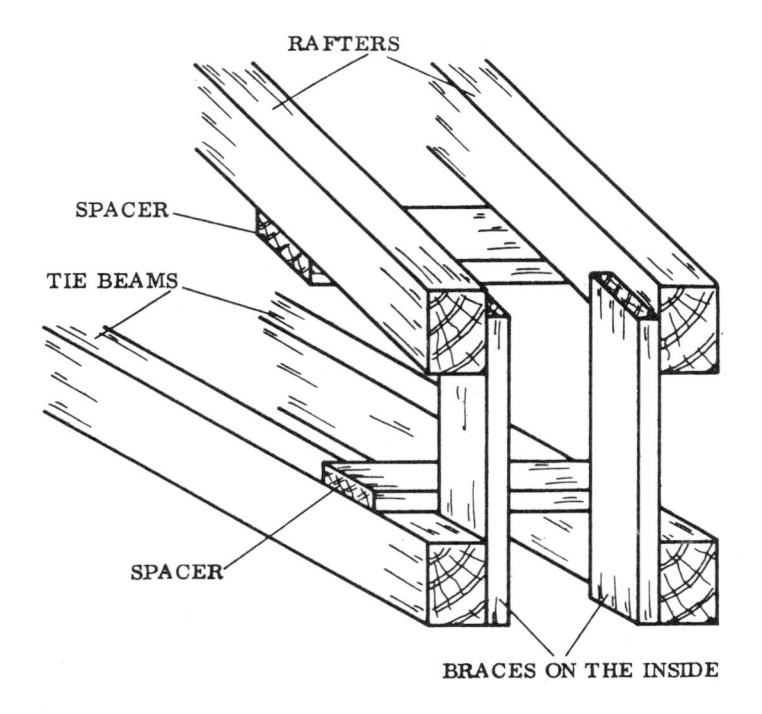

Fig. 4

HIP AND VALLEY ROOF.

NPVC
CON. 223

COVERING A HIP AND VALLEY ROOF

First fix the fascia boards and the purlins. Fix a line at eave level and cut the rafters according to this line. Then nail the eave purlin and the fascia boards.

At the place where the fascia boards meet at the valley, cuts have to be made in the fascia boards according to the shape of the gutter (Fig. 1).

Now fix the gutter in the valley. Lay a strip of sheeting metal lengthwise in the valley in the form of the required gutter. Fix it there temporarily with nails. The final fixing will be done as the other sheets are laid, since they have to be laid on top of the gutter. The overlap for the gutter sheets should be about 30 cm in the length.

Now lay the first sheet into position at the valley, mark and cut it. Nail this sheet and place the next sheet in position. Continue in this way until the hip and valley are covered.

Install the hip cap as explained for the hip roof.

The ridge cap on the main ridge is always covered last, since it has to overlap all the rest.

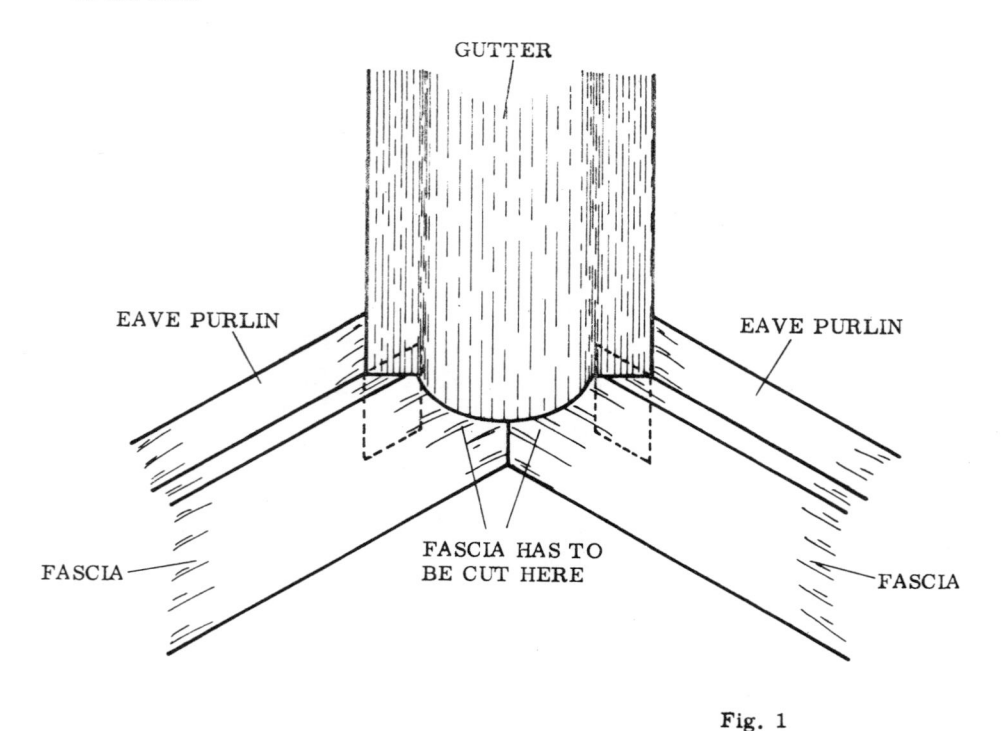

Fig. 1

N P V C		
224	CON.	HIP AND VALLEY ROOF.

PLASTER AND RENDER

Plaster or render is a mortar coating over the blockwork. We call the coating on the inside walls "plaster" and the coating on the outside walls "render". The main difference between the two is that render is generally richer in cement than plaster, because it has to be weather resistant. In the following pages we will refer to the application of either plaster or render as "plastering".

The function of the plaster inside is to make the walls smooth so that they are easier to clean, free of insect hiding places, and have a better appearance. Also, if the house is constructed out of wood or bamboo closed up with mud or clay, the plaster acts as a protection against fire.

The render on the outside surface of the walls is essential to protect them from the influence of the weather, especially if landcrete blocks or mud blocks are used. When these blocks are exposed to moisture for long periods, they gradually become soft, expand, and finally crumble away. The render must therefore be water resistant.

- All external walls should be covered with render. The sides of the building that face the heaviest weather (in north-western Ghana, this comes from the northeast) should be covered with a somewhat richer mix (Fig. 1, a, next page).
- All so-called wet rooms such as bathrooms, toilets, kitchens, and washing areas should be plastered and painted. (Fig. 1, next page: K = kitchen; S = shower; and W = washing area.)

COMPOSITION OF PLASTER OR RENDER

In Rural Building, plaster or render consists of cement, sand, and water. Lime can be added to the mix, but this is not always available.

Sharp and relatively coarse sands are preferred for render, although they produce a mortar which is not easy to work with. The size of the grains should be no greater than 5 mm; the sand should be well graded (Reference Book, pages 148 and 149); and it should not contain more than 20% very fine particles (powdered sand).

The sand for plaster should be well graded, but less coarse than sand for render (see Reference Book, page 159).

- MIX PROPORTIONS: The mix proportion for plaster need not be better than 1 : 8, and may be as low as 1 : 12. The exception to this is the plaster for an

Fig. 1

Fig. 2

JOINTS RAKED OUT

PREVAILING WIND

N

N P V C		PLASTER AND RENDER.
226	CON.	

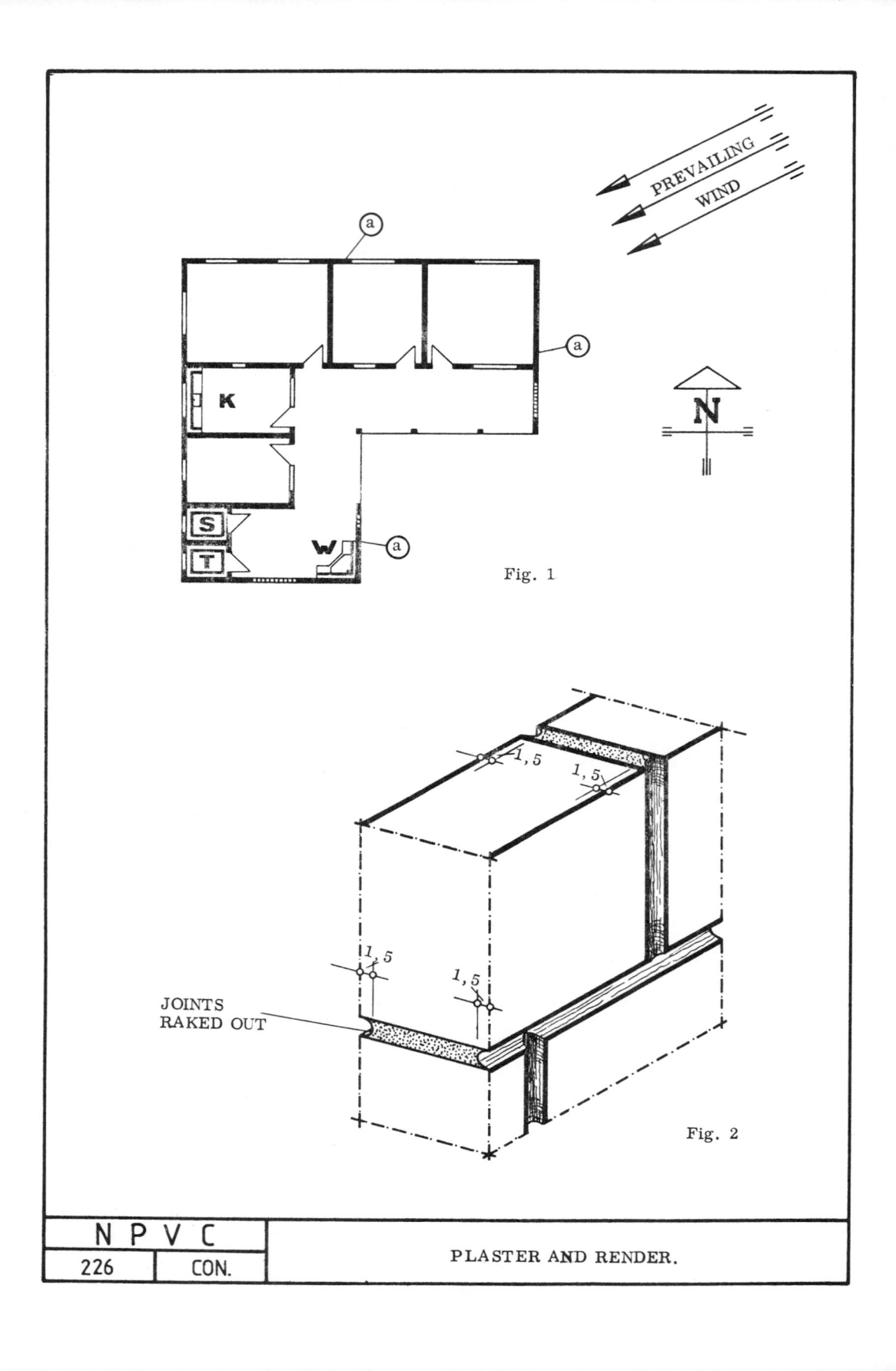

area which is often wet, such as a bathroom or washing area. These should be plastered with a mix of about 1 : 6.

The mix proportion for render should never be better than 1 : 6; this is for the north and east sides of the building (or whatever direction the heaviest rain comes from). The other sides of the building can be covered with a mix which is at least 1 : 10 (see the Reference Book, Tables of Figures, page 234).

Exceptions are made for special structures which need waterproof plastering, such as water storage tanks, septic tanks, inspection chambers, manholes, etc. The mix proportion of the plaster used on these may vary from 1 : 3 up to 1 : 1, because the permanently moist surfaces keep cracks from developing.

- REMEMBER: The richer the mix, the more likely it is to develop cracks. Save materials and money by following the rules for correct mix proportions.

PREPARING THE WALL FOR PLASTERING

During the blockwork as the wall is built-up, the first measure is taken to prepare the wall for plastering: the joints are raked out to a depth of approximately 1,5 cm (Fig. 2). This is done immediately after each course is completed. Raking out the joints gives the plaster or render a better grip to the wall.

Before any plaster or render is applied, the wall must be brushed to remove the dust and dirt which would weaken the grip of the mortar. The masonry must then be thoroughly watered, especially during the dry season. This reduces the absorption of water from the freshly applied mortar, so that the mortar hardens well.

- NOTE: Some experiments at NPVC have shown that the landcrete surface can be improved by leaving the walls exposed to one or two short rains before plastering. The rain washes out the finer particles, leaving a rough surface which bonds to the plaster or render better. This method is risky however, because it is impossible to tell in advance how long the rain will last and how heavy it will be.

APPLICATION TECHNIQUES

- SPATTERDASH: In order to give a good grip on smooth surfaces such as sandcrete, or to reduce the amount of moisture absorbed from the freshly applied mortar, spatterdash may be applied before plastering. This is done with the spatterdash apparatus (see Basic Knowledge, page 178), using a very wet and rich mix (1 : 3, up to 1 : 1). The spatterdash layer is never more than 5 cm thick.

WALL
PLASTER WORKING JOINT

Fig. 1

RISING WALL
FOOTING
EDGE BOARD
RENDER
(a) 1st STEP

SLANTED SURFACE
EDGE BOARD
(b) 2nd STEP

(c) 3rd STEP

Fig. 2

N P V C		
228	CON.	PLASTER AND RENDER.

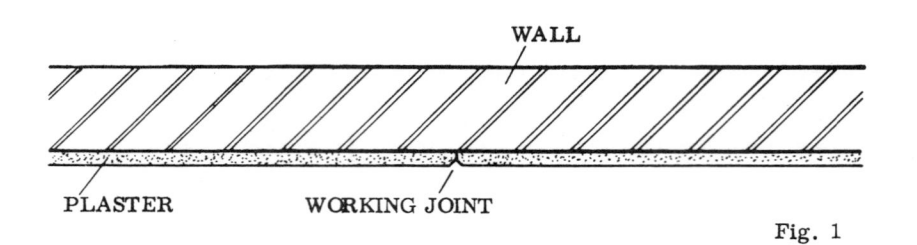

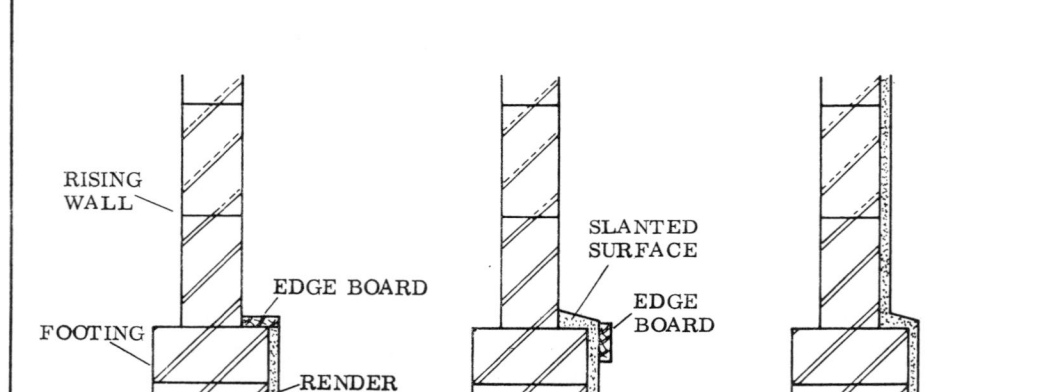

The plaster or render is generally applied in a single layer on top of the spatter-dash, or else directly on the blockwork surface. This layer will be 15 to 20 cm thick.

Sometimes a second coat or finish coat is applied on top of the first layer. This must be done before the first layer has hardened. The second coat is usually made where the plaster has to be very thick, for example where the wall is very uneven so that a thick coat of plaster has to be applied to get a smooth surface. The second coat should be about the same mix as the first coat, and about 5 cm thick.

Don't start plastering on a wall which is in direct sunlight. This would cause the render to dry out too quickly, resulting in a "burnt" render which is useless.

Instead, start plastering in the early morning on the west side of the building, going clockwise around the building to the north, east, and south sides so that the freshly plastered areas are always in the shade.

Areas which are too large to be completely plastered in one day are divided into smaller areas by introducing working joints (Fig. 1). The joints should always be vertical, so that water will flow away as quickly as possible from the joint.

In Rural Building, the inside of the building is generally plastered after the roofing is completed.

- PLASTERING THE FOOTINGS: Here we must add a note about the footings. These have to be plastered before the rising wall. This is done as shown on the left (Fig. 2; a, b, & c).

An edge board is laid on the ledge of the top of the footings, and the vertical surface below is plastered according to the procedure explained on the following pages.

Next the ledge is plastered. The render must form a slanted surface so that water runs away from the wall. It is important that this slanted surface is made before the rising wall is plastered, because otherwise the joint between the two would allow water to enter (Fig. 2; b & c).

NOTES:

PLASTER AND RENDER.

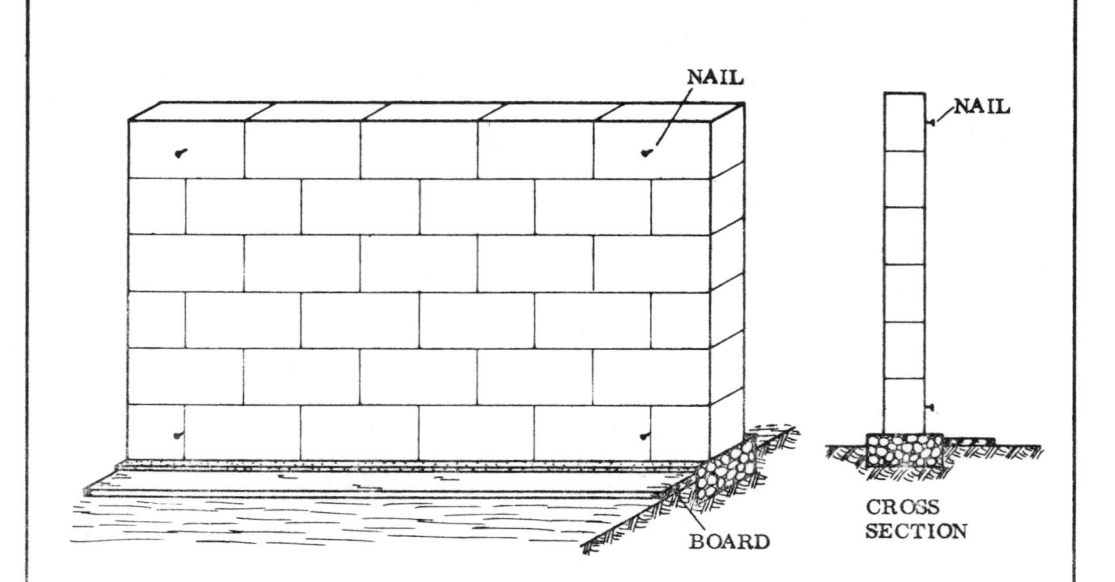

SEQUENCE OF OPERATIONS FOR PLASTERING A STRAIGHT WALL

a. Study the wall which is to be plastered; notice the holes and projecting parts.

b. Clean the wall with a hard brush and chisel off any projecting blocks.

c. Soak the wall thoroughly with water; try to wash off the loose particles. Clean the wall very well, especially the area near the footings and the footings themselves.

- NOTE: In the drawings here the footings have been left out, to make the drawing simpler. The plastering of the footing - rising wall connection is made as explained on the previous page.

d. Place boards against the foot of the wall to catch the mortar which is dropped.

e. Prepare some mortar which is of the same strength as the wall, and use it to fill up the larger holes in the wall.

f. Make a scaffolding to work from, if this is needed.

g. Insert four nails, one at each corner of the wall.

NOTES:

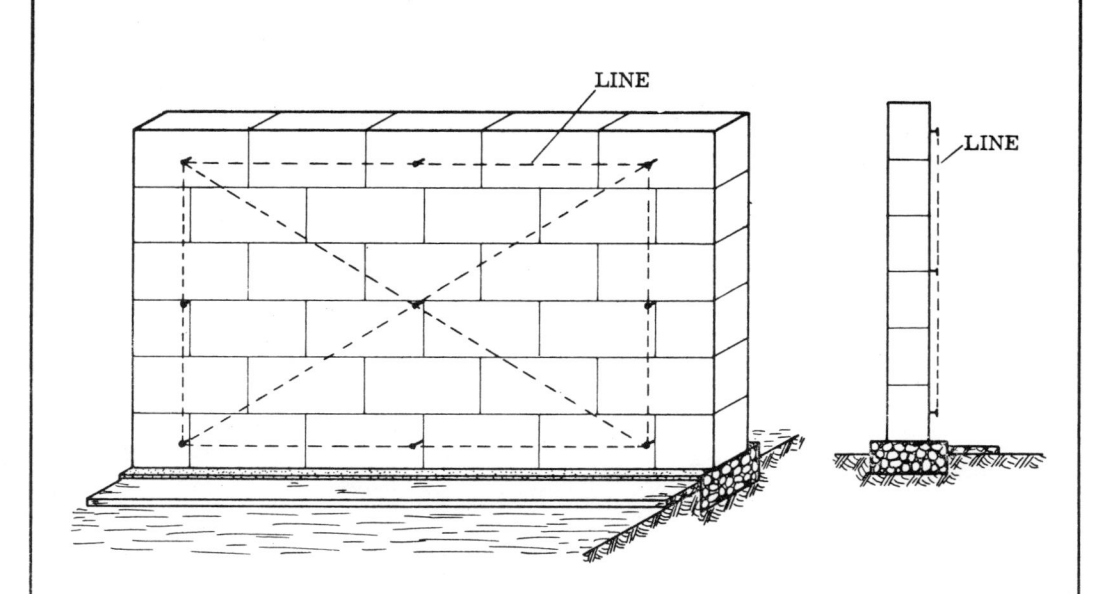

h. Fix lines around all four corners and across the diagonals.

i. The surface contained by the nail heads and the lines should be flat and plumb. Check this using the plumb bob and straight edge. Make corrections by knocking the nails in until they are all in the same plane; this will be the surface of the future plaster layer.

j. When the four corners have been adjusted, insert more nails along the mason lines, as shown above. The heads of the nails should be at the same height as the lines. The distance between the nails should be no more than the strike board can bridge.

NOTES:

PLASTER AND RENDER.

N P V C

CON. 231

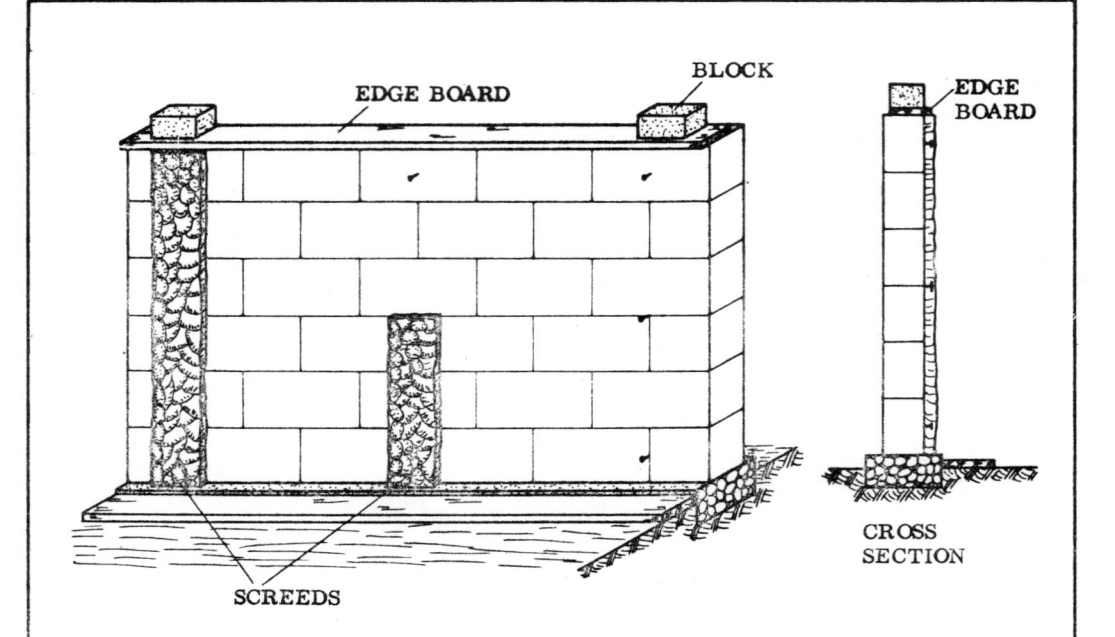

EDGE BOARD
BLOCK
EDGE BOARD
CROSS SECTION
SCREEDS

k. Fix the edge boards. The edge board at the top of the wall can be held in place with blocks on top of it. The edge boards should project by the same amount as the nails -- the thickness of the future plaster layer.

l. Now the plastering can start. Make sure that you have enough time to finish the whole area you want to plaster at once, otherwise wait until the next day.

m. Mix the mortar according to the required proportions. Take care that there are no stones in the plaster. Make up a dry mixture first and add water to this as it is needed. Do not mix too much at once, and keep the mortar covered with paper bags or other covering to keep out the sun and wind.

n. Sprinkle the wall with water again.

o. Build up screeds (guiding strips) by throwing mortar against the wall in vertical strips. The mortar should cover the nail heads completely.

NOTES:

N P V C   CON.

PLASTER AND RENDER.

STRIKE BOARD

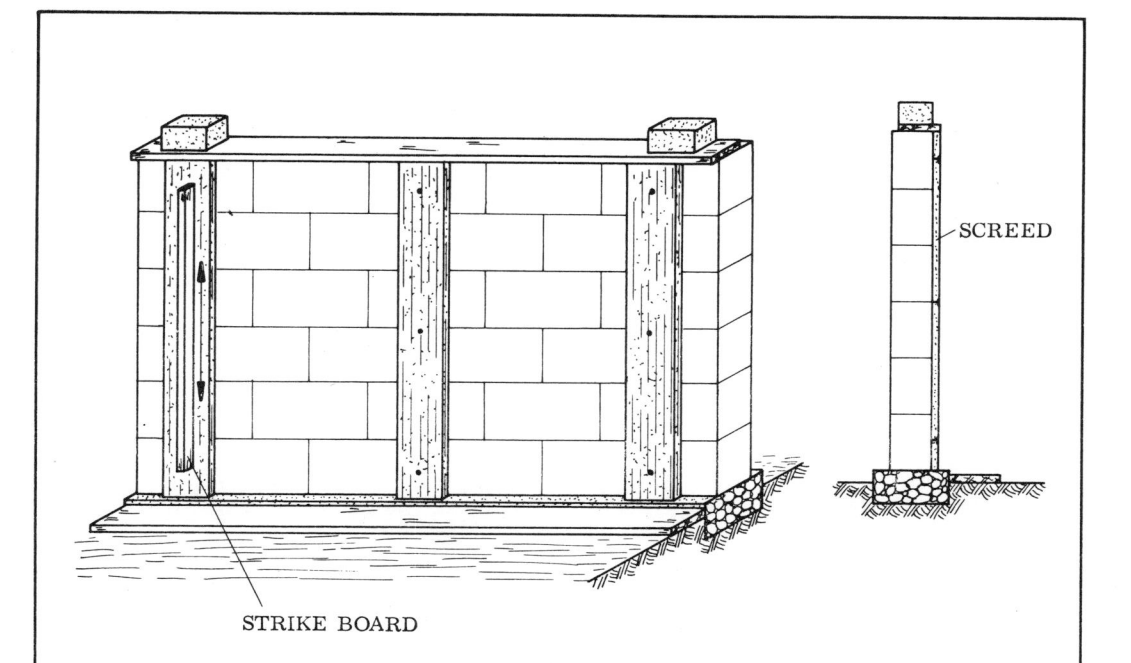

SCREED

p. Use the strike board to smooth off the screed flush with the heads of the nails. Return any mortar which drops down to the headpan. Give the strike board an up and down movement as you move it across the screed, to obtain a good surface.

q. Let the screeds set for a while, until they are hard enough so that no mortar will be taken off by the strike board during the next steps.

NOTES:

PLASTER AND RENDER.

NPVC
CON.   233

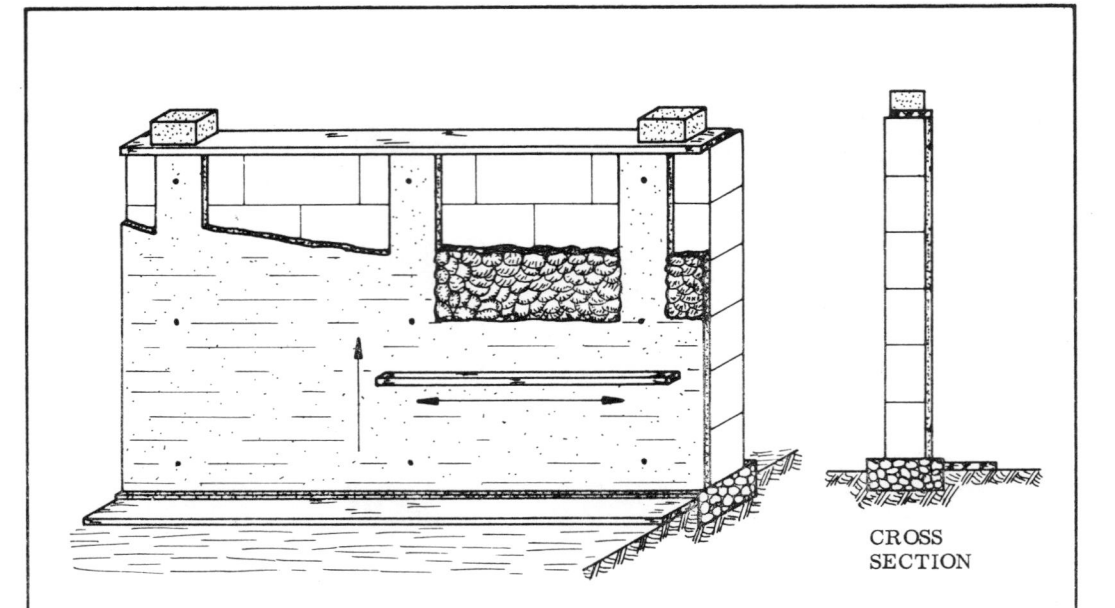

CROSS SECTION

r. When the screeds are hard enough, sprinkle the wall with water again.

s. Fill in between the screeds with mortar. Throw the mortar against the wall, so that it grips well to the wall. The mortar should be a bit thicker than the screeds.

t. Press the strike board to the screed and move it up the wall, holding it horizontally. The board should remove the excess mortar so that the surface of the mortar is flush with the screeds. Make sure that the strike board does not remove any of the mortar from the screeds. Return all surplus mortar to the headpan.

It is not necessary to fill up the whole section between the screeds before you start to smooth the plaster. Fill in the plaster as high as you can comfortably reach, then smooth this and fill in the rest with the help of a scaffold so you can reach the top of the wall. Smooth up to the top of the wall. However, you should never leave a section halfway finished, because it will not join properly when you continue the next day.

u. When the whole wall has been levelled off the nails should be removed and the last finishing work can be done.

N P V C		PLASTER AND RENDER.
234	CON.	

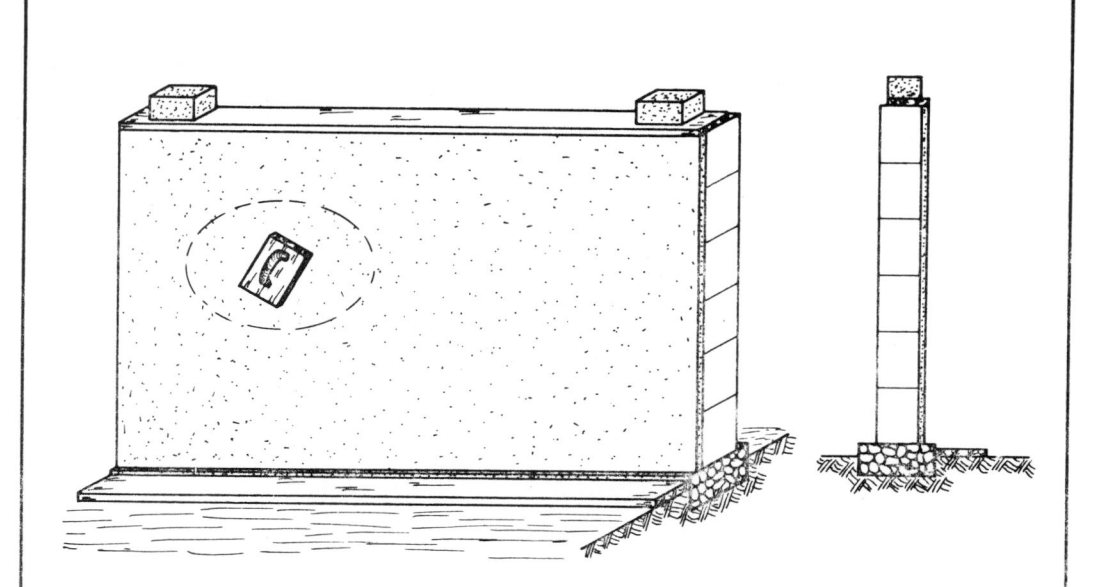

- FINISHING: If the wall should be finished with a textured surface, a wood float can do the job. Turn the float around systematically to obtain a regular texture.

If a wall should be smooth-surfaced, use a steel float and take care that the sharp edges do not remove mortar from the plastered surface.

Dip the wood float in water regularly, but don't over-wet the mortar because it could come down or "sag".

Take care that the plaster or render comes to the footings (here, to the foundations) and cut it off properly. The edges around openings and outside corners should be rounded or chamfered.

Before stopping work, make sure that the work is finished off properly and is clean. Someone should return in the evening and sprinkle water on the freshly made plaster or render. Water the plaster or render regularly in the next few days so that it cures properly.

The edge boards can be removed the next day.

- CURING: Watering the plaster or render is very important, especially during the dry season when the harmattan is blowing. Don't wait until the whole work is finished before you put water on the parts which were done first; the fresh plaster can dry out in a few hours.

If you started on the west side of the building in the morning, be sure to water those sections in the afternoon when the sun is shining on them.

PLASTER AND RENDER.

N P V C
CON. 235

POSITIONS OF EDGE BOARDS FOR PLASTERING CORNERS

Below are illustrations of the correct method for plastering around corners. In Fig. 1a, the first step is shown; in Fig. 1b the second step. The arrows indicate the areas which are plastered in each step. The boards project by the thickness of the plaster.

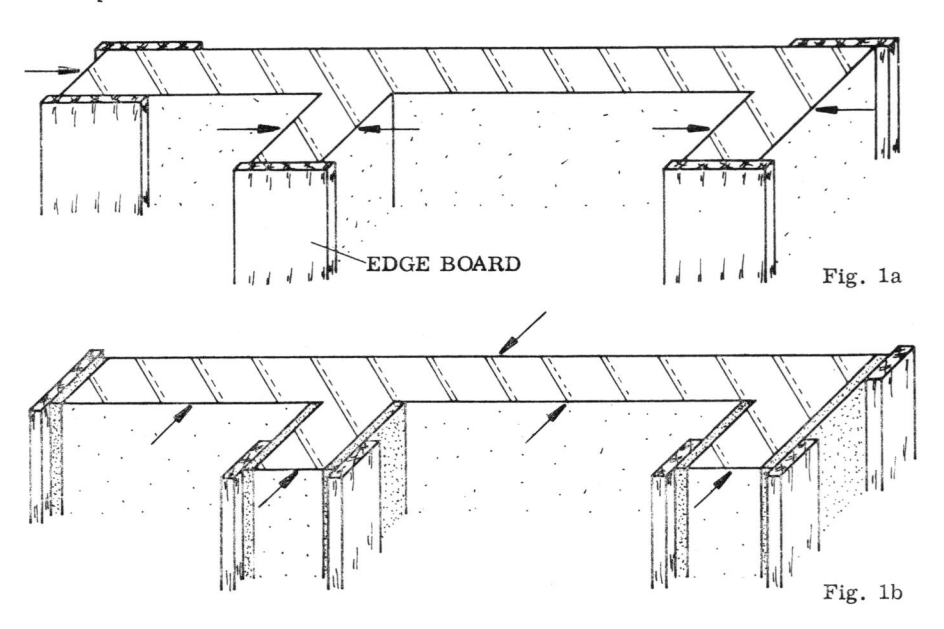

EDGE BOARD

Fig. 1a

Fig. 1b

POSITIONS OF EDGE BOARDS FOR PLASTERING AROUND FRAMES

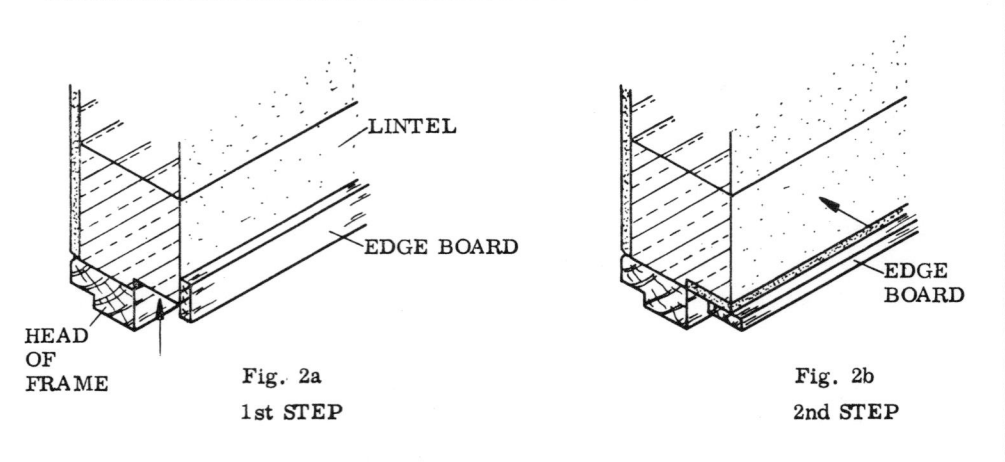

LINTEL

EDGE BOARD

HEAD OF FRAME

Fig. 2a
1st STEP

EDGE BOARD

Fig. 2b
2nd STEP

N P V C		PLASTER AND RENDER.
236	CON.	

Fig. 3a
1st STEP

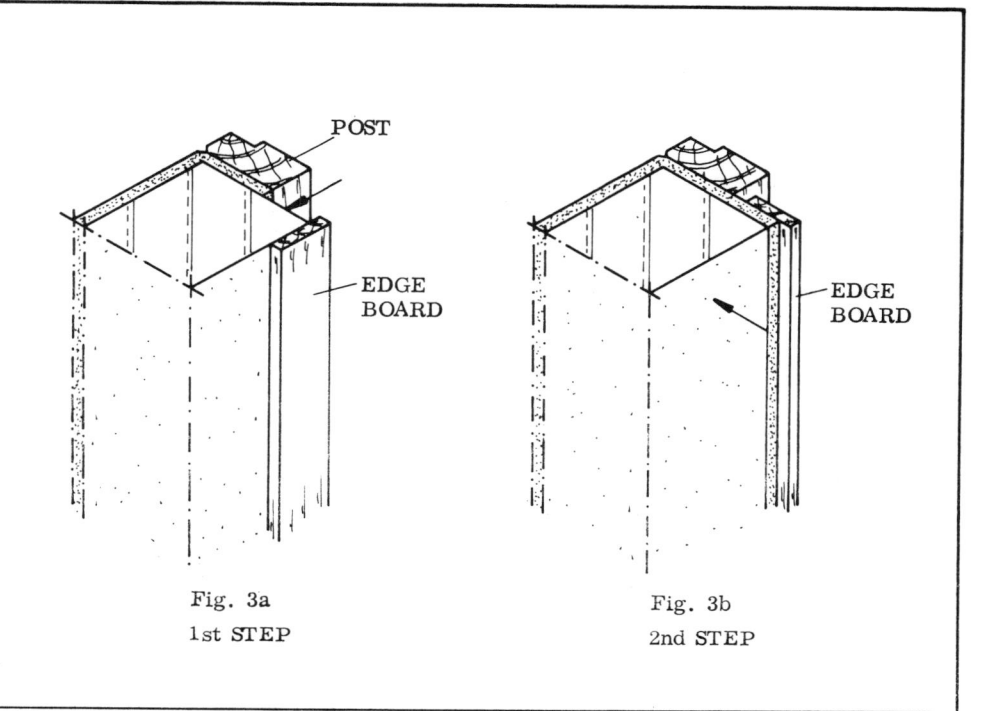

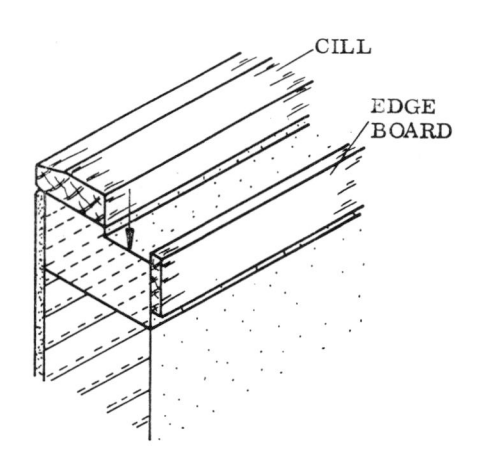

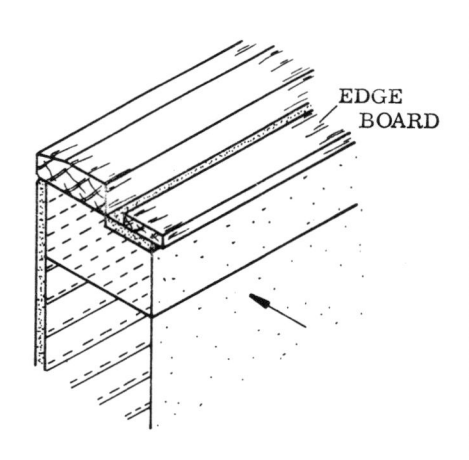

Fig. 3b
2nd STEP

Fig. 4a
1st STEP

Fig. 4b
2nd STEP

PLASTER AND RENDER.

NPVC
CON. 237

POSITION OF EDGE BOARDS FOR PLASTERING THE TOPS OF GABLES

Figs. 1a and 1b show the positions for the edge boards.

EDGE BOARDS

Fig. 1a
1st STEP

EDGE BOARD

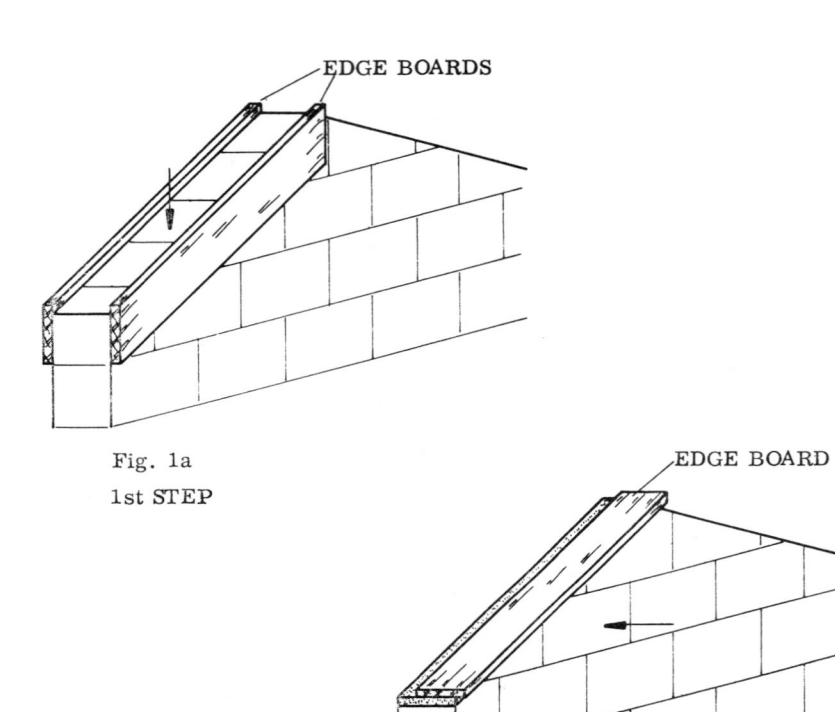

Fig. 1b
2nd STEP

NOTES:

N P V C
238 CON.

PLASTER AND RENDER.

POSITIONS OF EDGE BOARDS FOR PLASTERING THE TOP OF A PARAPETTED PENT ROOF

Figs. 2, 3, and 4 show the positions of the edge boards for plastering.

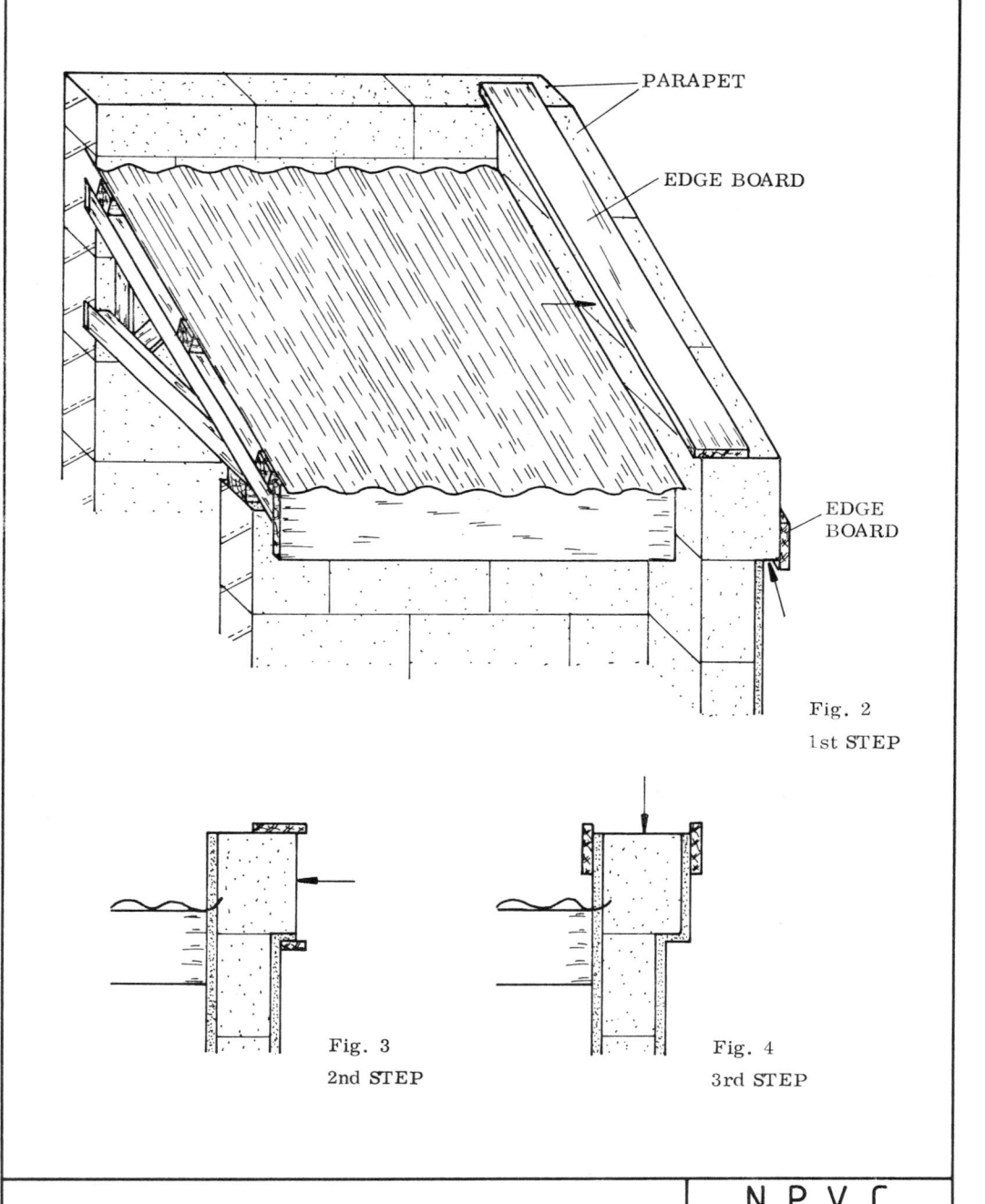

PARAPET

EDGE BOARD

EDGE BOARD

Fig. 2
1st STEP

Fig. 3
2nd STEP

Fig. 4
3rd STEP

PLASTER AND RENDER.

NPVC

CON. | 239

POSITIONS OF THE EDGE BOARDS FOR PLASTERING OPENINGS

Plaster the inside surfaces of the opening first (Fig. 1), then the rest of the wall can be plastered as usual (Fig. 2).

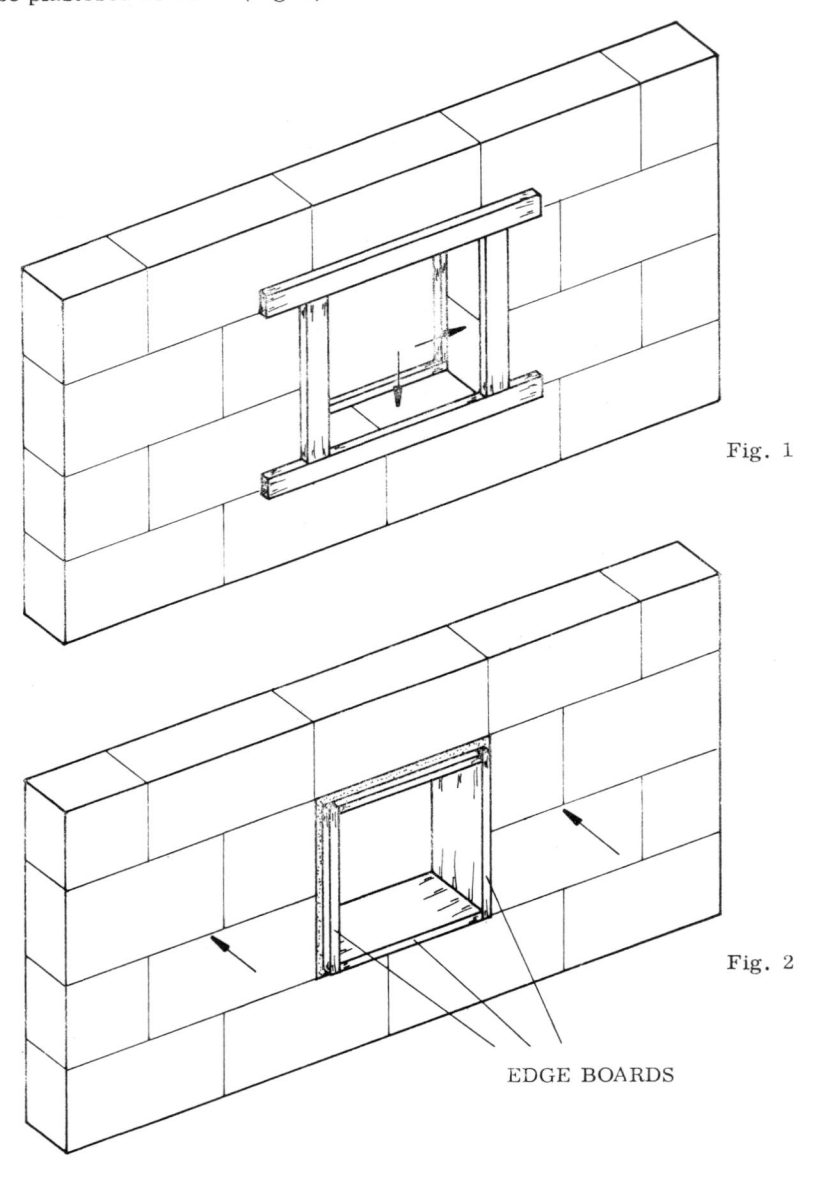

Fig. 1

Fig. 2

EDGE BOARDS

N P V C	
240	CON.

PLASTER AND RENDER.

FLOOR CONSTRUCTION

It is best to construct the floor after the roof covering has been completed. This makes it easier to cure the floor without problems caused by the concrete drying out too quickly and cracking.

Floors are constructed on top of the hardcore filling. There are two methods for floor construction used in Rural Building: one-course work and two-course work. One-course work means that the floor is made in a single layer; while in two-course work the floor is constructed in two layers, the base layer and the floor screed or finishing layer. For an explanation of the two methods and their respective advantages and disadvantages, see the Basic Knowledge book, page 179.

Any floor which is larger than 10 square metres should be divided into "bays" before it is cast (Basic Knowledge, page 180). The bays should be as nearly square-shaped as possible, so that the shrinkage on each side of the bay will be the same. The bays are separated from each other by edge boards during the casting. The smaller the bays, the fewer cracks will appear. In floors which are in the sun, the bays should be no more than 5 square metres in area, and expansion gaps should be made between them.

Expansion gaps and shrinkage gaps are explained in the Basic Knowledge book, page 183. Shrinkage gaps are made between the bays in either one-course or two-course work, to keep the floor from cracking as it hardens and shrinks a bit. Expansion gaps are larger gaps made in the base layer of two-course work when the floor is exposed to the sun or to large temperature changes, which can cause cracks when the floor expands with the heat and contracts as it cools.

The procedure for casting a floor is explained on the following pages.

NOTES:

SAND Fig. 1

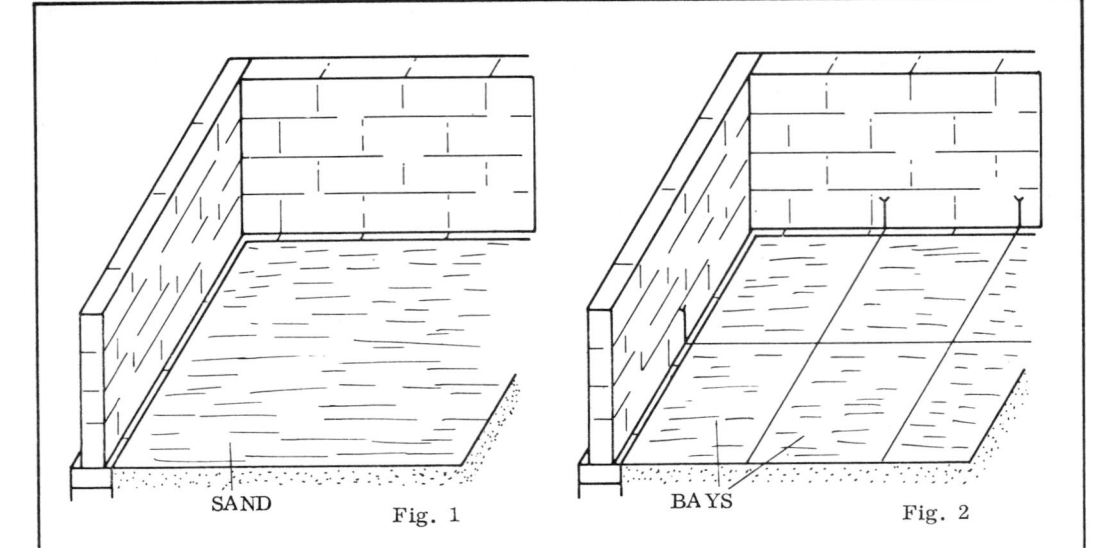

BAYS Fig. 2

SEQUENCE OF OPERATIONS FOR ONE-COURSE WORK

The hardcore filling is usually made well beforehand, to allow it to settle properly. It is constructed as explained on pages 59 and 60 in this book. The last 6 cm up to the top of the footings is filled with sand. The hardcore forms a firm support for the floor (Fig. 3).

a. Clean the whole area and level the sand surface (Fig. 1).

b. Divide the area of the room into bays and mark the bays on the walls (Fig. 2).

c. Mark the positions of the edge boards on the sand between the marks on the walls (Fig. 2). Lines may also be fixed between the wall marks at the height of the finished floor level (Fig. 3).

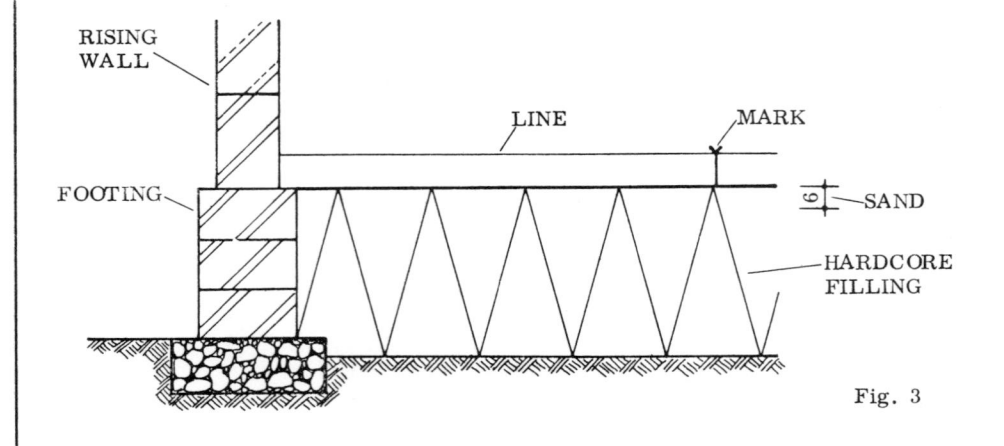

RISING WALL
FOOTING
LINE
MARK
SAND
HARDCORE FILLING

Fig. 3

NPVC 242 CON.   FLOOR CONSTRUCTION.

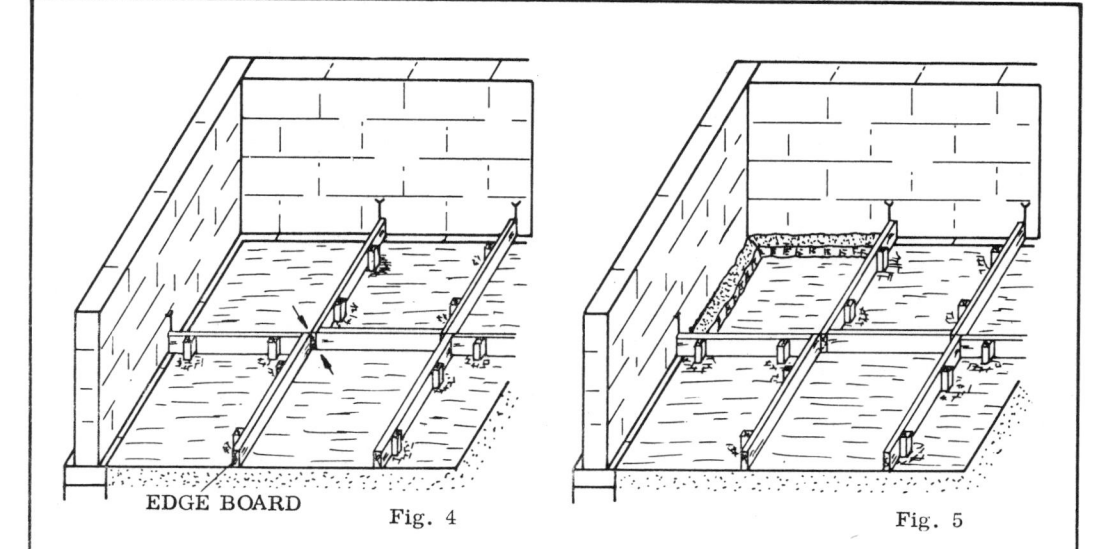

EDGE BOARD Fig. 4

Fig. 5

d. Cut the edgeboards to the proper length. Note the arrangement of the boards shown above in Fig. 4 and in Fig. 6 below, and cut the boards so that they fit in this pattern.

e. Nail the pegs to the sides of the edge boards.

f. Place the edge boards as shown in Fig. 4, along the marks. The corners of the boards should meet as shown in Fig. 6 below.

g. Knock lightly on the pegs until the edge boards are level (check with the spirit level) and at the correct height (the tops of the boards should be level with the finished floor surface).

h. Pour the guide strips to the height of the base layer (approximately 8 cm) (Fig. 5). Level the tops of the guide strips (Fig. 7).

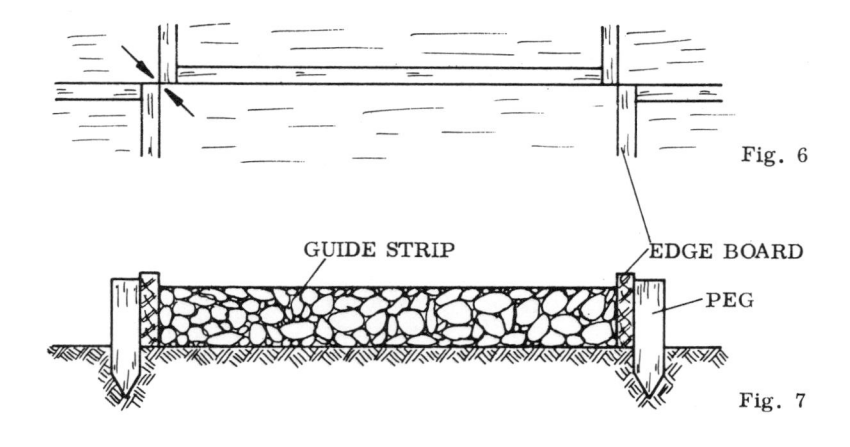

Fig. 6

GUIDE STRIP   EDGE BOARD

PEG

Fig. 7

FLOOR CONSTRUCTION.

NPVC
CON. 243

Fig. 1

Fig. 2

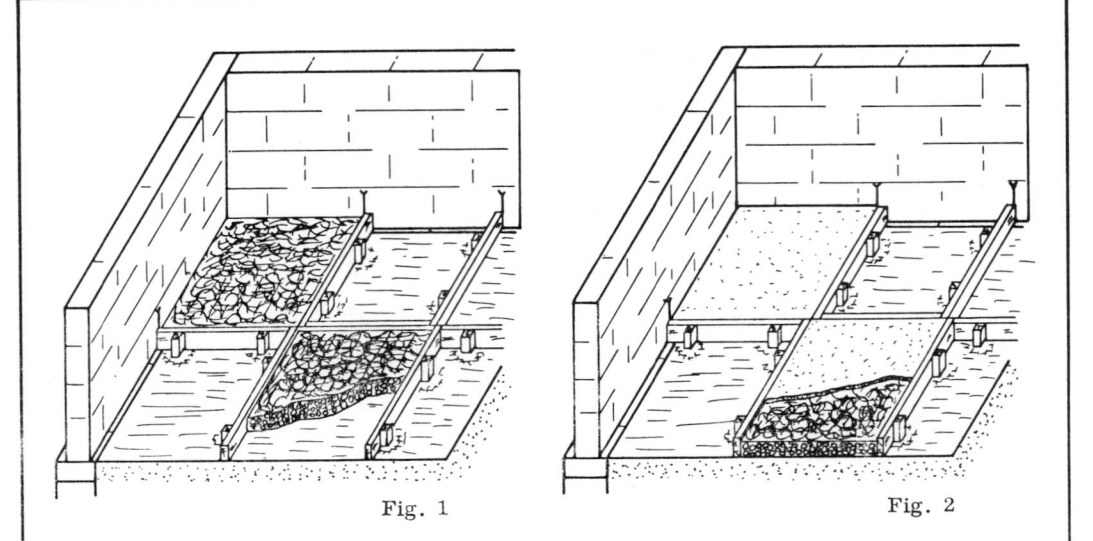

i. Pour the concrete for the base layer of one bay, making sure that it is well distributed near the walls and corners especially. Mix proportions can be found in the Tables of Figures, Reference Book, page 234 (Fig. 1).

j. Compact the concrete with a rammer, especially the corners and edges.

k. Use the notched strike board (Fig. 3) to level the surface by gently tamping until the layer is screed-thickness below the top of the edge boards.

l. Prepare the mortar for the screed layer. This should be moist but not wet.

m. Pour this mortar on top of the concrete. The screed is about 2 cm thick.

n. Tamp the screed down with a strike board until a level surface is obtained (Fig. 2).

o. Use the wood float to tamp the screed until the moisture from the concrete comes through. At the same time smooth the surface with the float. During this operation dip the float regularly in water and sprinkle water on top of the screed if necessary to make the work easier.

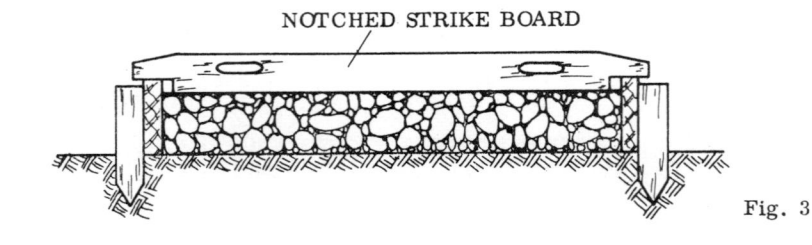

NOTCHED STRIKE BOARD

Fig. 3

N P V C	FLOOR CONSTRUCTION.
244	CON.

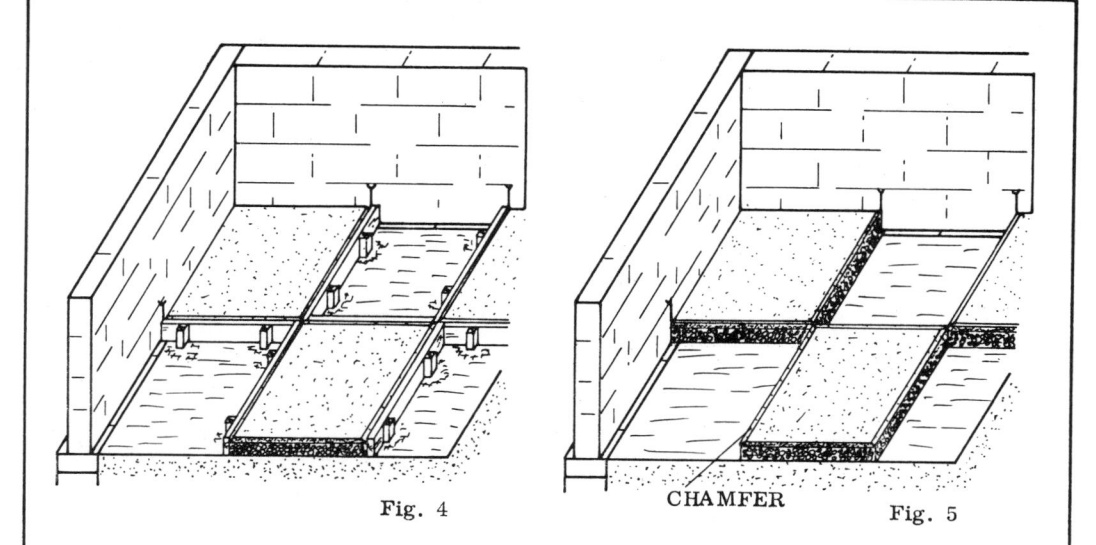

Fig. 4   CHAMFER   Fig. 5

p. If a smoother surface is desired, finish off with a steel float. Make a chamfer along each edge next to the edge board to form half of the "V" groove above the shrinkage gap (Figs. 4 & 6).

q. If blisters or bubbles appear when the concrete starts to harden, pop them and close the holes.

r. Let the floor harden for some hours, then sprinkle water over it. Keep the floor wet at all times.

s. The next day, put some clean sand on top of the floor and wet it thoroughly. Keep the sand wet until the hardening process is over in several days time.

t. Remove the edge boards and cast the remaining bays in the same way. Make sure that the bays are separate by putting pieces of plastic or paper between them before pouring the concrete in the other bays (Fig. 5). The tops of the finished bays act as guides for the strike boards, but be careful not to damage them (Fig. 7).

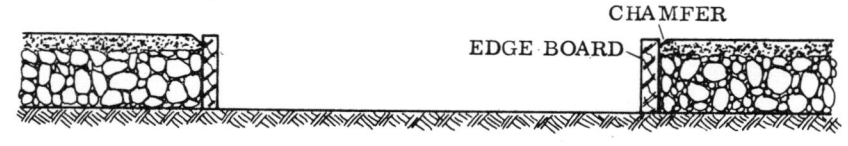

CHAMFER
EDGE BOARD
Fig. 6

PLASTIC
Fig. 7

FLOOR CONSTRUCTION.   NPVC   CON.   245

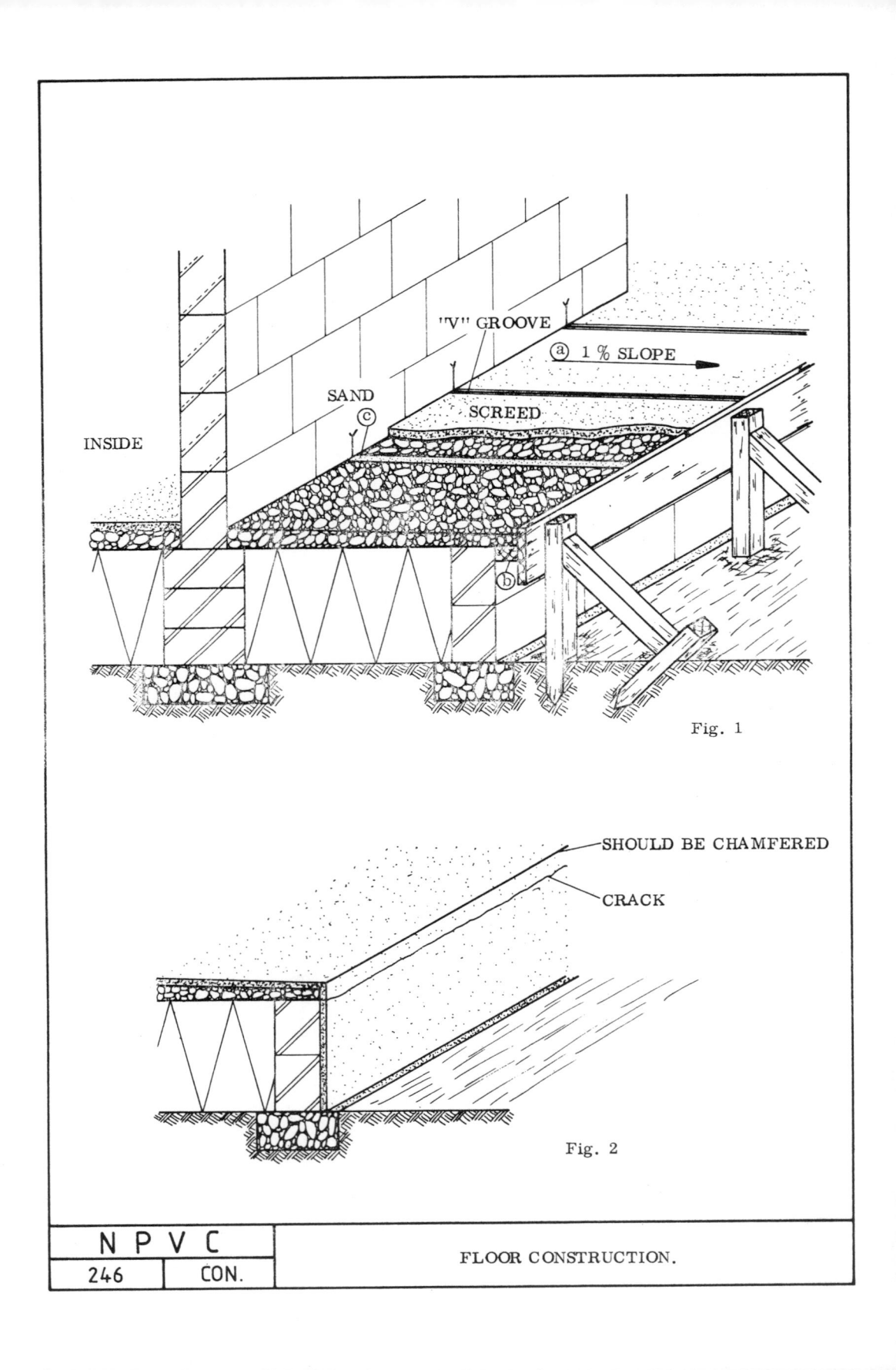

Fig. 1

Fig. 2

- "V" GROOVE
- ⓐ 1 % SLOPE
- SAND ⓒ
- SCREED
- INSIDE
- ⓑ
- SHOULD BE CHAMFERED
- CRACK

N P V C		FLOOR CONSTRUCTION.
246	CON.	

VERANDAH FLOORS

The construction of a verandah floor is different from that of an inside floor in three particular ways.

- Usually the verandah floor is built with a small slope towards the outside so that rain water can run off quickly (Fig. 1, a).

- The verandah floor can have a projecting outside edge (Fig. 1, b).

- Because the floor is exposed to the sun, expansion gaps have to be made in it. This means that the floor must be made with the two-course method. (Fig. 1, c, and Basic Knowledge book, page 183).

- SLOPING FLOOR: Verandah floors may have a slope which is about 1%; no more than 2%, because this could cause problems in walking and working on it. The greater slope increases the danger of slipping when the surface is wet.

  A slope of 1% means that the floor slopes 1 cm lower in every metre; in this case the slope is across the width of the floor.

  The construction of the slope is done by simply setting the edge boards between the bays at the required slope.

- PROJECTING OUTSIDE EDGE: If the outside edge of a verandah floor is kept flush with the render of the footing (Fig. 2), then cracks would soon appear along the edge. This is prevented by constructing the floor so it projects past the footing by no more than 7,5 cm. The shorter the projection the better, because the floor is not reinforced.

NOTES:

FLOOR CONSTRUCTION.

N P V C
CON. | 247

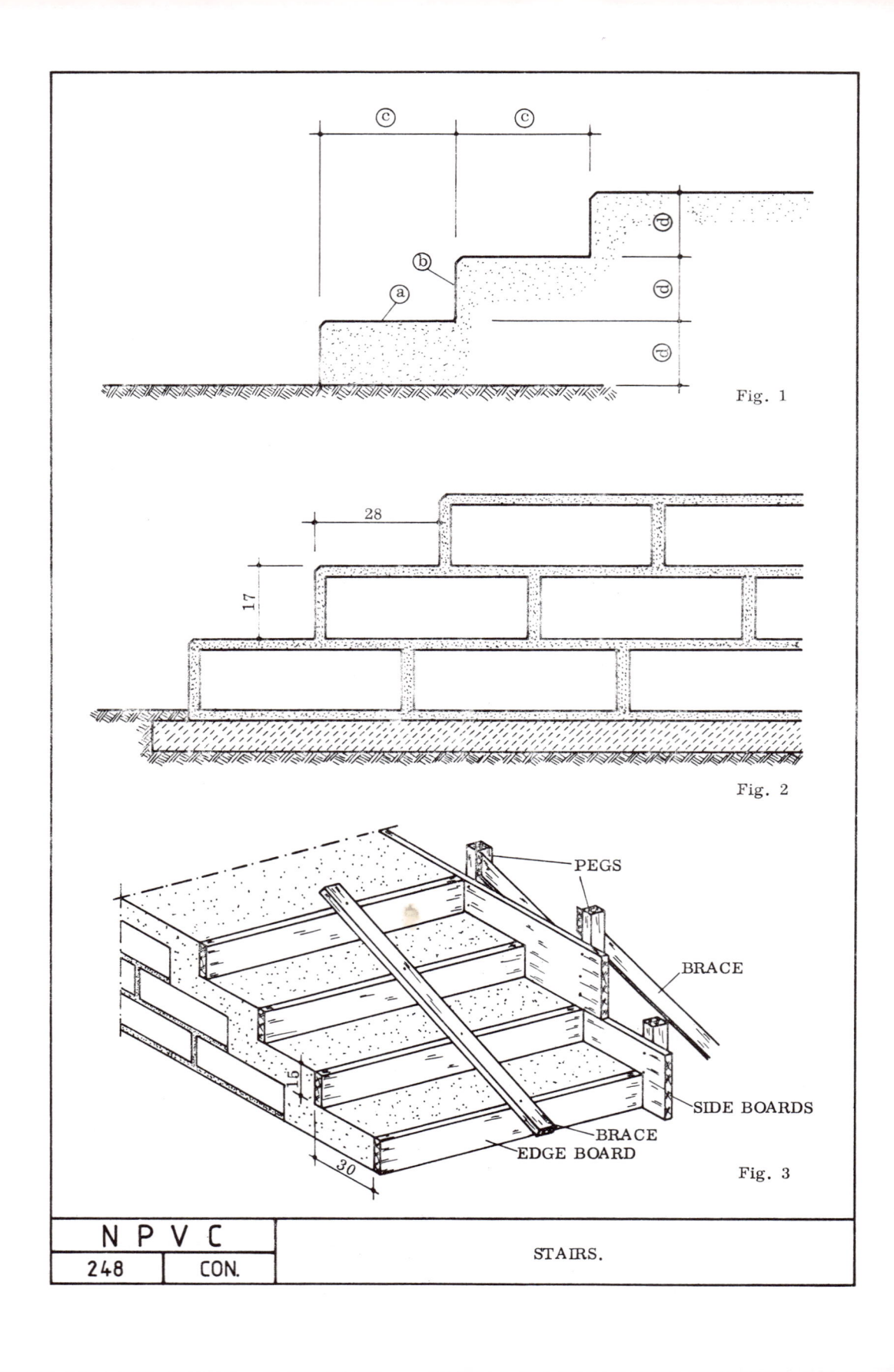

Fig. 1

Fig. 2

Fig. 3

PEGS
BRACE
SIDE BOARDS
BRACE
EDGE BOARD

N P V C		STAIRS.
248	CON.	

STAIRS

A stair means a series of at least three steps, also called a flight of stairs. A stair can form the link between two storeys of a building. In Rural Building we are only concerned with one storey buildings, so the stairs are usually only needed up to the level of the floor of the building, which is normally not much higher than 60 cm above ground level.

- TECHNICAL TERMS: Fig. 1 shows the parts of a staircase.

  - TREAD: The tread is the horizontal part of the step (a).
  - RISER: This is the vertical part of a step (b).
  - GOING: This refers to the width of the tread (c). The "going" of a flight of steps is the sum of the "goings" of all the steps of the flight.
  - RISE: This is the vertical distance between two treads, equal to the height of the riser (d).

REQUIREMENTS

The short flights of stairs we make in Rural Building are usually built with sandcrete blocks or are cast out of concrete. Regardless of the materials and the method of construction, the steps must be all the same height and the same width. A staircase with steps of different sizes is likely to cause accidents.

The rise of the steps is usually 15 cm (approximately) and the width of the tread, the "going", is usually about 30 cm for each step.

The steps in Fig. 2 are built with sandcrete blocks, so the rise of 17 cm is convenient because it fits in with the block size. The width of the tread can be about 28 cm, or slightly more than the half-block size.

Steps cast out of concrete can have any convenient rise, but normally the rise is about 15 cm high, and the width of the tread will be approximately 30 cm. The formwork for casting the steps is shown in Fig. 3.

Plan ahead. If the steps with a 15 cm rise do not reach the right level (measure the height before you start to make the steps) either change the height of the rise (to 14, 16, or 17 cm, etc); dig out some soil before to lower the height of the whole stair; or add a foundation slab to raise the whole construction by a few centimetres to the required height.

- REMEMBER: Never make steps of different sizes in the same stair.

WIRE
PIPE
Fig. 2a

DRIP
Fig. 1

Fig. 2

NAIL LINE
Fig. 3

HALF-ROUND BEAD
Fig. 3a

Fig. 4

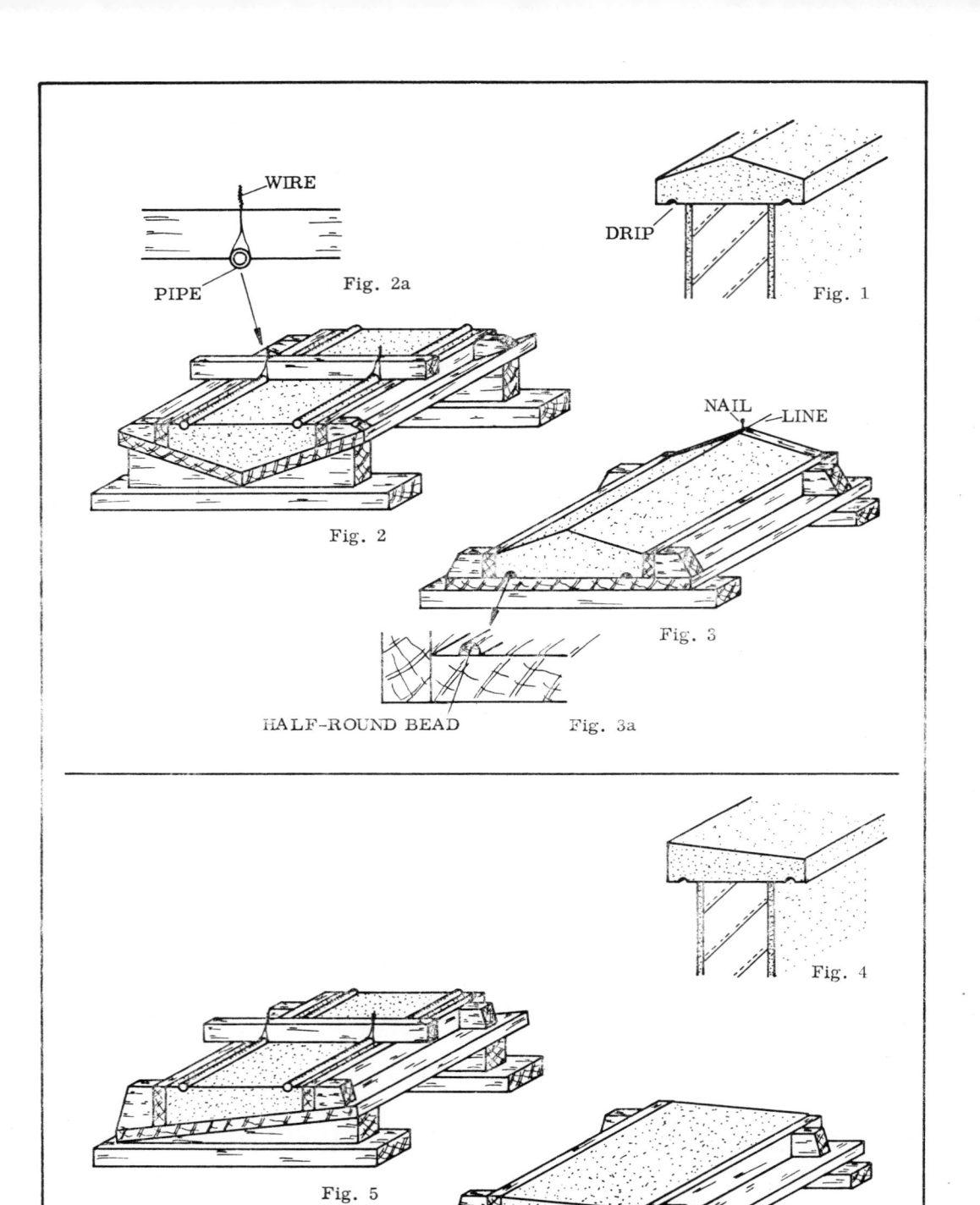

Fig. 5

Fig. 6

N	P	V	C
250	CON.		

FORMWORK.

FORMWORK

FORMWORK FOR COPINGS

A free-standing wall which is not part of the building (not under the roof) needs to be protected against rain, which otherwise might penetrate at the top of the wall and damage it. This is done by adding a "coping" at the top of the wall.

- SADDLE-BACK COPING: This is placed on top of the wall as shown in Fig. 1. The formwork for the coping is shown in Fig. 2. The drip or throating is made by setting two pipes or smooth iron rods in the wet concrete. The correct position can be assured by using wooden spacers tied to the pipe (Figs. 2 & 2a).

  A second way of making the coping is shown in Fig. 3. A line is fixed at the height of the ridge and the surfaces are shaped with the steel float. The drip is made by fixing half-round beads onto the soffit board of the formwork (Figs. 3 & 3a).

- SPLAYED COPING: The splayed coping in Fig. 4 can also be made in two ways, in a fashion similar to the saddle-back coping above (Figs. 5 & 6).

FORMWORK FOR PILLARS

The formwork for a concrete pillar is shown in Fig. 7 below.

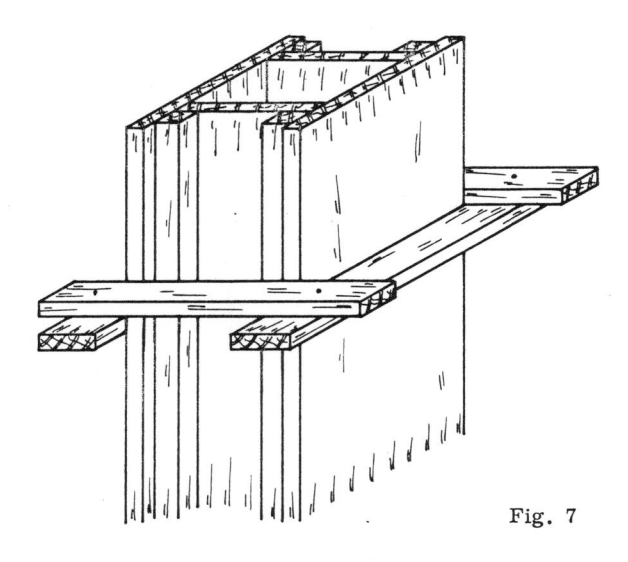

Fig. 7

FORMWORK. NPVC CON. 251

BRACE

PAPER

Fig. 1

BRACE

PAPER

Fig. 2

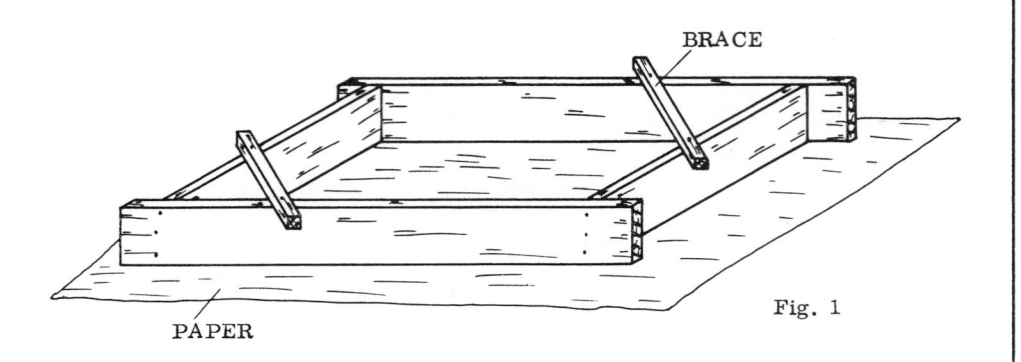

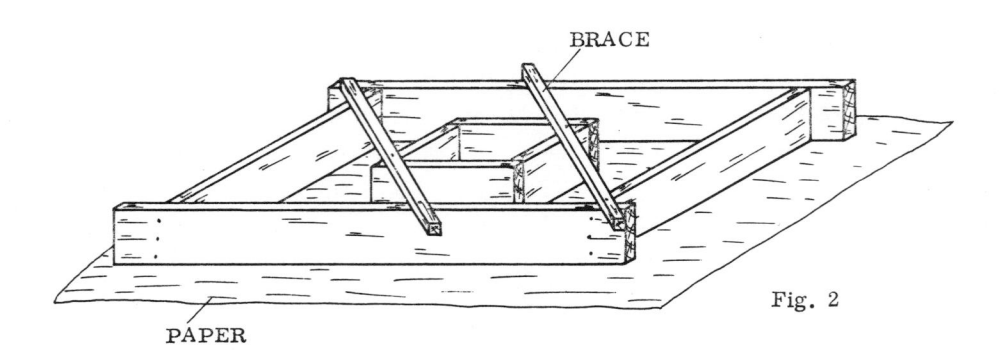

NOTES:

N P V C	
252	CON.

FORMWORK.

FORMWORK FOR SLABS

Fig. 1 shows an example of formwork for a slab. Paper should be set underneath before the concrete is cast.

Fig. 2 shows an example of the formwork for a slab with an opening in it.

GENERAL HINTS FOR FORMWORK

- Use good, straight-grained wood.

- Make a sketch of the formwork before you start to make it.

- Design the formwork so it can be easily removed from the piece after curing.

- Use enough supports and braces to make the formwork rigid and strong.

- Be aware that the wood swells when it is in contact with the wet concrete.

- If the concrete will not be plastered later, plane the formwork members smooth where they are in contact with the concrete.

- Do not overnail the structure, and fix the nails so that they can be taken out easily.

- Oil the form lightly where it is in contact with the concrete.

- For precast members, put the form on the ground in a level spot, and put paper underneath it (old cement bags).

- Remove the formwork only when the concrete has hardened sufficiently (the time depends on the member which is cast).

- Be careful not to damage the edges and corners of the concrete piece when you remove the formwork.

- Clean off the formwork with a steel brush and take out all nails when you finish.

NOTES:

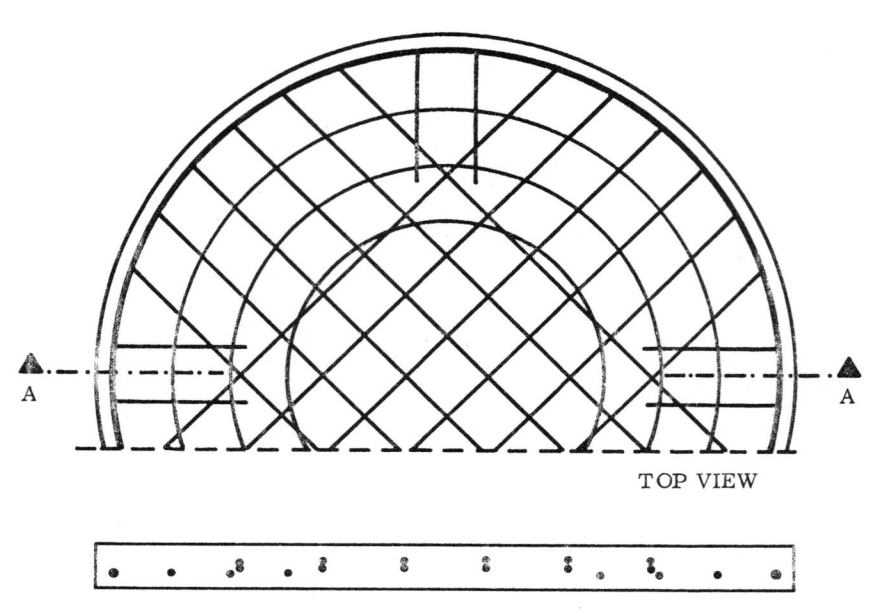

TOP VIEW

CROSS SECTION A-A   Fig. 1

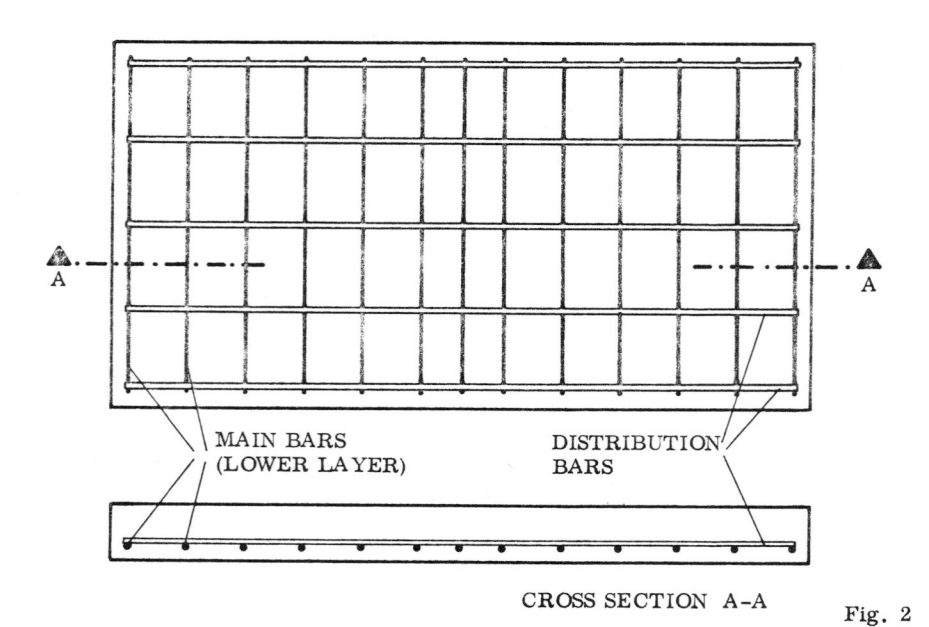

MAIN BARS
(LOWER LAYER)

DISTRIBUTION
BARS

CROSS SECTION A-A   Fig. 2

N P V C		REINFORCED CONCRETE SLABS.
254	CON.	

REINFORCED CONCRETE SLABS

In the northern part of Ghana a typical traditional house has a flat roof made of mud plastered over sticks, and supported by tree trunks. The flat roof is very appropriate for the area, but sometimes these traditional constructions have problems like leaking or even collapse during heavy rains.

In Rural Building we do not deal with the construction of flat roofs, even though these would be more appropriate to the local customs. The reason is because the materials to construct a safe, strong flat roof are expensive, and the construction of a reinforced concrete roof is very difficult and complicated.

Therefore, in this course we only treat the construction of relatively small reinforced concrete slabs such as manhole covers, latrine slabs, septic tanks, etc.

CONSTRUCTION

Reinforcement mats are a simple way of reinforcing slabs (Reference Book, page 173) but they are sometimes hard to obtain and they are expensive. The Rural Builder usually has to rely on single reinforcement bars to reinforce slabs.

Fig. 1 shows a typical arrangement for a circular slab, and Fig. 2 shows that for a rectangular slab.

The distance between the 12 mm main bars is never more than 10 cm. The thickness of the slab can vary between 7,5 cm for smaller slabs like manhole covers, and up to 15 cm for covers of septic tanks, or slabs on which people may walk.

- NOTE: This book is not meant to train engineers; it merely gives basic information. If you need to construct a larger reinforced concrete member, you are advised to seek help from a building expert.

NOTES:

Fig. 1

Fig. 2

Fig. 3

Fig. 4

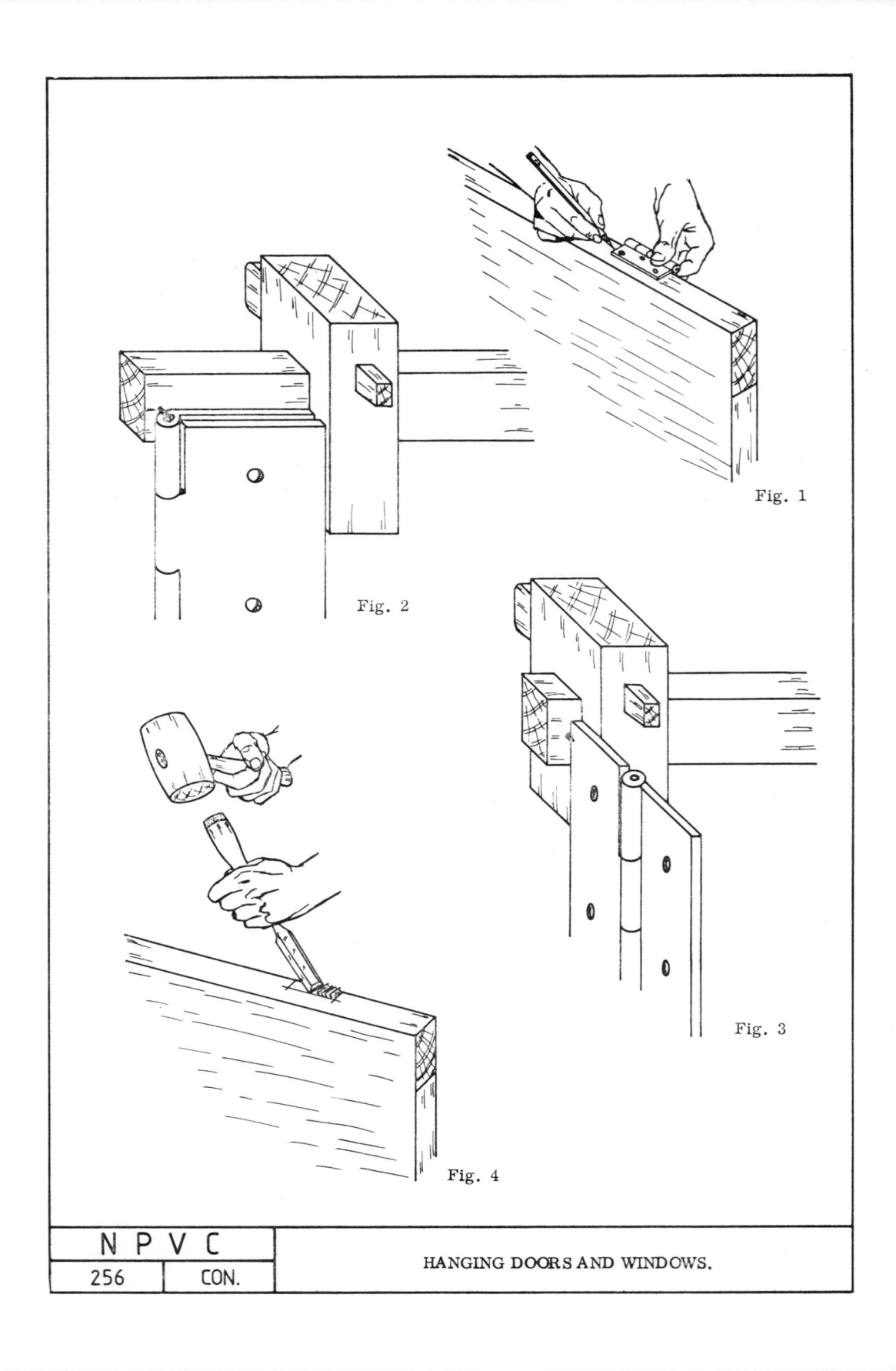

NPVC 256 CON.

HANGING DOORS AND WINDOWS.

HANGING DOORS AND WINDOWS

The first step in hanging a door is to make sure that the posts of the door frame are plumb. If the posts are not plumb the door will not hang properly. A door should remain in position, either open or closed, without the aid of a stop. It should not swing open or closed by its own weight. The same applies for window casements.

If the door posts are not plumb, the only solution is to fix the hinges in positions such that the door itself hangs plumb.

HANGING A DOOR WITH BUTT HINGES

Before you read the following section, look up butt hinges in the Reference Book, pages 219 to 221.

- SEQUENCE OF OPERATIONS:

  a. Set the door into the door frame.

  b. Check whether there is a sufficient clearance all around the door (at least 3 mm, more in the dry season).

  c. Mark the positions of the hinges, then take the door down and mark the length of the hinges on the door (Fig. 1).

  d. Set the marking gauge from the middle of the pin of the hinge to the edge of the hinge leaf, and mark the distance on the edge of the door (Fig. 2).

  e. Set the marking gauge to the thickness of one hinge leaf, and mark this on the face of the door (Fig. 3).

  f. Chisel the recesses for the hinges in the door, and fix each hinge with one screw (Fig. 4).

  g. Place the door in the door frame so that there is equal clearance around all the edges. A few thin pieces of wood between the door and frame will help to keep it in place.

  h. Mark the positions of the hinges on the post. Remove the door.

  i. Mark the depth and width of the hinge recess on the post, using the marking gauge.

  j. Chop the recesses in the post.

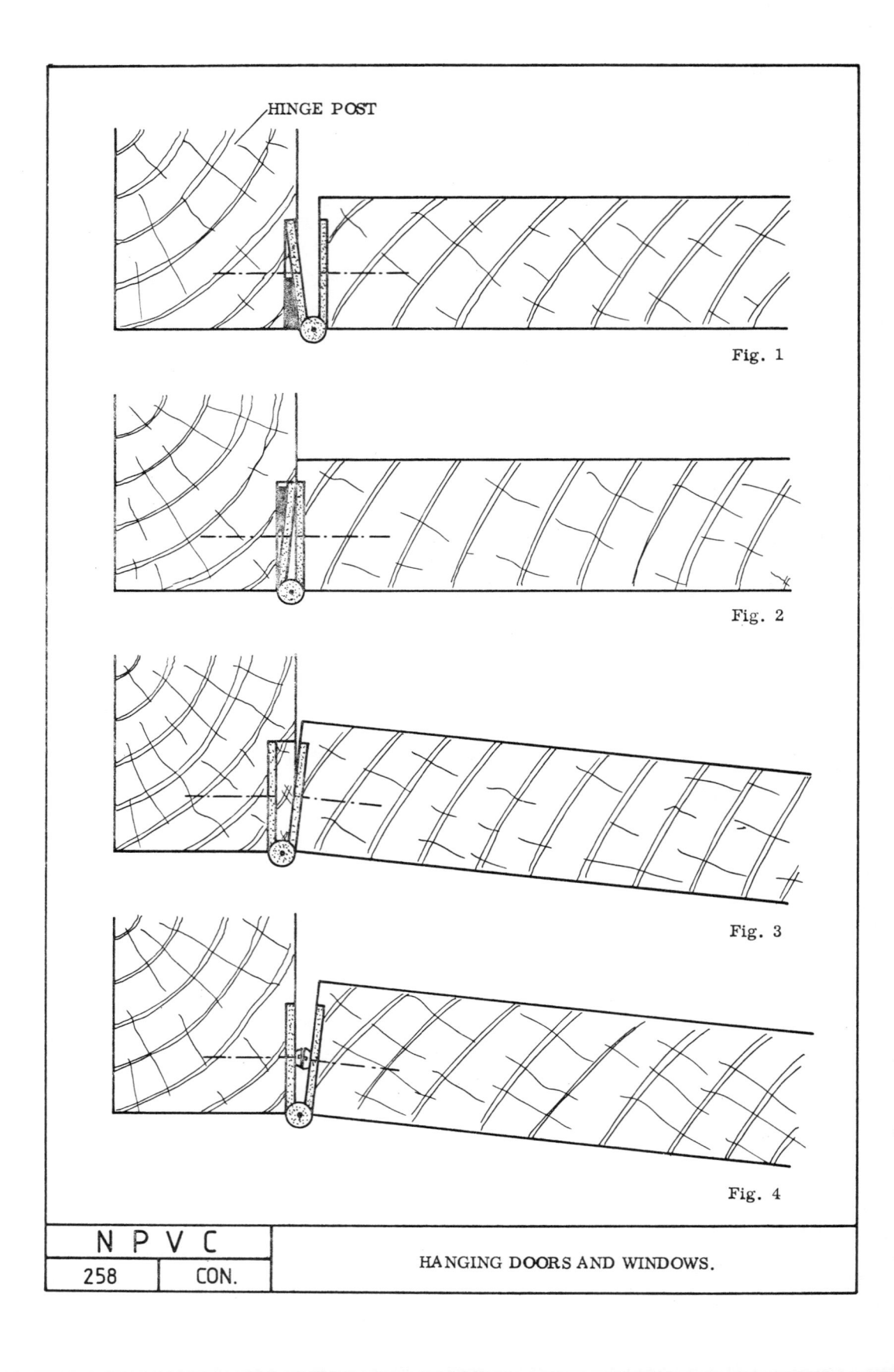

HINGE POST

Fig. 1

Fig. 2

Fig. 3

Fig. 4

N P V C	
258	CON.

HANGING DOORS AND WINDOWS.

k. Place the door in position and fix it to the door post with one screw in each leaf of the hinges. Close and open the door several times to see if it works smoothly. Notice in particular whether the door closes completely and does not scrape either post. It should stay closed by itself.

The same sequence as above also applies for casements.

- ADJUSTMENTS OF THE HINGES: Below are some tips for getting a door or casement to close properly.

  - If one hinge recess is deeper than the other, loosen the screws and pack the deeper recess with thin strips of wood, until the depths of the recesses are the same.

  - When a door binds (catches or rubs) on the hinge post, insert thin pieces of wood at the face edge of the hinges (near the pin) until the problem is relieved (Fig. 1). A strip about 1 cm wide and as long as the hinge will do.

  - When a door binds on the locking post, insert the wood strip at the back of the hinge, to force the door away from the lock post (Fig. 2).

  - If the wood strips are not enough to relieve the problem, you will have to remove the door and plane something off, alter the hinge recesses or even move them a bit up or down.

  - One possible reason that a door won't close well is that it is hinge bound. This means that the leaves of the hinges are set too deep. To remedy this, you must remove the hinge and place a thin piece of wood between the leaf and the post (Fig. 3).

  - When the hinge is screw bound, the heads of the screws obstruct the door so that it can't close properly. This is often the case when the screws are too thick for the hinges, or the heads are too large. The only remedy is to fix the correct screws (Fig. 4).

NOTES:

HANGING DOORS AND WINDOWS.

(b) to (e)   (f)

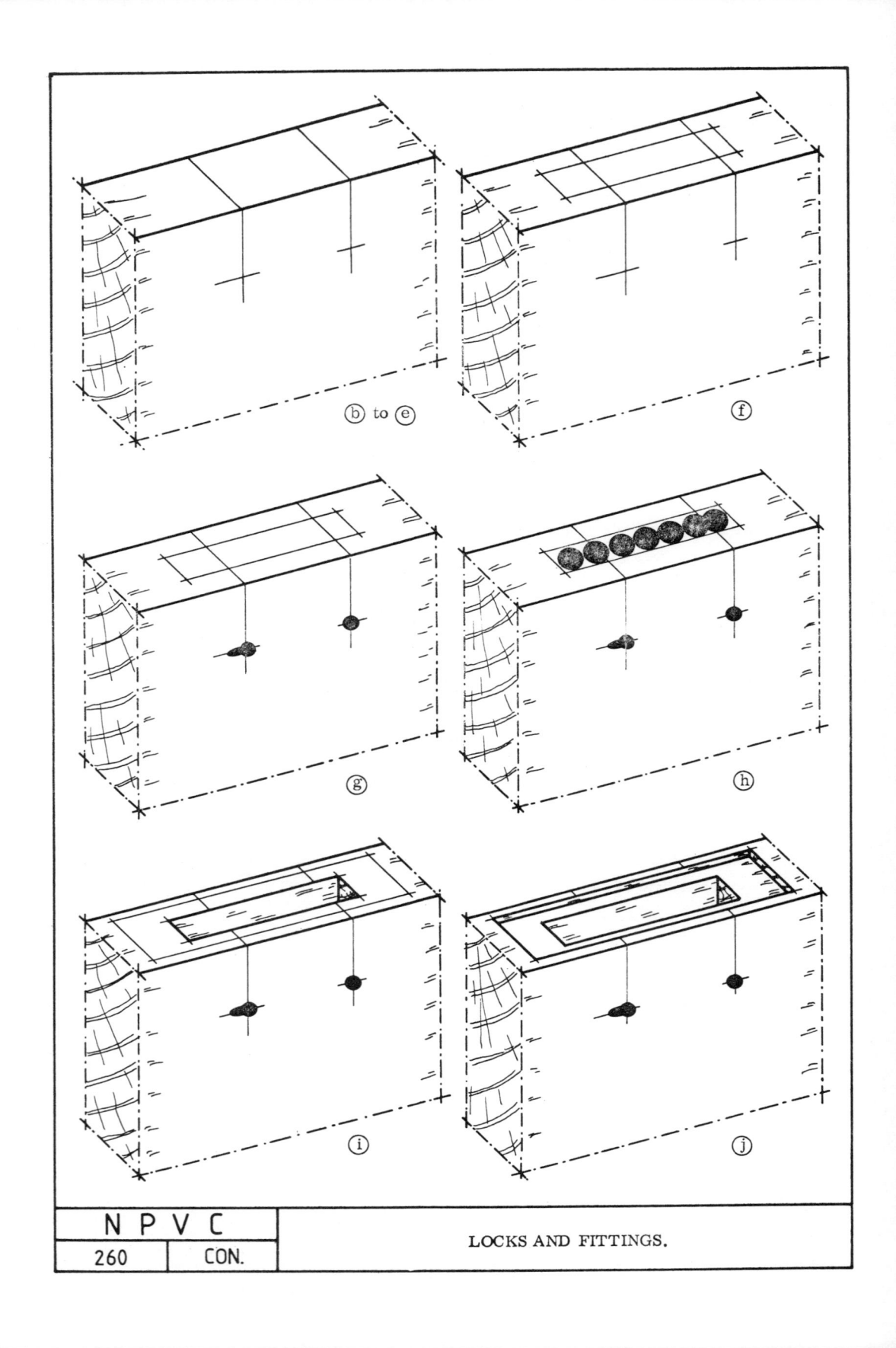

(g)   (h)

(i)   (j)

N	P	V	C		
260		CON.		LOCKS AND FITTINGS.	

LOCKS AND FITTINGS

MORTICE LOCK

Before you read this section look up mortice locks in the Reference Book, pages 221 to 223. Pay attention to the names of the parts of the lock.

- SEQUENCE OF OPERATIONS FOR INSTALLING A MORTICE LOCK:

  a. Place the door on edge against the side of the work bench, with one face against the bench and with the shutting stile upwards.

  b. Mark the distance from the bottom edge of the door to the centre of the "bush" (the bush is the hole where the handle fits through, normally it is 105 cm from the bottom edge of the door).

  c. Square the positions of the key hole and the bush on the face and edge of the door.

  d. Set a marking gauge with the distance from the outside edge of the lock face plate to the centre of the key hole. Add 2 mm allowance for planing off later.

  e. Mark this gauge setting on the door.

  f. Mark the position of the lock stock onto the shutting edge of the door.

  g. Clamp a piece of waste wood under the door to prevent splintering. Drill the bush hole and the keyhole. Complete the slot for the key hole with a chisel.

  h. Chop the recess for the stock with a mortice chisel. Much of the waste can be bored away with a drill. Take care that the stock fits tightly against the sides of the mortice.

  i. Drop the lock in the mortice and mark the position of the face plate on the door edge. Remove the lock.

  j. Chop the recess for the face plate with a mortice chisel.

  k. Drop the lock in again and insert the key and the handle with the fixed spindle. Check that there is nothing blocking the key and handle.

  l. Attach the lock with screws.

  m. Hold one leaf plate in position on the outside of the door. Insert the key and the handle with the spindle.

  n. Attach the leaf plate to the door with screws.

  o. Fix the leaf plate on the other side of the door in the same manner.

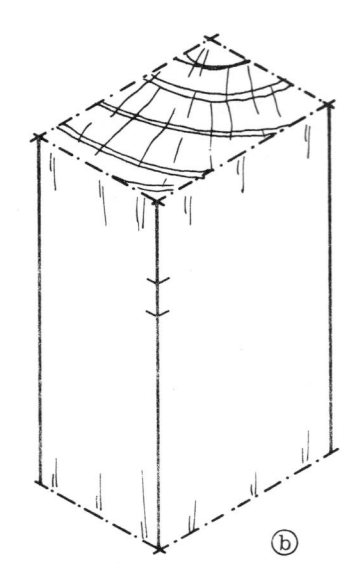

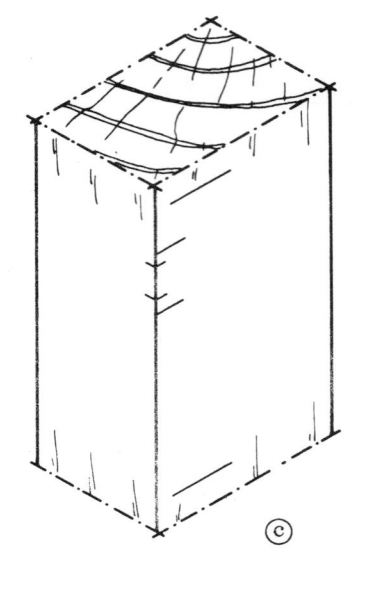

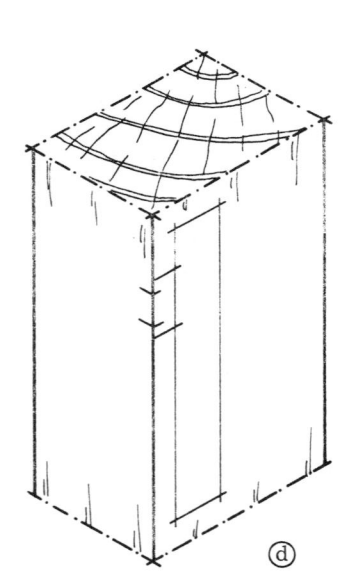

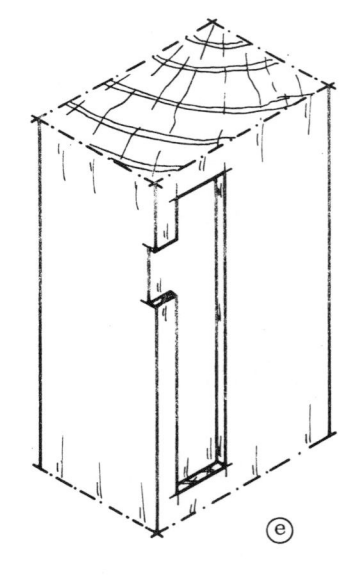

N P V C	
262	CON.

LOCKS AND FITTINGS.

p. Secure the loose handle onto the spindle, with the pin.

The installation of a mortice lock with a cylinder is similar to the above procedure; but instead of a key hole, a hole is drilled for the locking cylinder.

- SEQUENCE OF OPERATIONS FOR INSTALLING THE STRIKING PLATE:

a. Set the door with the mortice lock into the door frame.

b. Mark the position of the latch bolt onto the door frame.

c. Hold the striking plate to the post in the correct position and mark the position on the post.

d. Set the gauge and use it to mark the vertical lines.

e. Cut the recess for the striking plate.

f. Fasten the striking plate to the post with screws.

g. Chop the recesses for the latch bolt and the lock bolt while the striking plate is in position.

h. Cut off the projecting lug of the striking plate (arrow).

i. Close the door and check whether the latch bolt and the lock bolt can move freely and fit into their recesses.

In the case of a door frame with beads, the beads can be fixed after the door is properly hung and the locks installed.

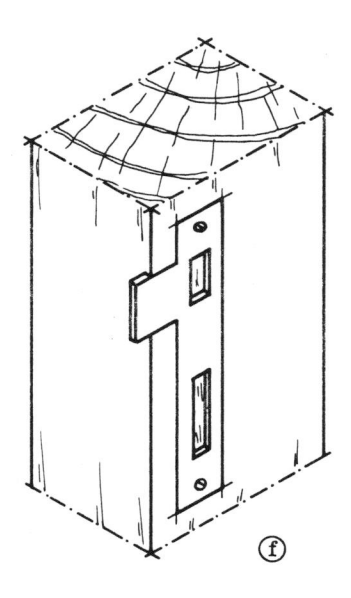

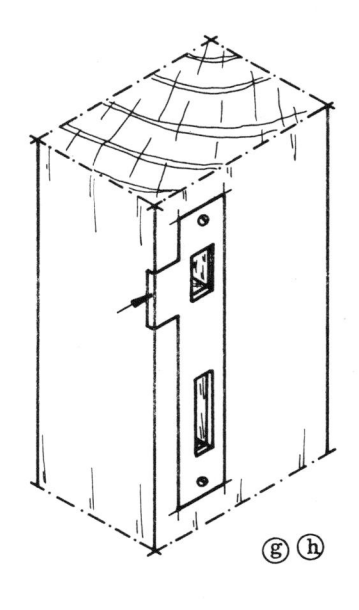

(f)

(g) (h)

LOCKS AND FITTINGS.

NPVC
CON. 263

RIM
CYLINDER
ⓐ ⓑ ⓒ

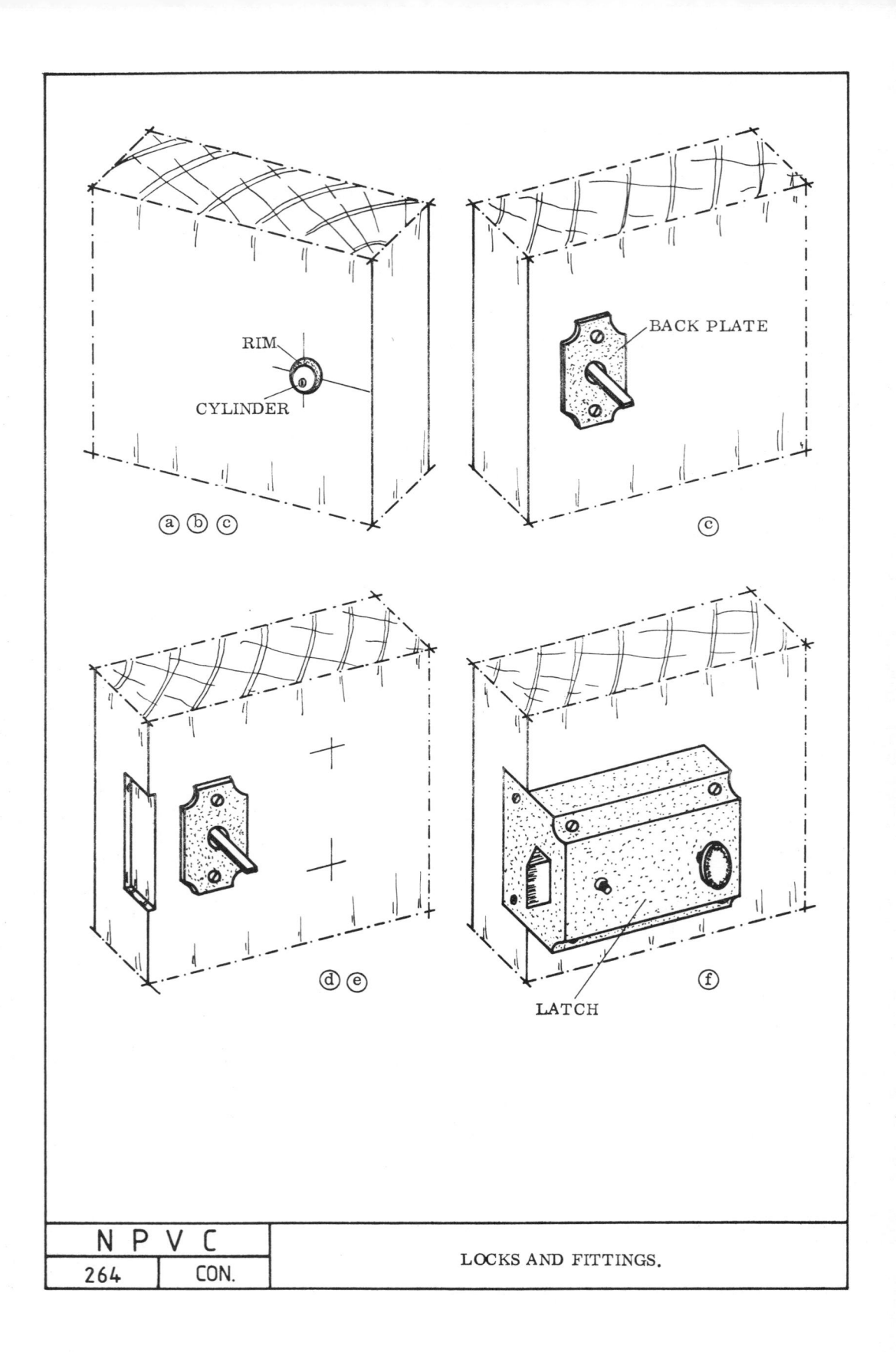

BACK PLATE
ⓒ

ⓓ ⓔ

LATCH
ⓕ

N P V C		LOCKS AND FITTINGS.
264	CON.	

RIM LOCK

Look in the Reference Book, page 225, for information about this type of lock and its parts before you read on.

- SEQUENCE OF OPERATIONS FOR INSTALLING A CYLINDER RIM NIGHT LATCH:

    a. Mark out the position of the cylinder on the door (normally 105 cm from the bottom of the door).

    b. Drill a hole with the diameter of the shell.

    c. Insert the cylinder with the rim into the hole, and attach the back plate to the opposite side of the door with screws.

    d. Mark the position of the latch and its face plate.

    e. Chop the recess for the face plate.

    f. Attach the latch to the door with screws. Also use two screws on the face plate to attach the lock securely.

    g. Turn the knob and check that the latch operates smoothly.

    h. Mark the staple on the post and chop the recess for the staple face plate.

    i. Attach the face plate to the post with screws.

    j. Check that the bolt of the latch fits in the staple.

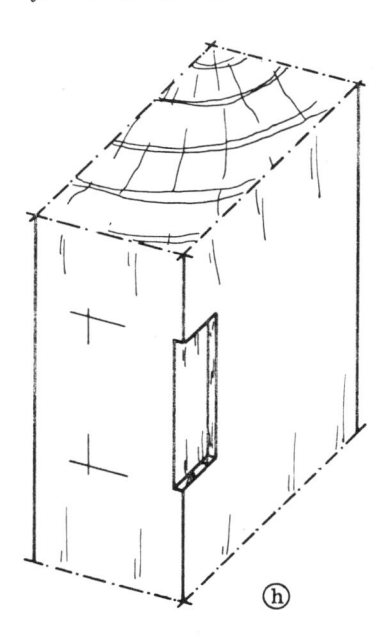

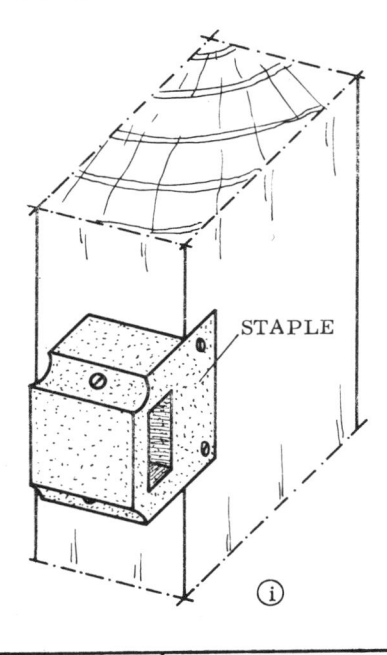

LOCKS AND FITTINGS.

N P V C

CON.   265

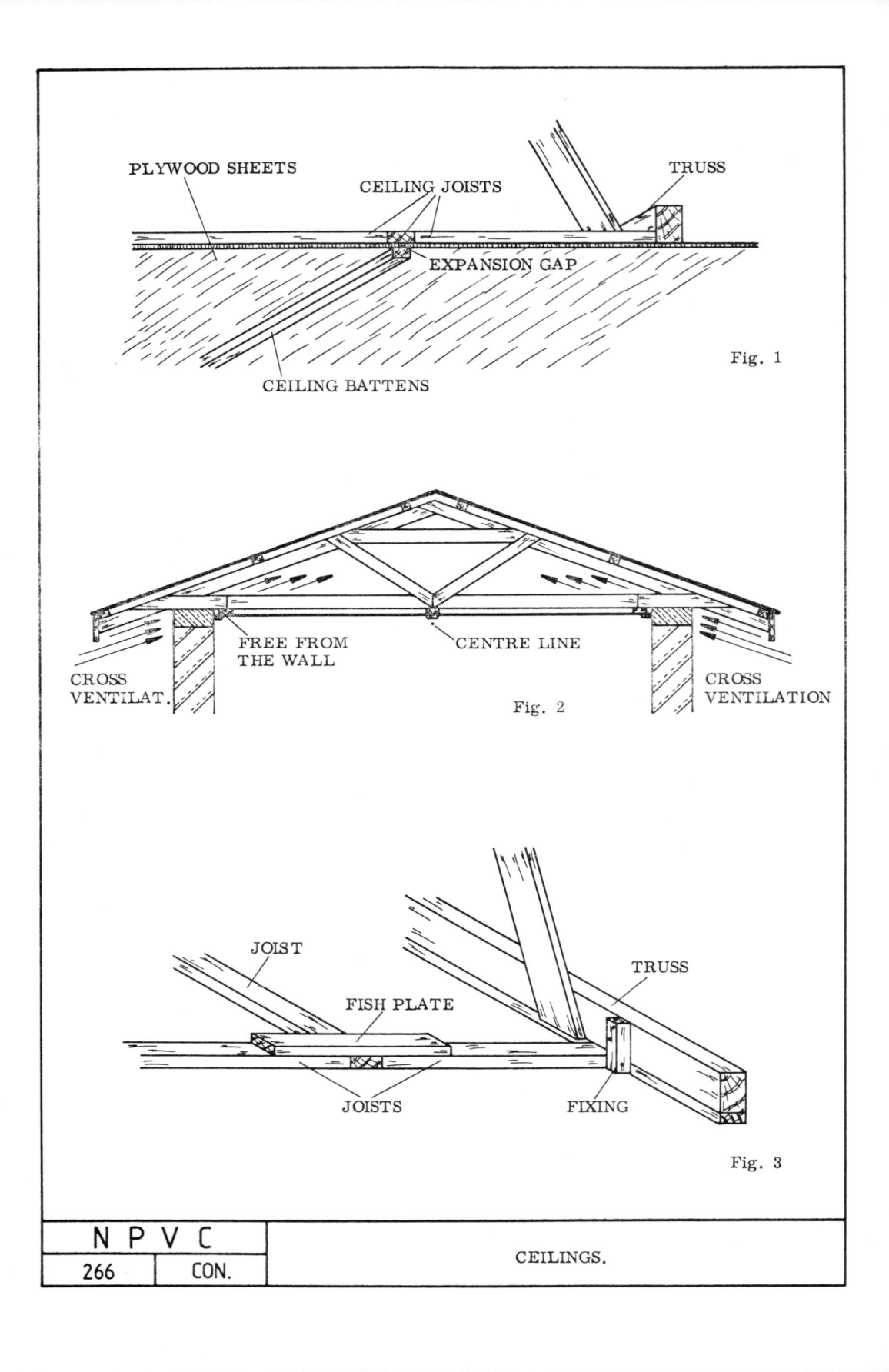

Fig. 1 — PLYWOOD SHEETS, CEILING JOISTS, TRUSS, EXPANSION GAP, CEILING BATTENS

Fig. 2 — CROSS VENTILAT., FREE FROM THE WALL, CENTRE LINE, CROSS VENTILATION

Fig. 3 — JOIST, FISH PLATE, TRUSS, JOISTS, FIXING

N P V C		
266	CON.	CEILINGS.

CEILINGS

The best method of keeping the building cool is to separate the rooms from the roof by a ceiling. The ceiling also keeps out dust and dirt, gives the room a nicer appearance, and makes it quieter.

In Rural Building, we deal only with plywood ceilings. These consist of sheets of thin plywood nailed to ceiling joists (the wooden beams which carry the ceiling). The edges where the plywood sheets join are covered by the ceiling battens (Fig. 1).

- NOTE: Cross ventilation is extremely important. Every ceiling should be constructed so that there is cross ventilation between it and the roof. If this is not done, bats and other animals will make their homes above the ceiling; and also any dampness will soon cause the ceiling sheets to rot.

Cross ventilation keeps heat from building up between the roof and the ceiling, and thus the rooms are cooler (Fig. 2).

CONSTRUCTION, PARTS AND SIZES OF PLYWOOD CEILINGS

It is easiest to construct a ceiling under a roof with built-up trusses. If the soffits (undersides) of the trusses are level, the ceiling joists can be nailed between the tie beams. If the tie beams are not level, use battens nailed to the tie beams to build up until the surfaces are level (Fig. 3).

If you want to fix a ceiling in a building with an ordinary pent roof (without trusses), a separate construction must be attached to the roof construction or to the wall, to provide support for the ceiling joists. Be sure to provide plenty of cross ventilation by some means (ventilation blocks could be used).

The distance between the ceiling joists should be no less than 122 cm. The joists can be left rough, because they are out of sight behind the plywood sheets.

When possible, design the rooms with the plywood sheet size in mind, just as the roof plan is made according to the size of the sheet material. Plan the size of the verandah and the positions of the inner walls so that the rooms are a convenient size and the plywood sheets fit without unnecessary waste.

NOTES:

- INSTALLING THE PLYWOOD CEILING SHEETS: Always start by fixing a line to show the exact centre of the room (Fig. 2, previous page). Start fitting sheets from this line. In case the sheets will not fit exactly in the room without being cut, start by putting one whole sheet in the centre of the ceiling, then divide the remaining space into equal areas.

There should be expansion gaps between the sheets to allow for changes in size with the changes of humidity. These gaps are covered by the ceiling battens. The battens are about 1 by 2,5 cm in size. Paint the ceiling and the battens before fixing the battens.

PAINTING

Paint preserves building materials from rot, rust, and general decay. Timber especially needs some finishing treatment, whether it is used inside or outside the building. Paint helps keep the wood from swelling or warping.

Protective finishes such as oil paints or varnish (Reference Book, pages 200 to 201) cover the wood with a protective "skin". In order to be effective this skin must be undamaged, and so we have to repair and maintain these finishes periodically.

Which type of finish is used depends on whether the wood will be used outside or inside the building and on the particular function of the wood piece.

The selection of the colours for the inside and outside of the building is important because it affects the temperature of the building. Light colours keep the building cooler.

PREPARATION OF SURFACES FOR PAINTING

Surfaces to be painted should be dry and clean; free from mud, dust, dirt, grease, rust, and old scaly paint.

All the boring, cutting, and shaping should be finished before the paint is applied.

Timbers should be well seasoned to prevent cracking. Paint with cracks in it is worse than no protection at all; water can enter the wood through the cracks, but it cannot evaporate off through the skin of paint so it causes the wood to rot.

If cracks appear in painted wood, sandpaper the wood before you apply new paint, to prevent the cracks from coming back. When you cut painted wood, don't forget to repaint the ends.

HOW TO PAINT

Paint when the weather is good for drying, and when there is little or no wind or dust in the air. Paint should not be applied in wet weather or when the wood is damp. In new buildings first make sure that they are not damp. New masonry must be thoroughly dry before paint is applied to it. Drying, especially of floors, can take up to 6 months.

Mix the paint thoroughly before you apply it.

Painting is done with a brush. Dip the brush into the paint up to about 1/3rd of the length of its bristles, and remove the excess paint. Never dip the whole brush into the paint because the excess paint will drip out and be wasted.

Use long sweeping strokes and brush the paint well to form an even coating. Start at the top of a surface at one edge, and work across and down. Try to finish each day's work at a corner of the building or at a window. If you stop in the middle of a wall the mark will show where you resume painting the next day.

All new work, either of wood or masonry, requires three coats of paint. Surfaces which have been painted before need only two coats.

Before you apply a new coat, the previous coat has to be thoroughly dry. With most paints, the first two coats may be diluted with thinner to allow a better distribution and penetration. Only the final coat must be undiluted.

If wood is unprotected from the rain and sun, it will be necessary to repeat the application of finish from time to time. Take care that the end grain absorbs as much paint as possible.

Never allow a brush to rest upright on its bristles. If you stop work for a few minutes, remove the excess paint from the brush by wiping it on the edge of the tin, then lay it flat across the top of the tin or on a smooth clean surface.

If work is stopped for a longer time (overnight or for a few days), put the brush in a tin of kerosene. Here is an illustration of a good way to keep the paintbrushes; note that the bristles are covered with kerosene but they do not rest on the bottom of the tin. Simply drill a hole in the brush for the stick to pass through.

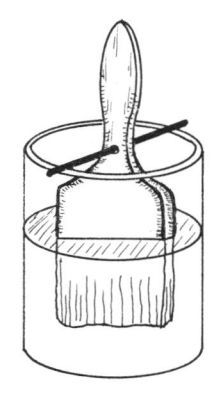

PAINTING.

NPVC CON. 269

HOW TO APPLY TIMBER PRESERVATIVES

There are two practical methods for applying timber preservatives:

- By painting (as described in the previous two pages)

- By soaking or dipping.

Different types of timber preservatives are required for different situations. The different kinds are described in the Reference Book, page 145.

- SOAKING OR DIPPING: Soaking or dipping is a much more efficient method than painting for applying any kind of timber preservative, because the chemical can penetrate deeper into the wood. This method can also be used for protective finishes such as oil paint.

    Use a bucket or other suitable container to soak the wood for several hours or days. This method is only practical for smaller pieces like the ends of fence posts or frame pieces.

In cases where you are repeating the application of a preservative, use a preservative from the same group as the first application. Remember that the waterborne preservatives cannot penetrate over oil preservatives, but it is possible to apply an oil preservative over a waterborne one.

Only waterborne preservatives should be used if the wood will be painted later.

Take care that the end grain and splits absorb as much preservative as possible.

- NOTE: Be careful; most finishes are poisonous and can be fatal to human beings. Paints and paint thinners are often flammable.

NOTES:

APPENDIX

The following pages contain information and designs for constructions which do not necessarily fit under the topic of Rural Building, but which are nevertheless important for the Rural Builder to know about, at least in general. One of the purposes of this course is to prepare the Rural Builder to become a part of the development process in the rural areas, and this includes not only the construction of safe, healthy and comfortable buildings, but also the provision of safe water supplies, sanitation methods, and even the storage of the agricultural produce which is indispensable to rural life.

The following simple designs can be adapted and should be useful to almost any rural community.

NOTES:

NOTES:

WATER PURIFICATION

A water source is important in planning any building. The source can be a well, deep bore hole, spring, stream, or rain water collected from a roof and stored in a tank. The importance of water needs no explanation: it is worth spending both time and money to make sure that pure water is available at all times.

Many diseases come from drinking impure water. The danger of disease can be reduced by taking such precautions as boiling and/or filtering drinking water. Water obtained from a clean source such as a properly dug well or a borehole should carry no disease in it.

There is much that the Rural Builder can do to help the people in his community to have clean water to drink.

SMALL WATER FILTER

One way to improve the quality of water is to use a water filter. A simple design for a filter which can be made by anyone is shown below. It is used for small amounts of water. The filter consists of a locally made pot which hangs from wires or ropes.

The pot is partly filled with layers of fine sand, gravel and stones. The water sinks down through this and drains off through holes in the bottom of the pot. The diameter of the holes should be small enough so that the stones can't pass through.

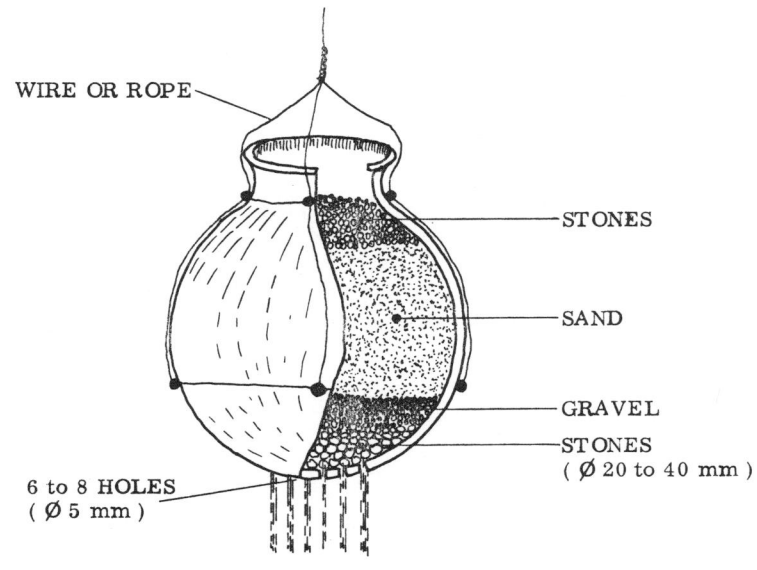

WIRE OR ROPE

STONES

SAND

GRAVEL

STONES ($\emptyset$ 20 to 40 mm)

6 to 8 HOLES ($\emptyset$ 5 mm)

WATER PURIFICATION.

COVER

WATER INLET

CONCRETE SLAB

DRAIN GROOVE

Fig. 1

WATER — 30

SAND — 55

GRAVEL — 15
STONES — 15

INSIDE CORNERS ROUNDED

7

120

8

60

15 100 15

Fig. 2

NPVC		WATER PURIFICATION.
274	APP.	

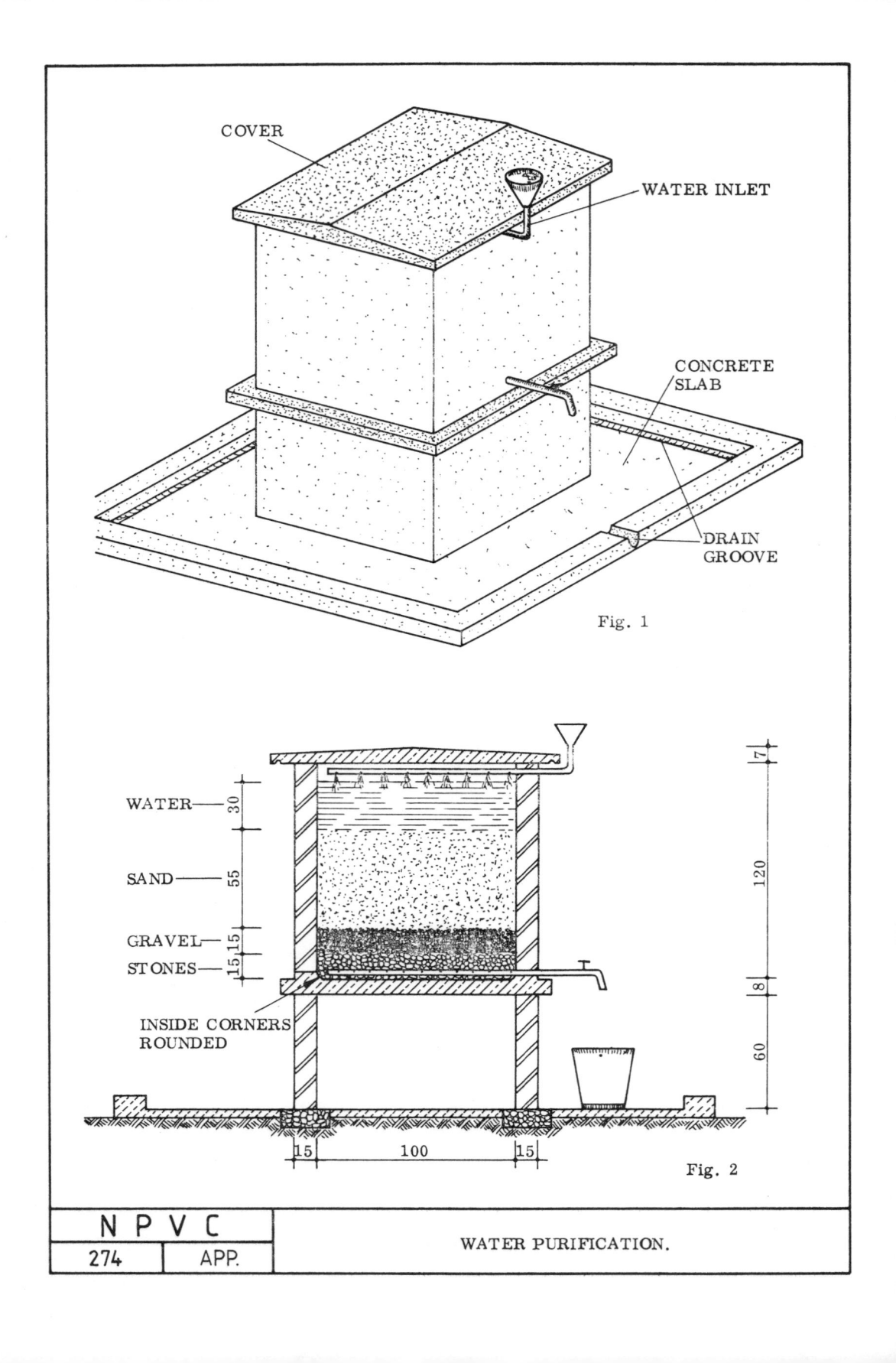

LARGE WATER FILTER

A built-up water filter (Fig. 1) can be used to purify larger amounts of water. Its size depends on the amount of purified water that is needed.

The cover can be removed to allow the user to clean the filter materials and the entire inside of the tank. If the cover is very heavy it can be made in two halves.

The water inlet also acts as a ventilation opening. The top of the funnel should be covered with copper or aluminium mosquito wire to keep insects out of the tank. The distribution pipe inside (Fig. 2) has many fine holes to make sure that the water is spread evenly over the sand. If necessary install more distribution pipes so that the whole top surface of the sand can be soaked with water; thus using all of the filter materials inside the tank rather than overloading a small area.

The space of 30 cm at the top for water storage may be increased if necessary. The depths of the different layers of filter materials should be no less than indicated. If the layers are made deeper still the water will be filtered more thoroughly, but generally the indicated depths are sufficient (also see Drawing Book, page 120).

At the bottom of the tank a piece of pipe with holes in it allows the filtered water to drain off. There should be plenty of holes, and the diameter of the holes should be small enough to prevent the fine gravel from entering the drain pipe. At the end of the pipe a valve can be used to stop or start the flow of water. It drains into a container which is placed under the pipe.

The floor of the tank should be constructed with a rim all around (Fig. 2), which carries the walls and projects at the base of the walls, so that there is no crack between the walls and the floor. The whole inside should be plastered smooth to make it easy to clean.

The filter should be built off the ground so that the water can drain easily into buckets. It is advisable to concrete the area surrounding the filter and to provide a drain for spilled water (Fig. 1).

When the filter needs cleaning the water flow will become very slow. To clean the filter it is only necessary to take off the top 2 cm of sand and replace it with clean sand. Depending on how clean the water is at the start, the filter may need cleaning only after several weeks or even months.

NOTES:

Fig. 1

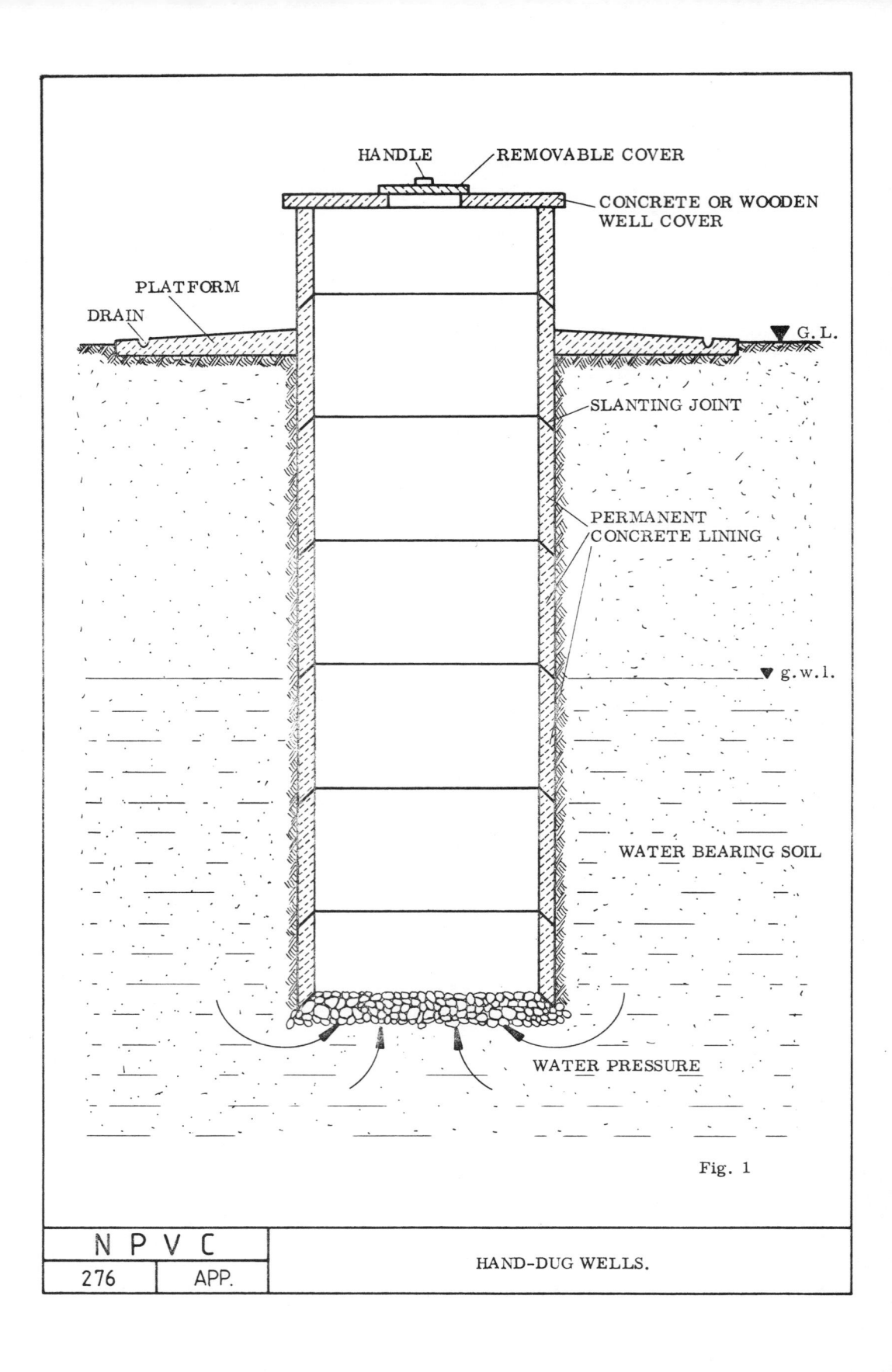

NPVC 276 APP.

HAND-DUG WELLS.

HAND-DUG WELLS

The hand-dug well is the most common kind of well. Unfortunately, most wells are dug according to very basic "hole-in-the-ground" methods, and they can become health hazards as they are easily infected by parasites and bacterial diseases. With modern methods and materials, hand-dug wells can safely be made up to 60 m deep and will give a permanent source of good water.

Untrained workers, if they are properly supervised, can safely dig a deep well with simple light equipment. The basic method is outlined here. This information is intended to help you to dig more safely, and also to make a well that will not become contaminated by surface water. However, there are other problems concerning well digging that are not covered here, and you are advised to seek the help of an experienced well digger when you make a well by this method.

LINING WELLS

Masonry or brickwork are widely used in many countries to line the walls of wells. These can be very satisfactory in the right conditions. In bad ground however, unequal pressures can make the lining bulge or collapse. Building with these materials is slow, and a thicker wall is required than with concrete linings. There is also the danger of the blocks moving during the construction in loose sandy soil or shale, before the mortar has set firmly between them.

This danger is prevented with concrete linings because the form is left in place to support the lining until the concrete has set hard.

Lining (Fig. 1) prevents the hole from collapsing; keeps out contaminated surface water; and supports the pump platform if there is a pump.

It is usually best to build the lining as the well is dug, since this avoids the need for temporary supports and also reduces the danger of cave-ins during digging.

There are two methods for doing the lining:

- As the hole is dug, the sections of lining are built into their permanent positions. This is the method explained on the following pages.

- Precast sections of lining are added at the top and the whole lining moves down as the hole is dug and earth is removed from under it.

NOTES:

Fig. 1

HAND-DUG WELLS.

NPVC 278 APP.

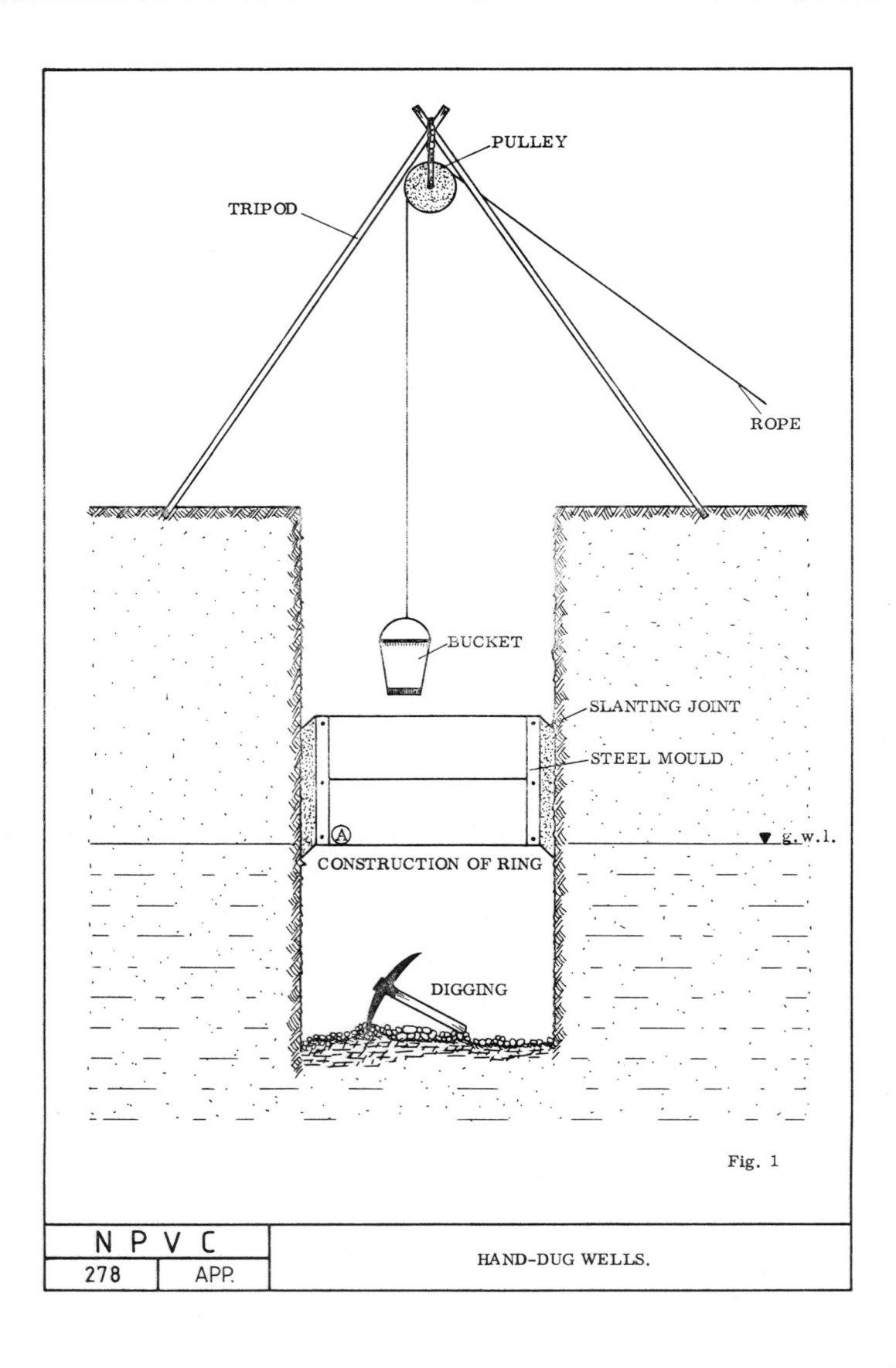

TOOLS AND MATERIALS NEEDED

- Shovels
- Pick-axes
- Buckets
- Ropes and pulley
- Steel forms and bolts
- Wooden or steel tripod
- Cement
- Reinforcement bars
- Sand
- Gravel
- Oil

CONSTRUCTION

Experience has shown that when one man is digging, he will be able to dig most efficiently if the well is about 100 cm in diameter. However if two men are digging together, a 150 cm diameter is best, and the two can dig more than twice as fast as one man. Thus two men in the larger hole is usually best, also for safety reasons.

- NOTE: Do not construct a well near a latrine, septic tank, or other source of contamination. The well should be as far away from these as possible; at least 30 m, and it should never be downhill from a source of contamination.

- SEQUENCE OF OPERATIONS:

    a. Erect a wooden or steel tripod to support a pulley and bucket for bringing out soil from the hole.

    b. Dig down until you reach ground water level (g.w.l.).

    c. Make the first concrete ring section just above the g.w.l. (Fig. 1, A)

    d. While the first section cures, proceed with digging the hole for the second section. The rings should be made with slanting edges as shown in Fig. 1, so that the joints between the rings slant downwards.

    e. Continue constructing ring sections until the well is deep enough.

    f. The other rings above ring A can be constructed during the digging.

    g. When the concrete lining is completed, the walls of the whole well should be waterproof, so that surface water can't enter the well. A concrete platform with a groove for drainage should be constructed, and a wall around the edge to keep people and animals from falling in. Either blocks or reinforced concrete can be used for the wall (Fig. 1, previous page).

    h. The well is closed off with a wooden or concrete cover, with a small opening for buckets to go through (Fig. 1, previous page).

HAND-DUG WELLS.

WOODEN LOCK PIECE

WOODEN LOCK PIECE

Fig. 1
STEEL MOULD

Fig. 2
WOODEN MOULD

NPVC 280 APP.

HAND-DUG WELLS.

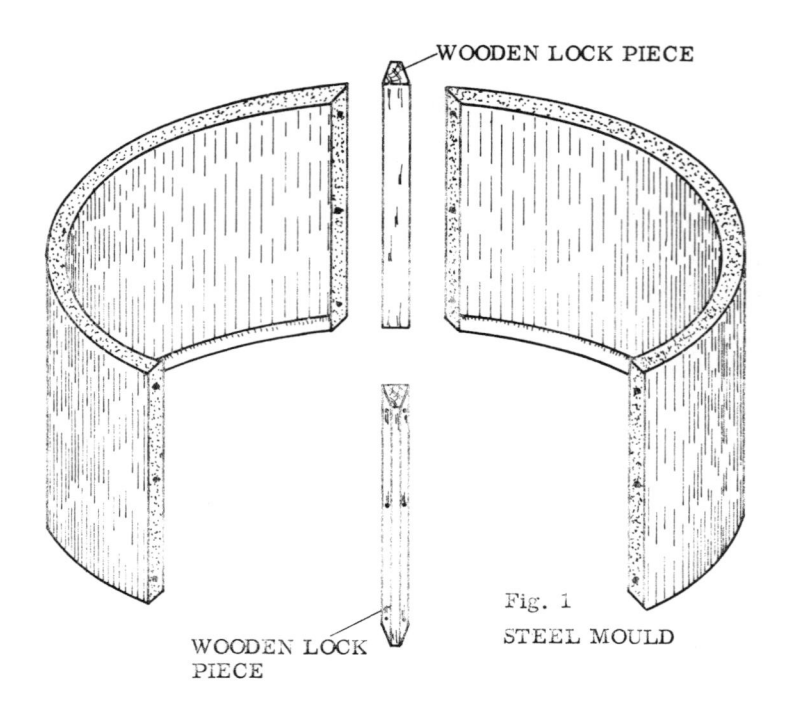

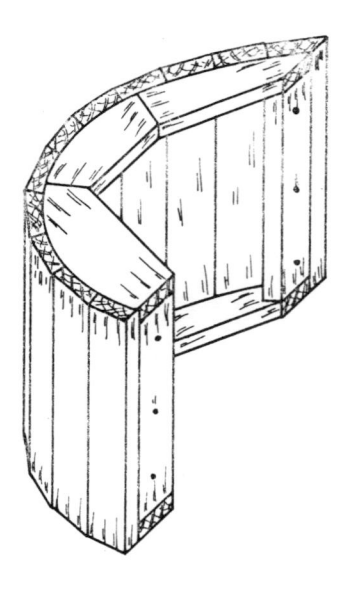

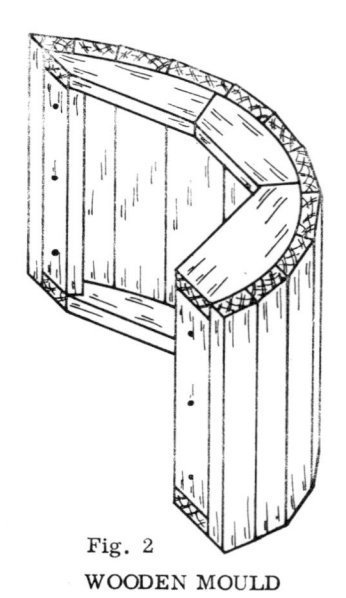

i. If possible, disinfect the entire well by adding a chlorine solution to it. Use about 1 heaping tablespoon of bleaching powder to 2 buckets of water and mix. Make up three buckets of the solution to wash the walls of the well thoroughly, and pour the remainder into the well. Make up another three buckets and put that into the well too. Leave it overnight before using the water.

- NOTE: Since there is a lot of water in the soil, don't use too much water in mixing the concrete for the lining. Be careful that there are no voids (empty pockets) in the concrete lining. These give surface water and soil a path to enter the well.

- MOULDS: A steel mould (Fig. 1) or a wooden mould (Fig. 2) is used to support the sections of concrete lining during construction and curing. After the ring has hardened sufficiently, the wooden lock pieces are removed and the mould is taken out and used for the next ring (Fig. 3).

When you assemble the mould, be sure that the wooden mould or wooden lock pieces are first soaked in water so that the wood does not expand when it contacts the wet concrete. If the pieces expand they will crack the concrete ring, and it will be very difficult to get the mould out again.

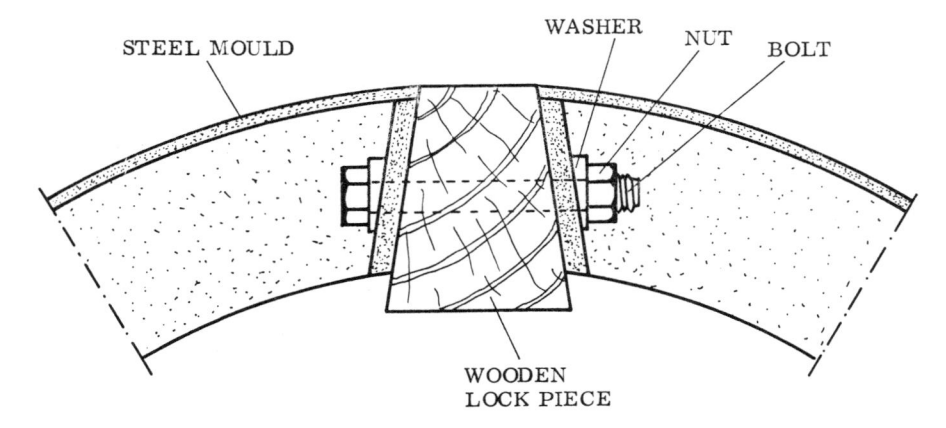

Fig. 3

HAND-DUG WELLS.

NPVC
APP. 281

Fig. 1

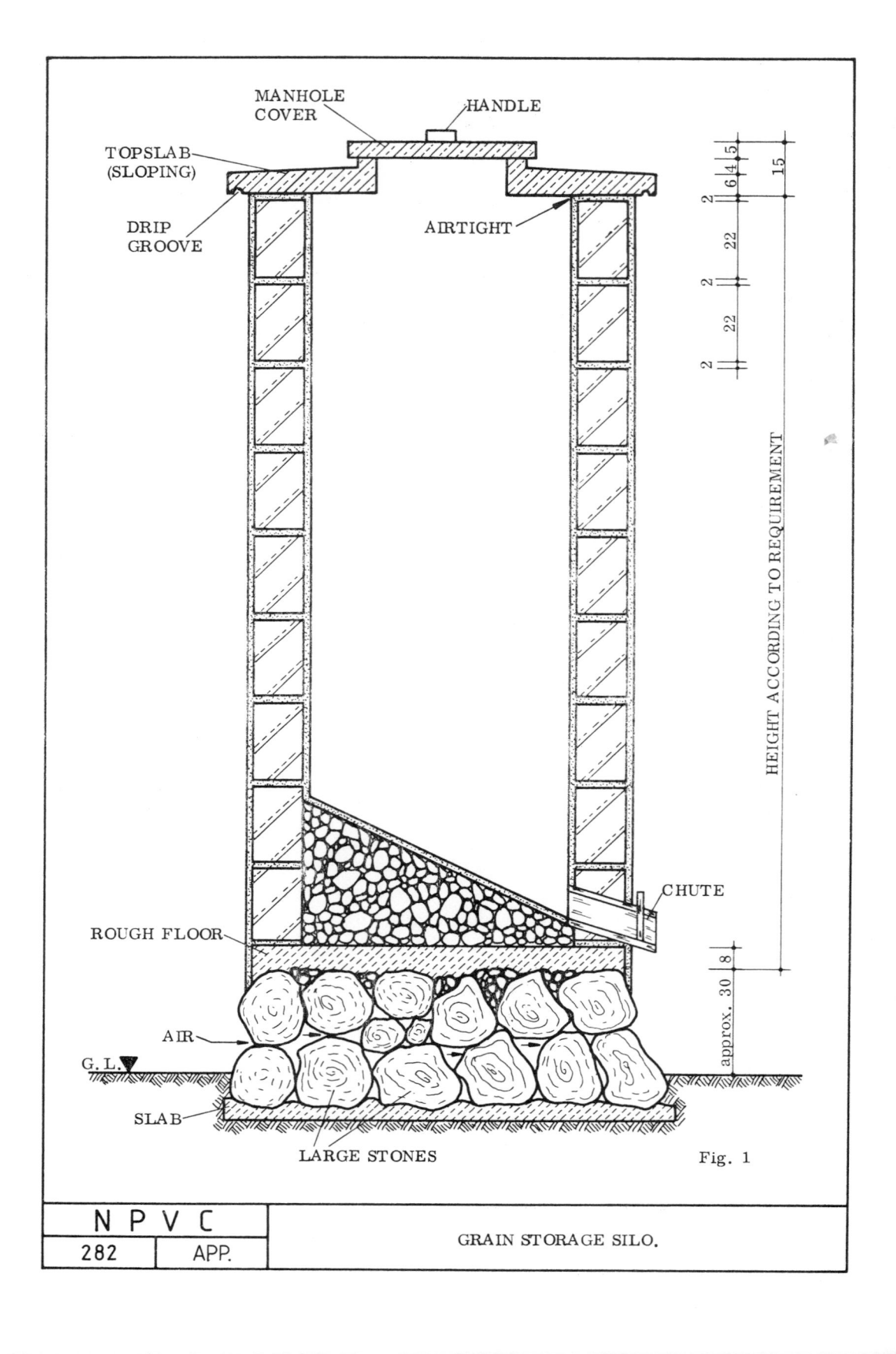

GRAIN STORAGE SILO.

NPVC 282 APP.

GRAIN STORAGE SILO

Often traditional storage methods result in the loss of valuable grain, because insects or mice etc. get into the grain. To help farmers with this problem the Garu Agricultural Station in the Upper Region has developed a new type of grain silo. This silo requires a few bags of cement plus stones or gravel to construct. It can hold between 9 and 16 bags of grain.

Compared to some types of traditional storage, the silo gives better protection to the grain against mice and insects. When the silo is closed so that no air gets in, CO_2 gas develops from the grain and stops the insects from developing.

CONSTRUCTION OF A SILO

Sun-dried laterite blocks are used for the walls of the silo. The blocks are made in the same way as any other mudblock.

Cast the topslab before starting to construct the rest of the silo, so that it will be cured and ready to use when you need it. Construct a frame for casting the slab, with an inner frame for the manhole (Fig. 1, next page). Reinforce the slab with some rods. For measurements, refer to the Drawing Book, page 118. Note that the top of the slab slopes to the outside. Also note that the edge of the slab is fitted with a groove underneath so that rainwater drips down instead of running back under the slab (Fig. 1, left).

Also make a frame for the manhole cover and cast this piece. Use a piece of iron rod set in the concrete for a handle (Fig. 1).

Place wet paper under the frames before you begin to pour the concrete. The concrete has to remain in the frame for three days. Water it twice daily so that it hardens well without cracking. After the concrete hardens (cures) remove the frames carefully so that they can be used again.

- FOUNDATION AND FOOTING: Dig out the top soil and make a concrete slab if needed (if the soil is soft). The slab will be 125 cm in diameter and about 6 cm thick. The footing is made of large stones set on the slab or on the ground. The stones should be placed in such a way that air can circulate freely between them and they should be built to a height of 30 cm above ground level. On top of and in between the large stones, small stones are used to fill in the gaps to form a surface for the floor to rest upon (Fig. 1).

- FLOOR: The floor is poured directly onto the footing. This floor is a working surface only and the top should be left rough so as to give a good grip to the slanting floor (Fig. 1, previous page).

- FIRST COURSE: Place the chute in position. The chute is a kind of open-ended box made from wood, with a sliding closure that can be locked (Fig. 1, previous page). Now proceed to lay the first blocks. The section after this one describes the procedure for laying blocks in a round structure.

- SLANTING FLOOR: Build up four courses, then pack the slanting floor on top of the rough floor. The mixture for this slanting floor should not be too rich, so it doesn't crack. After the mixture is poured into the silo, pack it down well and plaster the top of the slanting floor smoothly. To do this apply some cement powder and add some water, then smooth with a steel float.

- WALLING: Build the walls up straight, using a plumb bob on all sides and for each course. The height of the silo depends on the required storage area. When the desired height is reached, plaster the inside of the silo. The inside plaster should be very smooth and without holes or cracks, so that the silo is airtight.

- TOPSLAB: Provide a mortar bed for the topslab on top of the wall. Then place the slab into position. The slab will be heavy, so several people will be needed for this operation. One man should be inside the silo to direct the operation.

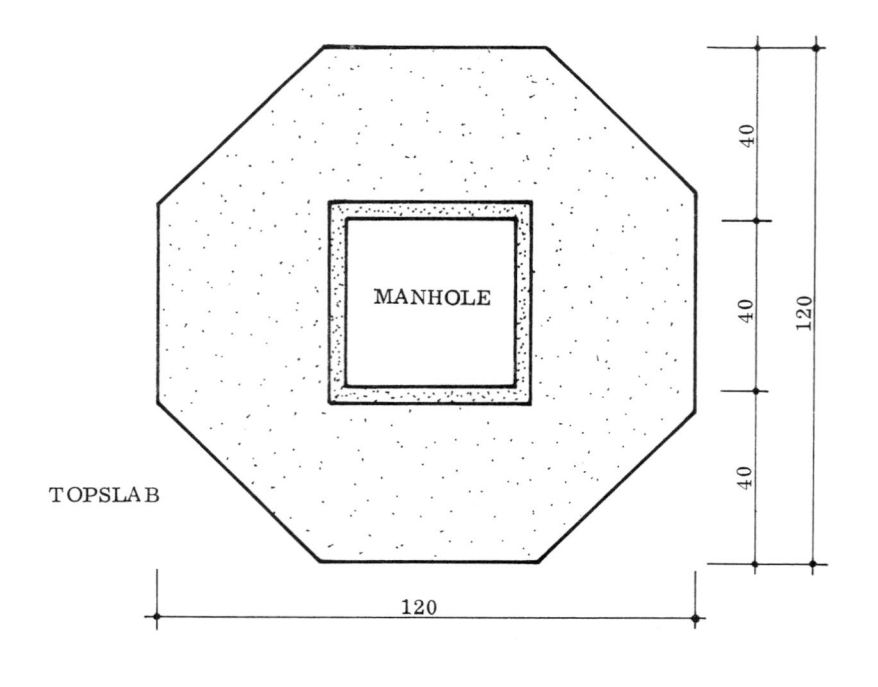

TOPSLAB

MANHOLE

40 / 40 / 40 / 120

120

N P V C   284   APP.

GRAIN STORAGE SILO.

- PLASTERING OUTSIDE: When the slab is set in position you can do the plastering on the outside of the silo. This has to make the silo waterproof, so the finish should be smooth and without cracks.

USING THE SILO

Dry the grain very well before you put it in the silo. Moist grain will rot, the silo will crack, and all the food will be lost. Distribute the grain well and fill the silo to the top.

Put the cover on the manhole and seal the edges with mud or cow dung. Also seal the chute so that nothing can enter the silo.

Keep the footings and the area around the silo clean.

Check the silo often for cracks, especially when it is full. Replaster any cracks immediately. The wooden chute should always close tightly and the plaster around it should not be cracked.

Paint the silo white or whitewash it. This makes it cooler inside and gives it a nice appearance.

When the silo is emptied, take care to empty it completely. Left-over grain attracts insects and rodents.

Before the silo is filled up again next year, it should be clean and insect free. Light a small bundle of grass on fire inside the silo. The smoke and heat will kill any insects or insect eggs. Sweep out the ashes and dust, and clean the entire silo thoroughly.

CIRCULAR MASONRY WORK

A section on circular work is included here because in Rural Building circular work applies mostly to structures like silos or protective walls around wells.

Here we describe a sequence which might be used to construct a silo. It may be adapted for other structures as well.

Regular blocks are used. Wedge-shaped blocks can be specially made for the job, but the shape of the wedge has to be calculated exactly according to the size of the circle.

NOTES:

Fig. 1

Fig. 2

NPVC		GRAIN STORAGE SILO.
286	APP.	

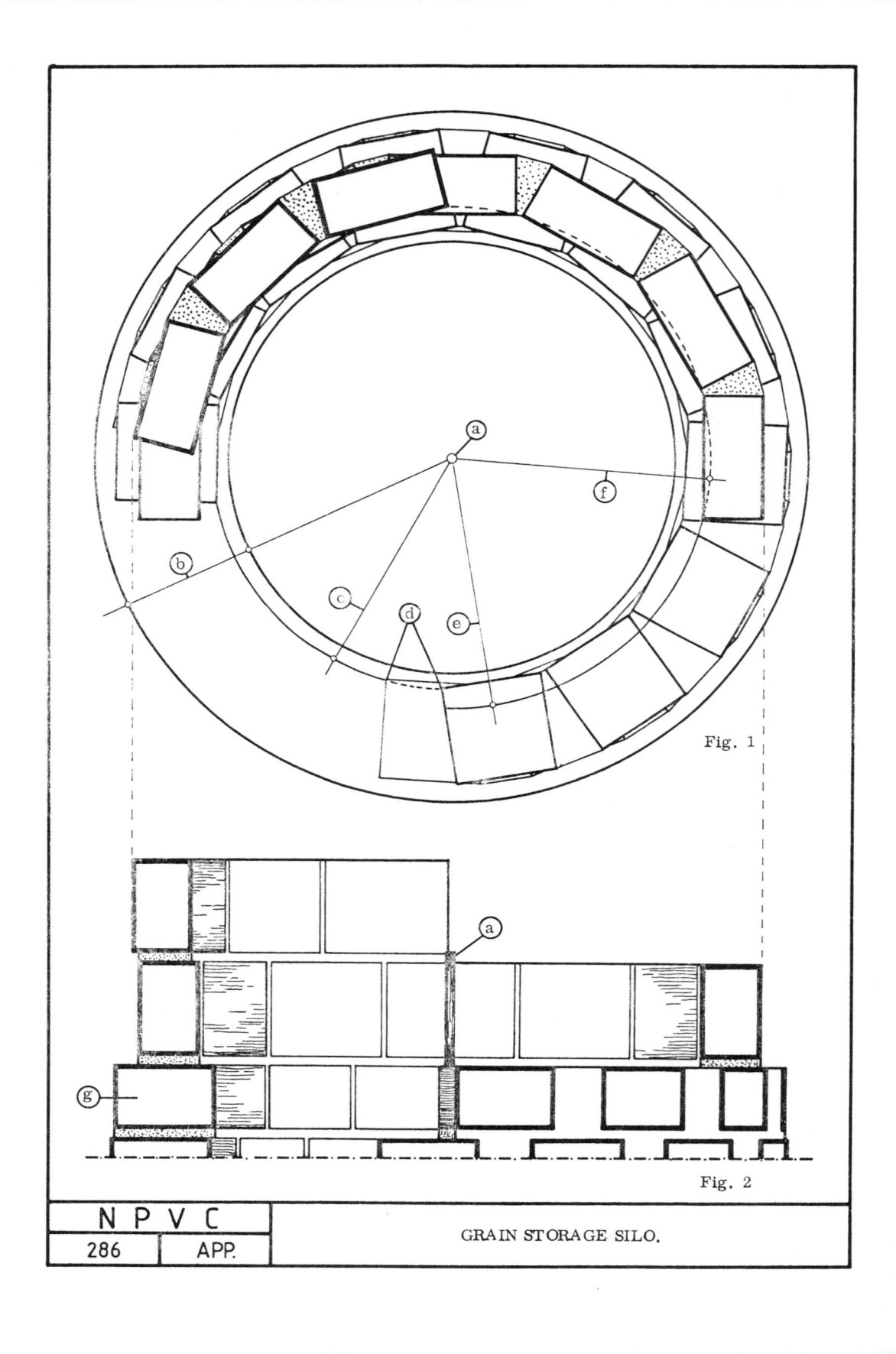

SEQUENCE OF OPERATIONS

a. Determine the centre of the circle. Drive an iron rod into the ground at this point (Figs. 1 & 2, a). The rod should be a little higher than the top of the first course of the future rising wall.

b. The inner and outer curves of the foundation are marked on the ground by fixing a mason line to the rod and describing circles with this at the desired radii (Fig. 1, b).

c. The foundation is completed, then the inner curve of the footings is marked on the top of the foundation using the same method (Fig. 1, c).

d. The footing blocks are laid so that the inside corners of each block touch the marked circle (d). The next course is marked on top of the first course and laid in the same manner.

NOTE: In this construction the footings are made with blocks, unlike the silo described in the preceeding pages, where the footings where made of large stones. If a silo is to be built on top of the footings, the cross joints between the blocks are not filled in; this permits cross ventilation. The space up to the top of the footings is filled in with large stones and the floor is made on top of these, as in the silo described before. Remember to set the chute in place on top of the footings before you continue with the rising wall construction. Only two footing courses are necessary to reach the required height of 30 cm for the silo floor.

e. When the footings are completed, the rising wall is laid out on top of the footings with the line and peg as before (Fig. 1, e). The blocks are positioned with the corners touching the circle as before.

f. The second course of the rising wall is laid out on top of the first course (Fig. 1, f), and the blocks are laid as before.

g. The following courses are laid with the aid of the plumb bob. All cross joints are 1 cm thick along the inner curve. Each course of the rising wall is plumbed according to the course two courses below it (3rd to 1st, 4th to 2nd, 5th to 3rd, etc.).

Fig. 2 shows a side view of Fig. 1, but only the last footing course is shown (g).

NOTES:

GRAIN STORAGE SILO.

Fig. 1
ORDINARY PIT LATRINE

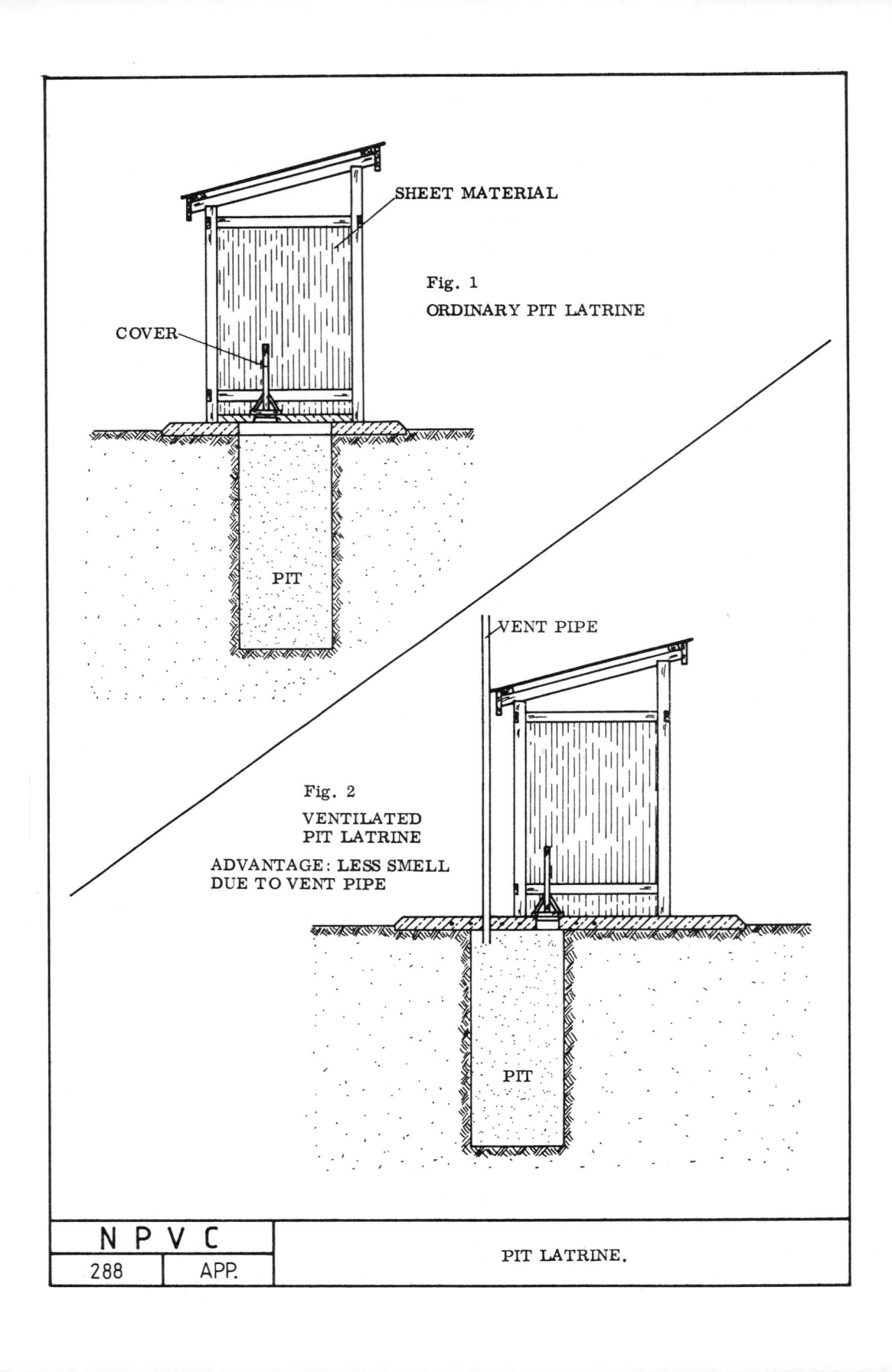

SHEET MATERIAL
COVER
PIT

Fig. 2
VENTILATED PIT LATRINE

ADVANTAGE: LESS SMELL DUE TO VENT PIPE

VENT PIPE
PIT

N P V C		PIT LATRINE.
288	APP.	

PIT LATRINE

Latrines, and especially pit latrines, are the most common and simple sanitation systems. A pit latrine is practical for rural areas because it is the cheapest possible system and the easiest to build.

One major problem with pit latrines is the limited capacity: only a certain number of people can use the latrine, or else it becomes full too quickly and another pit must be dug. Another problem is that the latrine, if it is improperly made or made in the wrong place, can contaminate nearby wells or surface water. Also, if the ground water level is near the surface it may not be possible to dig a deep pit.

CAPACITY

The volume of the pit may be planned by using the rough figure of at least 0,06 cubic meters per person per year. Thus a pit which is 1 m square and 3 m deep may serve a family of five for about 6 years, before it becomes two-thirds full and a new pit has to be dug.

Pit latrines of the type shown on the opposite page (Figs. 1 & 2) should not be used by large numbers of people unless there is space available to dig several pits. If many people use the same pit it will quickly become full, and the ground around it can become contaminated.

It is a good idea to dig two pits at once, so that when one is 2/3rds full it can be closed, filled to the top with soil, and left there for 6 months to a year; by which time the harmful, disease-causing organisms in it will have been destroyed. The sludge can be taken out and mixed with compost, making a very valuable (and non-imported) fertilizer for gardens or farms.

- NOTE: Human wastes which have not been composted for at least 6 months should never be used as a fertilizer for food crops, because disease can be spread that way.

LOCATION

The pit latrine should be located at least 6 m away from any house, and at least 30 m away from any water source: well, bore hole or stream. The latrine should never be located uphill from a water source -- this is extremely important.

NOTES:

	N P V C
PIT LATRINE.	APP. 289

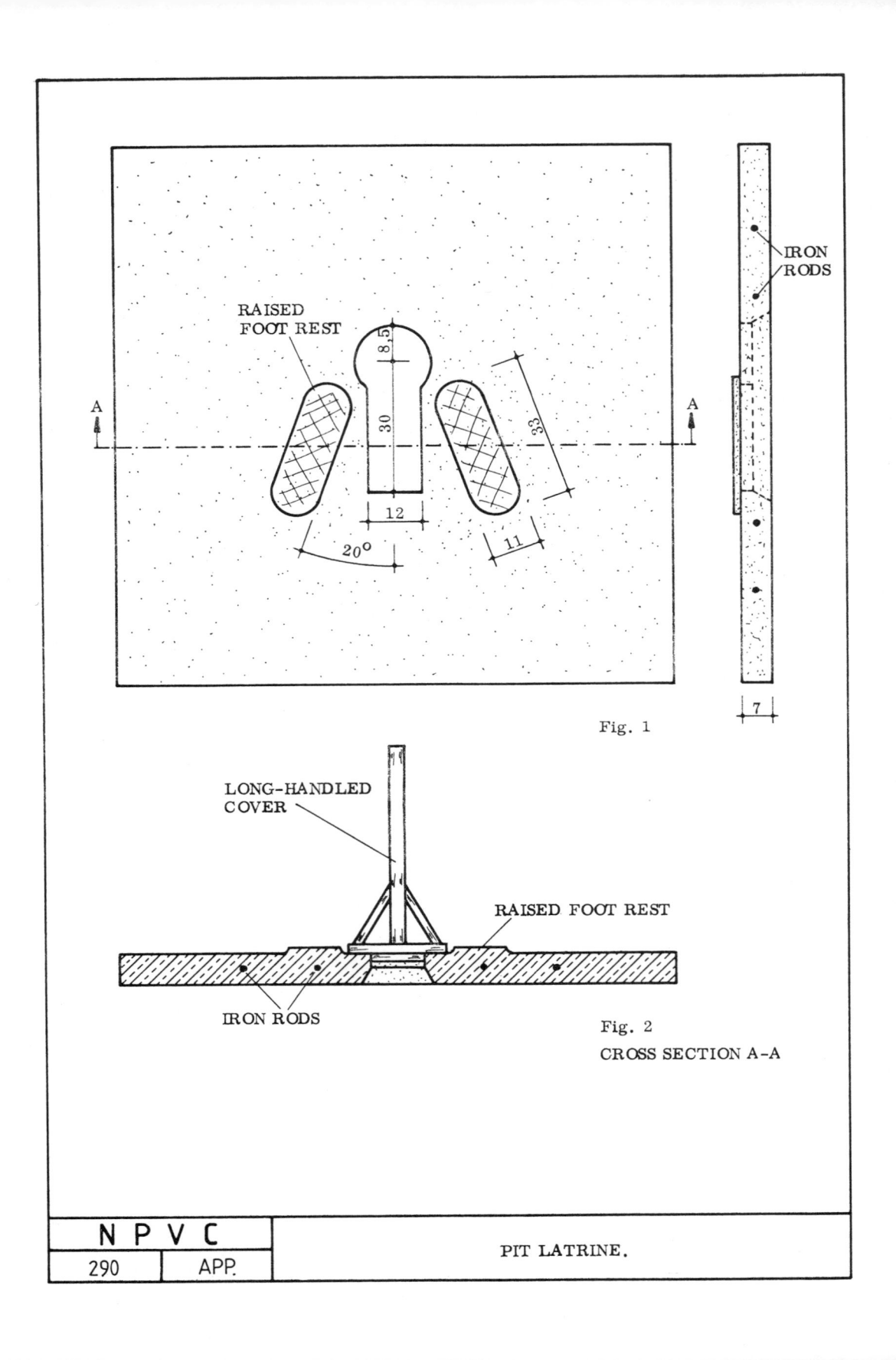

Fig. 1

Fig. 2
CROSS SECTION A-A

N P V C		
290	APP.	PIT LATRINE.

DESIGN AND CONSTRUCTION OF THE SQUATTING SLAB

The squatting slab should be designed with easy cleaning in mind. It is made out of concrete reinforced by iron rods as shown in Fig. 1. The rods should be about 8 mm in diameter.

The slab may be square, round, or rectangular in shape. A common size is 100 by 100 cm. The whole slab will weigh about 140 kg. The advantage of a round slab is that it can easily be rolled around the site and into place. The slab should be about 7 cm thick. The raised foot rests make the slab easier to clean. The shape of the hole and foot rests may vary; a typical arrangement is shown here. The hole should be at least 36 cm long to prevent soiling of the slab, and less than 18 cm wide so that small children can't fall through.

The distance from the back of the hole to the back wall of the latrine should be at least 15 cm, so that it is not necessary to lean against the wall when squatting.

Make a cover for the latrine hole out of wood, with a long handle (Fig. 2).

NOTES:

PIT LATRINE.

NPVC
APP. 291

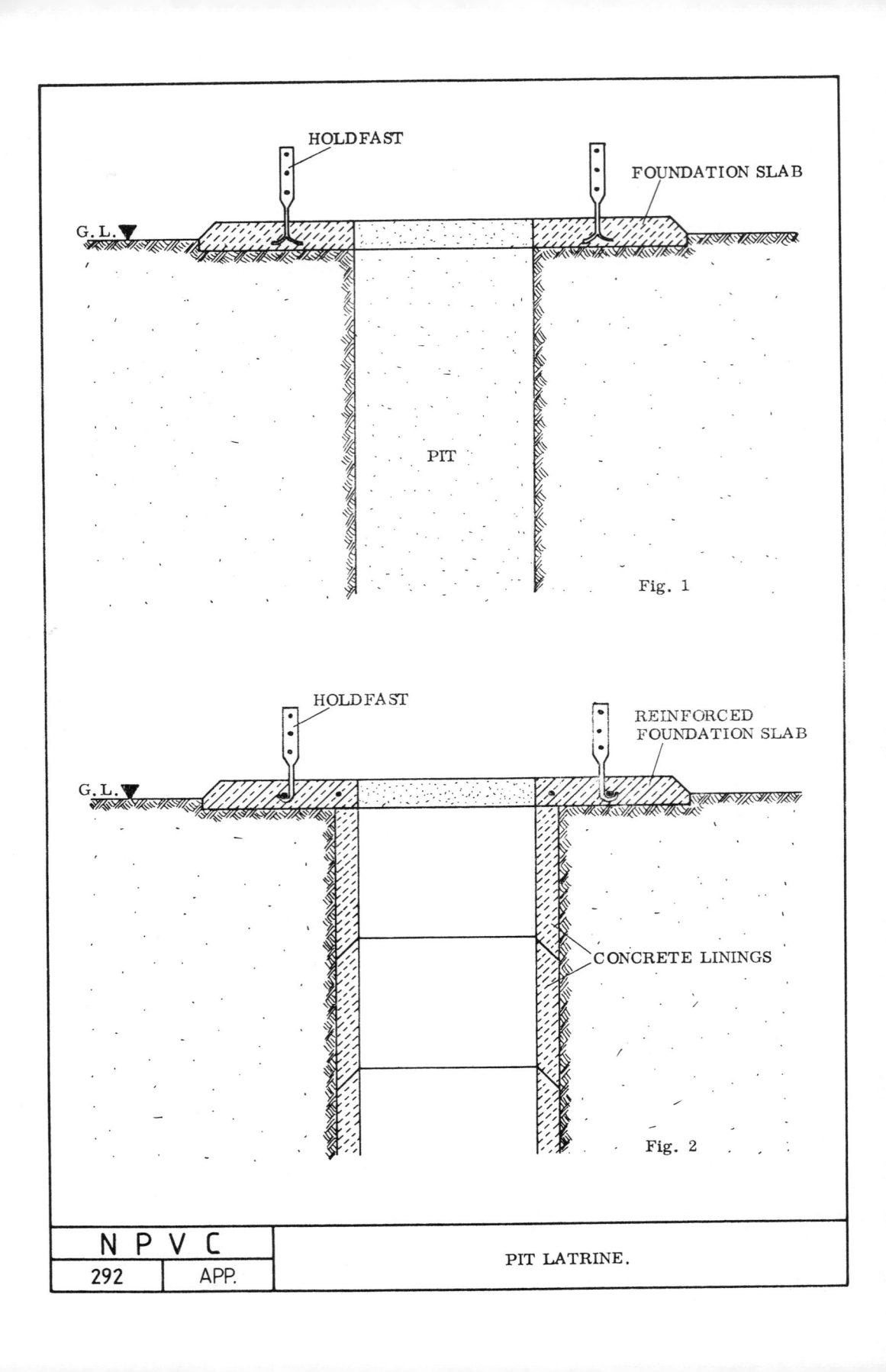

Fig. 1

- HOLDFAST
- FOUNDATION SLAB
- G.L.
- PIT

Fig. 2

- HOLDFAST
- REINFORCED FOUNDATION SLAB
- G.L.
- CONCRETE LININGS

N P V C		PIT LATRINE.
292	APP.	

CONSTRUCTION

The first step is to precast the squatting slab, as described on the previous page.

Make the hole for the latrine large enough so that it won't become full too quickly. The soil should be firm enough so that the weight of the latrine structure won't cause the hole to collapse. If the soil is firm enough, the foundation slab can be poured around the edge of the hole (first remove the top soil) (Fig. 1). If the soil is a bit soft, reinforce the foundation slab.

If the soil is very soft, it will not be able to bear the weight of the construction. Then it will be necessary to construct a lining for the hole to be sure that it will not collapse. The lining can be made with blocks; or in a round hole with concrete rings, as described under the section on wells, page 279. The foundation slab then rests on top of the lining and cannot sink down into the soft soil (Fig. 2).

When you construct the foundation slab, fix some steel rods or holdfasts in the concrete, as shown in Figs. 1 & 2. These can be used later to anchor the wooden structure of the toilet.

When the foundation slab is cured, place the precast squatting slab on top of the foundation and secure it in place with a weak mortar (Fig. 3).

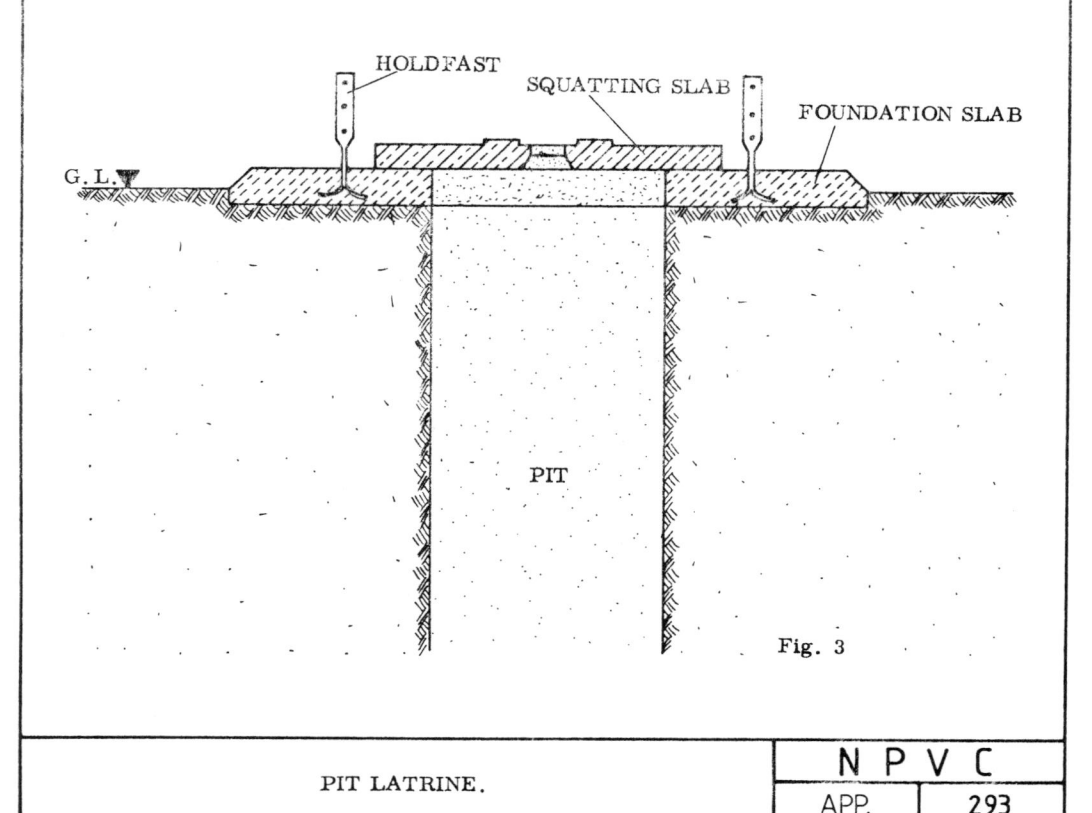

Fig. 3

PIT LATRINE.

NPVC
APP. 293

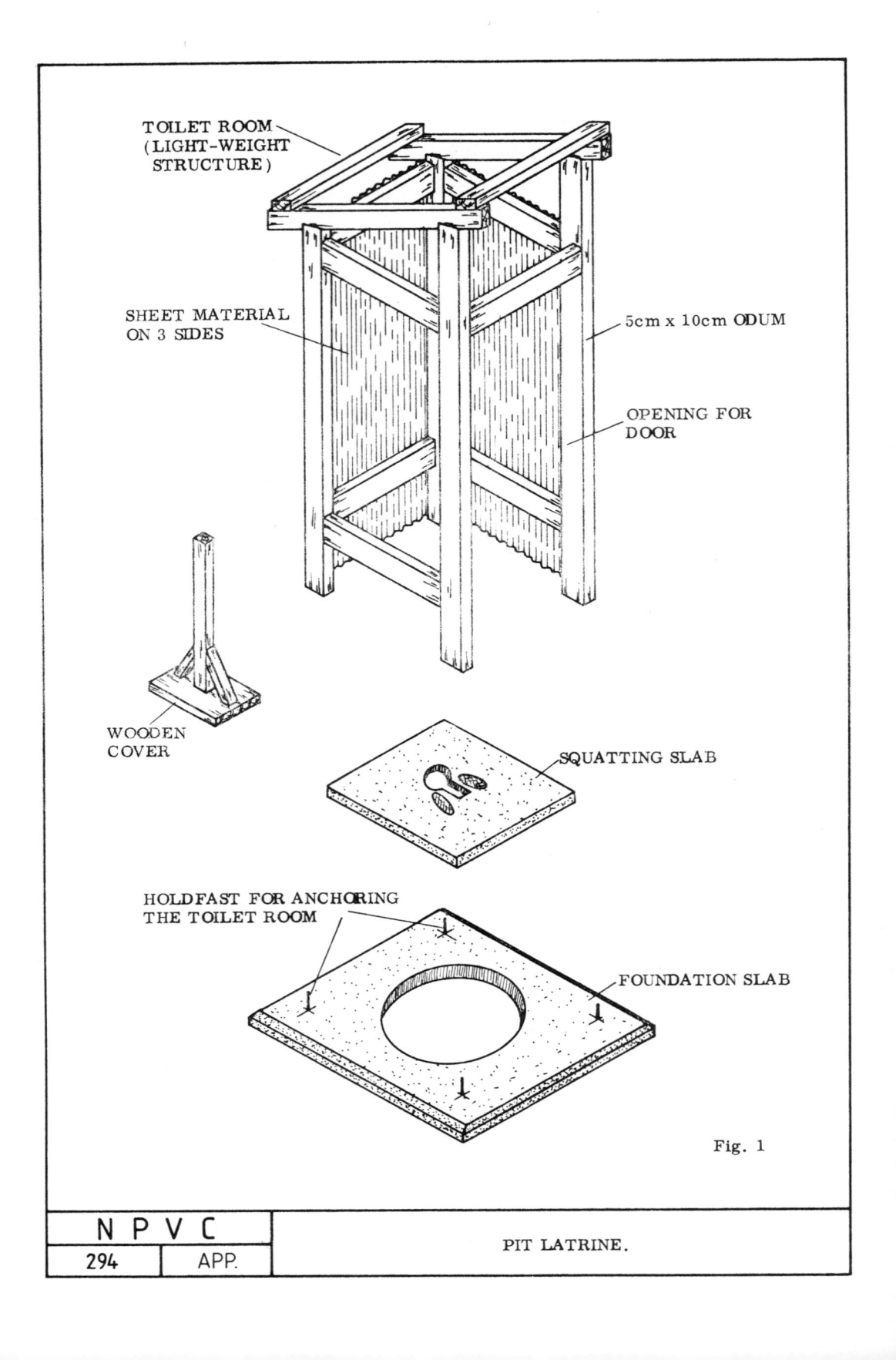

TOILET ROOM
(LIGHT-WEIGHT
STRUCTURE)

SHEET MATERIAL ON 3 SIDES

5cm x 10cm ODUM

OPENING FOR DOOR

WOODEN COVER

SQUATTING SLAB

HOLDFAST FOR ANCHORING THE TOILET ROOM

FOUNDATION SLAB

Fig. 1

NPVC 294 APP.

PIT LATRINE.

TOILET ROOM

The toilet room can be built from mud or landcrete blocks, and plastered with mud. However, be sure that the foundation can support the walls because they are heavy, and in soft soil they may sink down and collapse the hole. Also, when the hole is full and a new one has to be dug, the walls have to be built completely new.

Because of the above problems with block walls, it is advisable to construct a light-weight structure out of timber and metal sheets. The light-weight structure can be removed when the pit is full and installed again over a new pit. The squatting slab can also be re-used (Fig. 1).

The structure must be anchored well to the foundation slab so that it is not blown away in strong winds.

TOILET PARTITIONS

If a community latrine with several compartments is needed, use the system drawn on page 124 of the Drawing Book.

First do the setting out. Then dig the round pits and make foundation slabs around the pits. Construct arches or lay a reinforced concrete beam across the center of each pit.

There are four squatting slabs over each pit. Lay each one carefully so that one edge rests on the concrete beam and two other sides rest on the foundation slab. Use some weak mortar between the slab and the supports, to ensure that the slabs are flat and stable.

The partitions between the toilets should be light-weight structures of timber and metal sheets or waterproof plywood. The outside walls can be constructed from landcrete blocks or mud.

Many of the materials from the above structure can be re-used after the holes become 2/3rds full and new pits are dug. The light-weight partitions, the squatting slabs and the reinforced beams can all be taken out and used in the new construction.

NOTES:

PIT LATRINE.

NPVC
APP. 295

Fig. 1

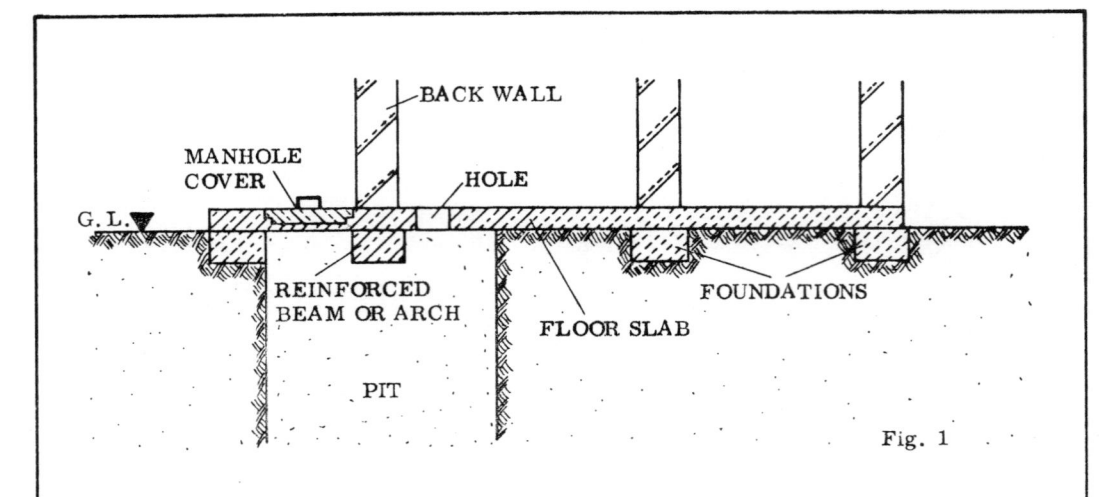

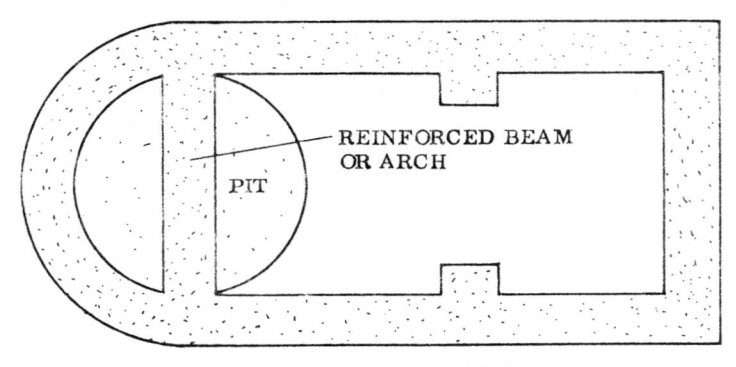

Fig. 2 FOUNDATIONS

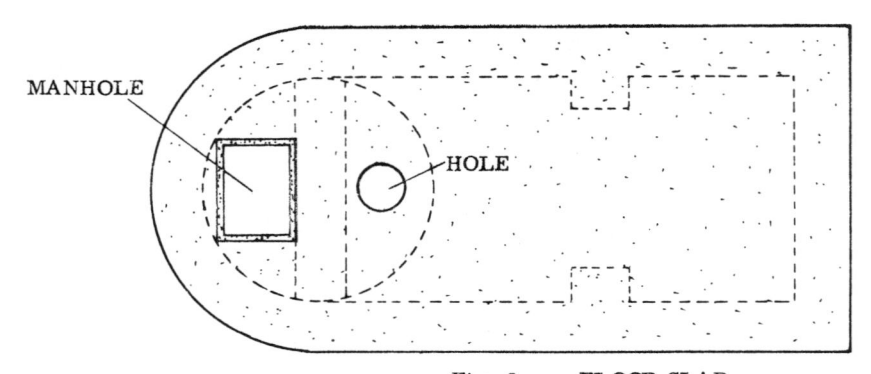

Fig. 3 FLOOR SLAB

N P V C	
296	APP.

PIT LATRINE.

PERMANENT LATRINE

If the latrine is constructed so that the pit can be cleaned out when it becomes two-thirds full, then it is no longer necessary to dig a new pit every few years. In the design explained here, the pit contents are removed through a manhole from the outside of the toilet room.

Make a reinforced concrete beam or an arch across the pit (Figs. 1 & 2) to support the back wall of the toilet, the floor, and the concrete manhole cover (see also the Drawing Book, pages 121 and 124).

Cast the foundations around the pit and under the future walls of the toilet (Figs. 1 & 2). On top of the foundations cast the floor slab with the openings for the manhole and toilet. The toilet can be made with the same squatting slab design as the other latrines, or a seat may be constructed above the hole (Fig. 3).

The rising walls are constructed directly on top of the floor slab (Fig. 4). They may be built out of mud, sandcrete or landcrete blocks, since the structure will be permanent. Make sure that the toilet room is well ventilated and the roof is securely anchored to the building. The wall in front of the entrance can be built up to about 180 cm high (Fig. 4).

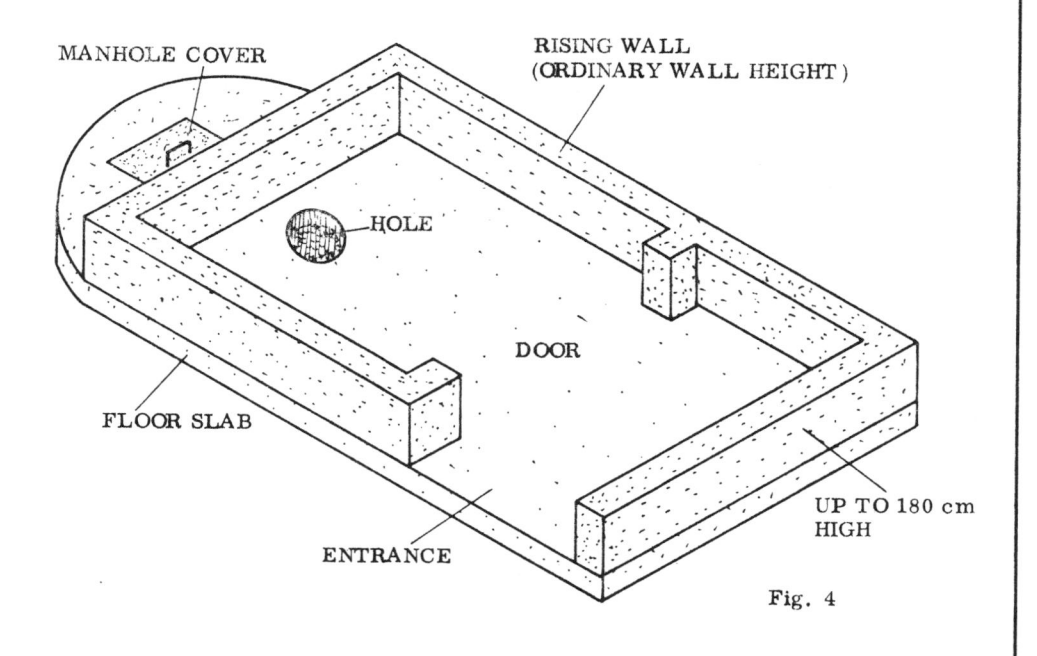

Fig. 4

PIT LATRINE.

Fig. 1

Fig. 2

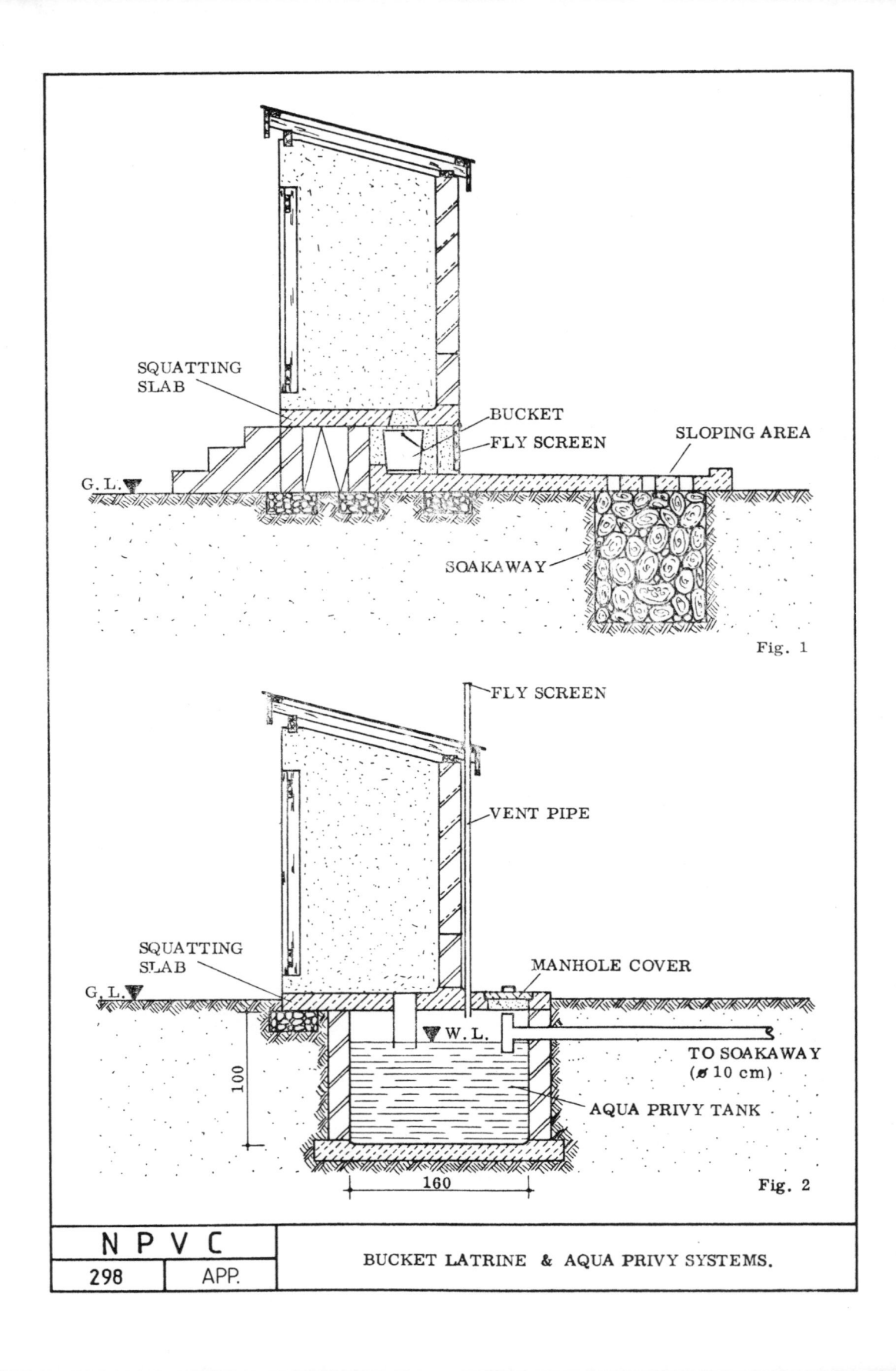

N P V C	BUCKET LATRINE & AQUA PRIVY SYSTEMS.
298 APP.	

BUCKET LATRINE

One of the oldest and generally least cleanly systems for waste removal is the bucket latrine. As in the cross section in Fig. 1, the squatting slab is set over a collection chamber with a bucket. The chamber should be closed off with a removable fly screen. The bucket should fit into a niche on the floor, so it is always returned to position directly under the hole in the slab when it is replaced after cleaning.

The area behind the latrine should be paved, and a drain and a soakaway should be made to get rid of the water used for cleaning the toilet and collection chamber.

The bucket latrine system is only possible where there is an organized system for regular collection and disposal of the bucket contents. This is usually not possible in rural areas.

AQUA PRIVY SYSTEMS

An aqua privy is basically a septic tank directly under a latrine. Its advantages are that it has fewer problems with smells, flies, and disease spreading than a regular latrine. However, at least two buckets of water must be poured into the privy every day for it to operate properly, and this can be a problem where there is no water source nearby.

An aqua privy for the use of one family may have a pit which is about 1 cubic meter in volume (about 0,15 cubic meters per person). The manhole should be large enough so that the sludge can be removed from the pit from time to time; whenever the tank is one-third full. If the tank is not emptied in time the drain pipe will become blocked and the system will fail.

Add some reinforcement to the squatting slab over the tank to support the back wall of the toilet. Make sure that the inside of the tank is waterproof, because if it leaks the water level cannot be maintained and smells can pass through the chute. If necessary paint the inside walls of the tank with waterproof paint. Make sure that there is a good ventilation system, with a fly screen over the end of the pipe which ventilates the tank (Fig. 2).

It is very important to maintain the water level within the tank. If possible, washing facilities can be made nearby and connected with the tank, so that the water from the sink or shower room goes into the tank and maintains the water level.

HANDLE MANHOLE COVER
G.L.
PIPE
Fig. 1a

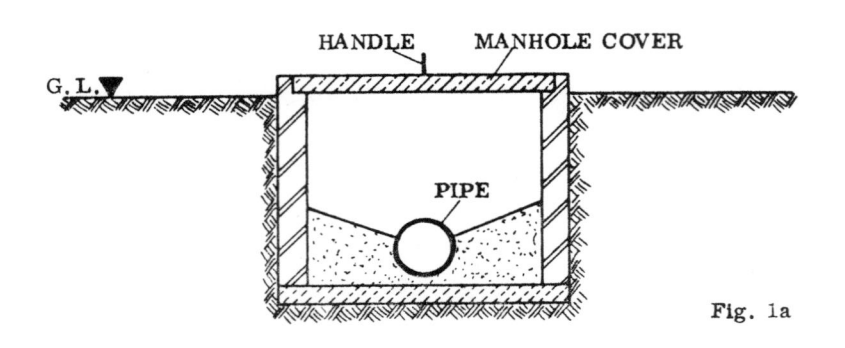

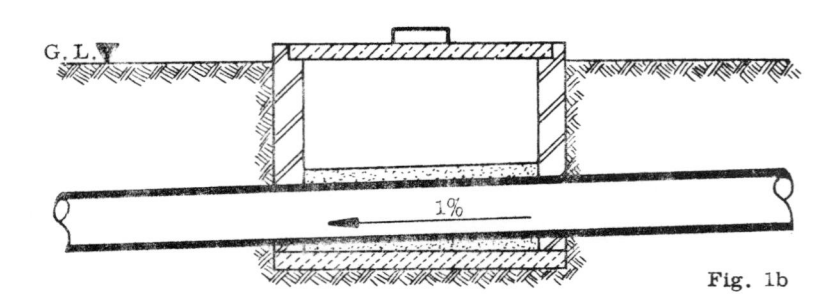

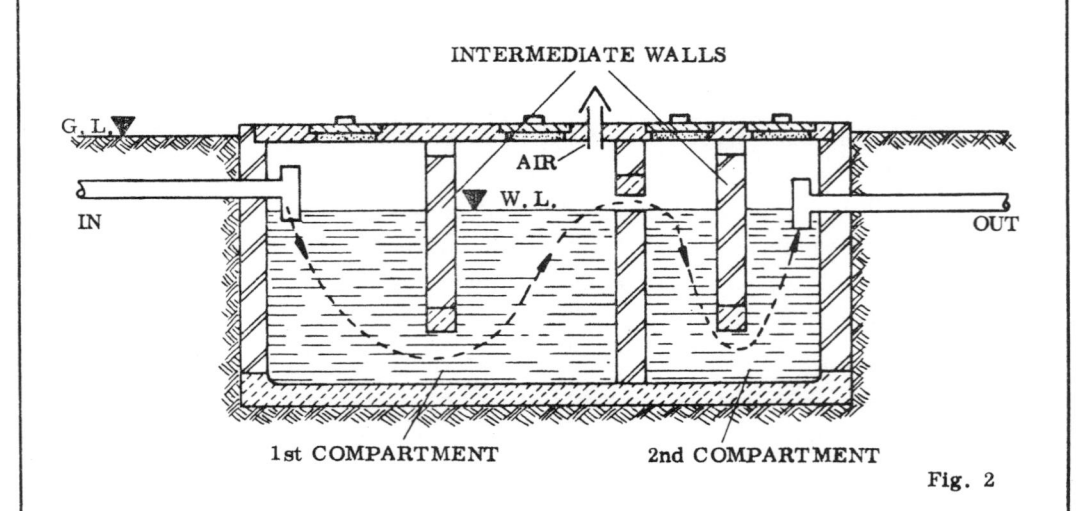

G.L.
1%
Fig. 1b

INTERMEDIATE WALLS
G.L.
AIR
W.L.
IN
OUT
1st COMPARTMENT   2nd COMPARTMENT
Fig. 2

N P V C	
300	APP.

SOAKAWAYS & MANHOLES & SEPTIC TANKS.

SOAKAWAYS

A soakaway is basically a hole in the ground filled with stones, through which water can seep away into the surrounding soil instead of forming a pool on the surface of the ground where mosquitos can breed.

Dig a hole and fill it with large stones at the bottom and smaller ones at the top. Drain pipes from bathing areas, etc. can lead into the pit; they should end up in the centre. Cover the soakaway with soil, but make sure that the soil does not enter between the stones and block the drainage pipe. The soakaway can be covered with a layer of concrete to keep the top soil out (Fig. 1, previous page).

Soakaways should be at least 30 m away from wells or streams, and never uphill from a water source.

MANHOLES

Sometimes manholes are needed where long pipes have to be inspected at intervals. Manholes give easy access to the pipe junctions. A typical manhole layout is shown in the Drawing Book, page 125. This manhole can also serve as a kind of junction hole, from which more than one pipe can cross. The slope should be the same for pipes entering and leaving the manhole (Figs. 1a & 1b).

SEPTIC TANKS

Septic tanks are designed in such a way that the water takes at least 24 hours to pass through the system. During that time the heavier solids will settle to the bottom, forming the sludge. In the tank the solids are gradually broken down and become much reduced in volume.

Septic tanks should have two compartments (Fig. 2). The first compartment is twice the size of the second compartment. Intermediate walls are sometimes made in order to reduce the speed of the water flow, and to make the distance that the water has to travel longer (Drawing Book, page 126).

The sludge has to be removed from the tank every few years, whenever it becomes 1/3 full. This sludge will not be safe to use as fertilizer until it has been composted for several months. The water flowing out from the tank will also be contaminated with bacteria and disease organisms.

SOAKAWAYS & MANHOLES & SEPTIC TANKS.   N P V C   APP. 301